Historical & Cultural Astronomy

The Historical & Cultural Astronomy series includes high-level monographs and edited volumes covering a broad range of subjects in the history of astronomy, including interdisciplinary contributions from historians, sociologists, horologists, archaeologists, and other humanities fields. The authors are distinguished specialists in their fields of expertise. Each title is carefully supervised and aims to provide an in-depth understanding by offering detailed research. Rather than focusing on the scientific findings alone, these volumes explain the context of astronomical and space science progress from the pre-modern world to the future. The interdisciplinary Historical & Cultural Astronomy series offers a home for books addressing astronomical progress from a humanities perspective, encompassing the influence of religion, politics, social movements, and more on the growth of astronomical knowledge over the centuries.

The Historical & Cultural Astronomy Series Editors are: Wayne Orchiston, Marc Rothenberg, and Cliff Cunningham.

More information about this series at http://www.springer.com/series/15156

Tomokazu Kogure

The History of Modern Astronomy in Japan

Springer

Tomokazu Kogure
Yawata, Kyoto, Japan

ISSN 2509-310X ISSN 2509-3118 (electronic)
Historical & Cultural Astronomy
ISBN 978-3-030-57060-6 ISBN 978-3-030-57061-3 (eBook)
https://doi.org/10.1007/978-3-030-57061-3

Originally the author wrote a different book in Japanese. He got an idea from a chapter of its book and reconstructed the table of content for this book project. The author cleared the copyright issue with this original Japanese publisher and it is considered a new book.

Cover illustration text: This image illustrates a region of bright and dark nebulae in the constellation Cygnus extending along the Milky Way. Bright part is the hot ionized-gas cloud, while dark part the cold dust cloud.

This Springer imprint is published by the registered company Springer Nature Switzerland AG
The registered company address is: Gewerbestrasse 11, 6330 Cham, Switzerland

Preface: A History of Modern Astronomy in Japan

Modern astronomy fascinates people with its challenging wide range of discoveries about the sun, stars, and the universe. Japan is now making exciting discoveries in a wide range of fields in astrophysics. While Japan has a long history of astronomy since the seventeenth century, remarkable developments in astrophysics belong to the latter half of the twentieth century.

I have long been interested in the history of astrophysics from its origin up to the present with special attention to the lives and works of pioneering astronomers. I planned to write the history of modern astronomy in Japan on the basis of the same idea and started the writing of history on this line. However, a difficulty has arose in the history of the last quarter of the twentieth century by the appearance of so many numbers of researchers and research subjects. I am not a professional historian of astronomy but an astrophysicist who has worked in several fields of stellar and galactic astronomy since the late 1950s. In this situation, it is almost impossible to cover all the research fields of astrophysics. That is why the subjects in this book have been selected based on my interest.

In this book, I divided Japan's timeline of astronomical history into the following four periods from the Tokugawa era to the postwar era until around the end of the twentieth century:

1. Tokugawa era: 1603–1868: age of calendar-making,
2. Meiji-Taisho era: 1868–1926: age of foreign studies,
3. Early Showa era: 1926 ~ 1945: age of self-efforts,
4. Postwar era (late Showa and Heisei period): 1946 ~ 2000: age of development of astronomy and astrophysics,

where Meiji, Taisho, Showa, and Heisei denote the names of emperors, as historical eras in Japan are casually named after the emperor who ruled during that particular era. The history in these periods will be presented in separate eight chapters. Besides, the names of Japanese people are given surname first according to the custom in Japan, which will be followed throughout this book.

Period 1, astronomy in the Tokugawa era, is presented in Chap. 1. The subject of astronomy in this era was confined to calendar-making in the frame of luni-solar calendar. In parallel, western cosmologies from Aristotle to Newton were mainly accepted by civilians, particularly by Dutch interpreters.

Period 2, astronomy in the Meiji-Taisho era, is described in Chap. 2. The Tokugawa government was abolished in 1868 by Meiji emperor's army and a new Meiji government was established. This is called the Meiji Revolution. The new government adopted the policy of modernization, accepting western systems in politics, military, and other affairs. Initially astronomy was introduced in the field of position astronomy to adopt the Gregorian calendar and to build a new astronomical observatory. The observatory mainly had geopolitical purposes such as time keeping, calendar making, and observations of latitude and longitude.

Astrophysics was introduced through foreign studies in Western Europe and the USA in the late Meiji era around 1905. In the Taisho era (1912–1926), astrophysics was still promoted similarly through foreign studies.

Period 3, astronomy in the early Showa era (1927~1945), is given in Chaps. 3, 4, and 5, covering the development of astronomy in three distinct areas: Tokyo, Kyoto, and Sendai, where astronomy was promoted separately and almost independently. In this era, Japan was under a military government and finally rushed into World War II. International collaboration in science was getting difficult, and astronomers were forced to work for military purpose. Nevertheless, some notable works were produced in some theoretical fields. For example, physics of the solar corona and chromosphere, and basic physical processes of planetary nebulae were carried out mostly independent of western astrophysics. This period may be cited as the age of self-efforts.

Period 4 presents the development of astronomy in the postwar era (1946 ~ 2000), described in Chaps. 6, 7, and 8. In 1945, World War II ended with the defeat of Japan. Social and economic confusion lasted for several years. Thereafter, the Japanese economy expanded at a stunning pace. This enabled the construction of large scientific facilities including telescopes and orbiting satellites. In parallel with the economic growth, the education system was greatly changed in the form that compulsory education was extended from 6 to 9 years, and the university system also changed. Seven Imperial Universities were restyled to National Universities, along with the establishment of 60 other new National Universities on the mainland. Under these economical and education-system backgrounds, astrophysics remarkably flourished in the postwar era.

Chapter 6 is devoted mainly to the development of astrophysical instruments in optical, radio, and space observations, along with some results of observations to show the capabilities of these instruments.

In Chap. 7, the development of astrophysical research is traced in the solar and stellar physics, galactic astronomy, cosmology, and theoretical astrophysics. The progress was slow in the 1950s and gradually accelerated by the establishment of large observational facilities. The style of research has changed from individual study in the 1950s to group study in the 1970s and thereafter. Even inter-university or international researches have been often organized.

In Chap. 8, the progress in the education and popularization of astronomy is briefly traced. General education in the lower-level schools and universities is considered in parallel with technical education in the graduate courses of astronomy. Progress in the popularization of astronomy was remarkable, with the construction of large planetariums and public observatories equipped with large telescopes.

In the appendix, history of the rise of astrophysics in Western countries is briefly traced. While astrophysics had no effect on astronomy in the Tokugawa era, it offered great influence to the progress of modern astronomy in Japan.

In this book, the history of astronomy in Japan has been traced up to around 2000 or so. The research of astrophysics in the twenty-first century is highly advanced to the international level in the world. For this reason, I have closed the history of astronomy around the end of the twentieth century.

Yawata, Kyoto, Japan Tomokazu Kogure

Acknowledgment

The author expresses sincere gratitude to Professor Kam-Ching Leung of the University of Neburaska, USA, and Professors Kato Shoji and Hirata Ryuko of Kyoto University for their kind critical reading of the manuscript.

The author is thankful to Mrs. Itoh Noriko, librarian in the Department of Astronomy, Kyoto University, for the extensive search of references.

The author's sincere gratitude is also due to Professor Seki Munezo of Tohoku University (on the life of Hitotuyanagi Zyuiti), Mr. Nakagiri Masao of the National Astronomical Observatory (on the life of Ichinohe Naozo), Dr. Hirabayashi Hisashi (on the VSOP and VERA) and Dr. Okuda Haruyuki (infrared astronomy) of the Institute of Space and Aeronautic Science, Professor Sawa Takeyasu of Aichi Kyoiku University (on the education of astronomy), Professors Kurokawa Hiroki and Shibata Kazunari of the Kwasan and Hida Observatories (on the DST observations), Dr. Tomita Yoshio of Kyoto University (on the lives of Shinjo, Shinzo, and Yamamoto Issei), Professor Miyajima Kazuhiko of Doshisha University (on the history of astronomy in the Tokugawa era), Dr. Izumiura Hideyuki (on the observation programs) of the Okayama Astrophysical Observatory, Dr. Kazu Tsugito and Dr. Watanabe Yoshiya (on the activity of planetarium) of the Osaka Science Museum, and Mr. Nariai Tatuo and Mrs. Hashimoto Eiko of Ashiya City (on the life of Nariai Hidekazu) for the data search on the respective subjects.

The original idea of writing this book is owed to Ms. Niko Hisako, editor at Springer, to whom I express my hearty thanks. I also thank anonymous reviewers for the helpful comments.

For the helpful editorial works – (to be continued).

For the editorial works, I sincerely express my gratitude to the Editors of Springer: Ms. Camilya Anitta, Mr. Werner Hermens, Ms. Hisako Niko, Ms. Rebecca Sauter, and Mr. Solomon George. The efforts of these editors made this a far better book than it would have been otherwise. Finally, I am thankful to my wife for supporting everyday life of an old author.

Contents

Chapter 1
Astronomy in the Tokugawa Period, 1603–1868

Abstract In the Tokugawa period, Japan was governed by a feudal government run by the Tokugawa family with some economic and cultural flourishing. The role of astronomers was the making of the lunisolar calendar, which occurred four times. In parallel, Western cosmology from Aristotle to Newton was introduced by civilians, particularly Dutch interpreters, in Nagasaki. The Tokugawa period is an age of prehistory of modern astronomy in Japan.

1.1 Tokugawa Period and Astronomy

In 1603, a new central government was established by the Tokugawa family in Edo (江戸, now Tokyo), apart from the Imperial Court in Kyoto. The country was divided into many domains of feudal clan and governed by respective clan lord. The Tokugawa government's stable reign lasted from 1603 to 1868 (Nakayama 1969; Gordon 2003).

Japan was geographically distant from Western countries, and the Tokugawa government adopted a policy of isolation from other countries to protect itself from the propagation of Christianity. The Netherlands was the only country permitted to have commercial trade at Nagasaki. The aim of this policy was the preservation of traditional Chinese and Japanese thought, as the principal philosophical background of Japanese culture.

In the Tokugawa period, astronomy was essentially limited to calendar reforms in collaboration with Japanese mathematicians. Since these calendars were of a lunisolar type, the addition of a leap month was required to reconcile the calendar with the actual seasons. This required lengthy and laborious work involving numerical calculations, and calendar reform was implemented four times in 250 years (Araki 1940; Uchida 1986).

During this period, the translation of Dutch books with their contemporary concept of Western cosmologies from the Aristotelian to the Newtonian was introduced into Japan. In 1720 the eighth Shogun Tokugawa Yoshimune (徳川吉宗), permitted the importation of Western technical literature into Japan, especially

T. Kogure, *The History of Modern Astronomy in Japan*, Historical & Cultural Astronomy, https://doi.org/10.1007/978-3-030-57061-3_1

Dutch works on physical and medical sciences. Dutch learning became popular and served as the basis of culture in the late Tokugawa period.

1.2 Calendar Reforms

The first calendar used in Japan was imported from China in 554 A.D. during the regime of Emperor Kinmei (欽明), and several successive emperors also adopted new Chinese calendars without any revisions. Finally, it was in 862 that the Hsuan-Ming calendar (*Senmyo reki*, 宣明暦) was imported from China and promulgated for yearly use.

This calendar was used for more than 800 years, despite the growing discrepancies between observed phenomena and the calendar.

In the sixteenth century, the situation surrounding Japan changed dramatically. In 1543, a Portuguese ship came shore and introduced Western civilization to Japan for the first time. In 1549, Francis Xavier, a Jesuit priest, also landed on the southwest shore, and Christian evangelism was received with enthusiasm in some Japanese quarters. In 1603, the Tokugawa government was established in Edo and lasted for around 250 years, accompanied by domestic peace and economic prosperity. These two factors, the impact of the West and national stability, made possible the calendar reform during the Tokugawa period. In Western countries, the Julian calendar was replaced by the Gregorian calendar in Catholic countries in 1582, but it took around 200 years for it to be accepted in Protestant countries. Since these calendars were of the solar type, the reform had no direct impact on lunisolar calendars in China and Japan. Impacts may have been confined to technical areas, such as mathematical and observational techniques.

The official astronomical office in the Tokugawa era was the Tenmonkata (天文方, Bureau of Astronomy and its personal position) in Edo and the On-yo Ryo (Yin-Yang Board, 陰陽寮, Office of Astrology) in the Kyoto Imperial Court.

1.2.1 Shibukawa Harumi and Jokyo Calendar

Shibukawa Harumi (渋川春海, 1639–1715) inaugurated Japan's first calendar reforms after many centuries based on Western astronomical ideas and techniques.

Harumi was born into the Yasui family, a professional *GO* (碁, board game) player. From his youth, Harumi had the reputation of being a prodigy in *GO* play and showed a remarkable understanding of astronomy, particularly calendar reform.

With his calendar works, Harumi was recommended to the Shogun by Clan Lords as the most qualified person to carry out calendar reform.

It was the Chinese *Shou-shih* calendar (*Juji reki*, 授時暦) that Shibukawa had taken as the basis of his calendar reform. This calendar was produced in 1281 by

Kuo Shoujing (郭守敬, 1231–1316) in the Yun (元) dynasty and was used for more than 360 years in China (Nakayama 1971).

The *Shou-shih* calendar was introduced to Japan in around 1670. However, in the long span of time since its production, there had arisen nonnegligible discrepancies in calendar time for the prediction of sky phenomena. Shibukawa began to construct a new calendar based on the *Shou-shih* calendar by taking into account the difference in longitude between Japan and China. After observations of solar and stellar motions for several years, he completed a new calendar system and called it the Yamato calendar (大和暦). On the other hand, the Tsuchimikado family (土御門家) of the Imperial Court in Kyoto proposed the adoption of the *Daito-reki* (大統暦*Ta-t'ung* calendar, the calendar in the Ming Dynasty, 明朝).

After a long dispute on the accuracy of these two calendars, the Yamato calendar was finally accepted and promulgated in 1685 under the name of Jokyo calendar (貞享暦).

With this achievement, Shibukawa was appointed for chair of Tenmonkata at the Surugadai Observatory (駿河台天文台) in the city of Edo. After Shibukawa's death, this observatory remained inactive for several generations since Tenmonkata was a hereditary position and no able astronomer was able to replicate Shibukawa's works.

1.2.2 Horeki Calendar

In the early eighteenth century the Eighth Shogun, Yoshimune (徳川吉宗, 1684–1751), himself an enthusiastic amateur astronomer, tried to modernize the calendar, appointing Nishikawa Seikyu (西川正休) as a Tenmonkata in charge of calendar making. Nishikawa had an outstanding reputation for his knowledge of Western astronomy.

Nishikawa and his college Shibukawa Noriyoshi (渋川則休) intended to formulate a new calendar based on the *Shou-shih* calender (授時暦). On the other hand, Tsuchimikado Yasukuni (土御門泰邦) and his family in Kyoto tried to develop a new calendar based on the traditional Jokyo calendar. There arose a bitter conflict between these two groups, and Nishikawa's group was finally defeated after the death of Shogun Yoshimune in 1751.

As a consequence, the new Horeki or Horyaku calendar (宝暦) was developed by the Tsuchimikado family in 1755. However, since Yasukuni did not have sufficient knowledge of calendar calculation, the essentials of this calendar were the same as those of the Jokyo calendar. The failure of this calendar became apparent a few years later when the prediction of a solar eclipse failed. Despite this failure, the calendar was used for 43 years up to 1797.

During the formation of the Horeki calendar astronomical observations in Edo were conducted at the Kanda Observatory, Edo. This observatory was shut down in 1757 and temporarily moved to Ushigome in Edo. To promote calendar reform, the government established a new observatory in Asakusa, in the city of Edo, in 1782.

Fig. 1.1 Asakusa Observatory. (「寛政暦書」より国立天文台蔵), (Nakamura 2012, p. 77) (Courtesy Library of the NAOJ)

This observatory, called Asakusa Observatory (浅草天文台), was equipped with a *Kanten-gi* (簡天儀, a kind of armillary sphere), *Shogen-gi* (象限儀, a kind of quadrant), and some small refracting telescopes (Fig. 1.1). The calendar reform at this observatory will be discussed in the next section.

1.2.3 Kansei Calendar

Asada Goryu (麻田剛立, 1734–1799) played an important role in reforming the calendar and in training students in the latter half of the eighteenth century.

Asada was a son of a Confucian scholar and physician in Kitsuki Province (杵築), Kyushu. He educated himself in mathematics, astronomy, and medicine, among which he was especially interested in astronomy since his youth. He was also gifted in technology and constructed a small reflecting telescope, with which he observed sunspots, the lunar surface, and planets. He successfully predicted the solar eclipse that occurred on September 1, 1763. Since this eclipse was not noticed in the then current Horeku calendar, the prediction garnered him fame at the time.

In 1777, Asada moved to Osaka, where, keeping up his daytime profession as a physician, he established a private school of astronomy, which he called Senjikan (先事館), where he studied and taught astronomy.

In the 1790s, the Tokugawa government launched a project to revise the current Horeku calendar, but no suitable astronomer able to carry out this work was found in

Edo. The government first invited Asada for this work, but, due to illness, Asada was unable to accept the post and recommended two of his students, Takahashi Yoshitoki (高橋至時, 1764–1804) and Hazama Shigetomi (間重富, 1756–1816). Asada himself stayed in Osaka and died in 1799 at age 65.

Takahashi was born the son of a lower-class official (同心, a policeman) in Osaka. From his youth, Takahashi was interested in mathematics. In 1779, he took over his father's position and studied in his spare time at Asada's Senjikan. Takahashi found common interest with a wealthy pawnbroker and instrument maker named Hazama, who was also a student of Senjikan. Hazama was talented in technology and built many astronomical and geodetic instruments, including a reflecting telescope and a spherical armillary.

Both Takahashi and Hazama moved to Edo in 1795 and started observations at Asakusa Observatory for carrying out calendar reform. The new calendar was completed in 1797 and was called the Kansei calendar (寛政暦). It was officially put into operation in 1798.

The Kansei calendar was based largely on the sequel to the *Li-Hsian K'ao Ch'eng* (暦象考成総編, Compendium of Calendrical Science and Astronomy), which was chiefly edited by the German missionary Ignatius Kögler (1680–1746) and made use of the elliptical orbit of the Sun without reference to the heliocentric system. Takahashi was really the first to apply the elliptical theory. For planetary motions, however, Takahashi and Hazama applied the Tychonic epicyclic system, following the original *Li-Hsian K'ao Ch'eng* (暦象考成).

Just after completing the Kansei calendar, Takahashi encountered the Dutch version of J. J. L. de Lalande's *Treatise on Astronomy* (five volumes). He was greatly impressed by this book and the comprehensive achievements of Western astronomy with respect to planetary theory and mathematical technique. Takahashi wrote a multivolume work entitled *Lalande Rekisho Kanken* (ラランデ暦書管見, *A Private Review of Lalande's Astronomy*, 1803) and began to translate this book, but the work remained unfinished at the time of his death in 1804 (Nakayama 1972). This work was taken over by his sons.

1.2.4 Tenpo Calendar

The complete translation of Lalande's book was completed in 40 volumes by Yoshitoki's two sons, Tekahashi Kageyasu (景保) and Shibukawa Kagesuke (渋川景祐). The translation was published in 1836 under the title *Shinko Rekisho* (新巧暦書, *Astronomy by the New Technique*) and represented the highest level of astronomy in China and Japan at the time. In 1841, the Tokugawa government ordered to Shibukawa Kageyasu and his group, including Yamaji Kaiko (山路諧孝) and Adachi Shinto (足立信頭), to undertake calendar reform. The calendar was completed in 1842 and named the *Tenpo* calendar (天保暦), which was promulgated in 1843 (Uchida 1986).

Following completion of the *Tenpo* calendar, Kagesuke wrote a book, *Shinpo Rekisho* (新法暦書, *Calendrical Treatise by the New Method*), in 1846. In its preface, Kagesuke criticized the concept of heliocentrism. He emphasized that the difference between geostatic motion and heliocentric motion of planets was merely a matter of coordinate transformation; the astronomical implication is all the same for the purpose of calendar making. Both theories have an equal logical claim to verisimilitude. Kagesuke's attitude reflected a traditional astronomer's approach, not open to new cosmological and physical concepts. Astronomers in his period remained tradition-bound within the framework of calendar reform and accepted Western astronomy in its technical aspects, but not in its cosmology.

1.3 Acceptance of Western Cosmology

In the Tokugawa period, Western knowledge in science and technology was transmitted via the Dutch language. Astronomy, mapmaking, calendar making, and medicine are among the main fields whose knowledge was acquired from Dutch texts. Dutch interpreters in Nagasaki were officers of the Edo government and promoted trade and cultural exchange. Dutch learning gradually spread among scholars in Edo and Osaka.

Gradually, Western cosmologies, Aristotelian, Copernican, and Newtonian, were introduced and successively accepted (Hirose 1964; Nakayama 1969).

1.3.1 Mukai Gensho and Aristotelian Cosmology

Mukai Gensho (向井玄升, 1607–1677) introduced Western astronomical ideas in Japan during the early Tokugawa period (Hockey et al. 2007; Nakayama 1969).

He was born in Hizen (肥前, now Saga Prefecture, Kyushu) and moved to Nagasaki, where he learned Confucian philosophy and medicine, as well as astronomy. In 1658, he moved to Kyoto, where he practiced medicine and remained until his death. Mukai was interested in Western astronomy based on Aristotelian cosmology and wrote a set of commentaries around 1650, entitled *Kenkon Bensetsu* (乾坤辯説, *Western Cosmography*, with critical commentaries). This book was written based on a translation by the apostate Jesuit missionary Christoph Ferreira (Japanese name, *Sawano chu'an* 沢野忠庵).

In *Kenkon Bensetsu* he introduced the geocentric Aristotelian system to Japan with some commentary. In this system, heaven consists of nine spherical shells surrounding the spherical earth at rest. The Sun, Moon, and planets are fixed in their respective shells, and fixed stars are placed on one shell. Heavenly bodies are made of perfect matter, composed of four elements: earth, water, fire, and wind.

Mukai made two comments in *Kenkon Bensetsu*. One was his opposition to the four-element hypothesis. He argued that terrestrial materials are composed of five elements: water, wood, fire, earth, and metal. The second comment related to a disagreement with the celestial-terrestrial dichotomy. He insisted that the composition of the heaven and the earth is the same, that both were composed of five elements. Mukai's arguments are mostly based on ancient Taoism in China. Thus, he defended traditional Chinese and Japanese philosophy.

1.3.2 Kobayashi Kentei and Aristotelian Cosmology

Kobayashi Kentei (小林謙貞, or Kobayashi Yoshinobu 小林義信, 1601–1684) was a surveying engineer and an astronomer who was interested in cosmological ideas (Hirose 1971a, b).

Kentei was born in Nagasaki and learned Western technology and astronomy from Hayashi Kichiemon (林吉右衛門), who was a Christian scholar later executed for his forbidden faith. As a student of Hayashi, Kentei was sent to prison for 21 years. After his release from prison, he became a surveying engineer in Nagasaki and introduced Dutch and Portuguese surveying techniques to Japan. In astronomy, Kentei wrote a book entitled *Nigi Ryakusetsu* (二儀略説, *Outline Theory of Terrestrial and Celestial Globes*) (ca. 1667), which was based on Pedro Gomez's book *On the Spheres* (ca. 1593), which dealt with cosmology, meteorology, and matter theory. Kobayashi introduced Aristotelian cosmology, as in the case of Mukai, but he restricted himself to a faithful translation of the core of cosmology without commentary.

1.3.3 Motoki Ryoei and Copernican Cosmology

Motoki Ryoei (or Motoki Yoshinaga, 本木良永, 1735–1794) was a translator of Dutch books and known as the first person to introduce Copernican cosmology to Japan.

Motoki was born the son of a physician, Nishimatsu Sen, and became an adopted son of the Dutch interpreter Motoki at age 13. Besides working as an interpreter, he studied natural science by reading Dutch books, almost entirely independently.

On astronomy, Motoki's first work, *Tenchi-Nikyu Yoho* (天地二球用法, *The Use of Celestial and Terrestrial Globes*, 1774), was based on a book by Wellem J. Blaeu (or Blaaw, 1572–1635, Dutch cartographer and atlas maker), which consisted of two volumes on Ptolemy's and Copernicus's cosmologies (1666).

Motoki's first book, *Tenchi-Nikyu Yoho*, was restricted to a treatment of Ptolemaic geocentric theory. After around 20 years, Motoki wrote a sequel to *Tenchi-Nikyu Yoho* in 1792–1793. This sequel was based on Copernican cosmology.

Motoki described the heliocentric solar system with elliptical orbits, but parts of dynamic theory and Kepler's laws were arbitrarily omitted. After all, Motoki was a faithful interpreter, avoiding to refer the new science.

1.3.4 Shizuki Tadao and Newtonian Cosmology

Shizuki Tadao (志筑忠雄, 1760–1806) was a translator and commentator of works on natural philosophy. He introduced Newtonian cosmology to Japan and attempted to reconcile its principles with Confucianism (Matsuo 2007; Hockey et al. 2007).

Shizuki was born into the Nakano family and later became an adopted son of the Shizuki family of a professional Dutch interpreter. Shizuki was a disciple of Motoki Ryoei. After learning Dutch, astronomy, and other sciences from Motoki, he was hired as an official interpreter in 1776. In the following year, however, he resigned his official position because of illness and began to work on translations and commentaries.

He completed his main book, *Rekisho Shinsho* (暦象新書, *New Treatise on Calendrical Phenomena*), in three volumes in 1802 (Ohmori 1971). This book was a translation of books on physics and astronomy written by John Keill, English mathematician and astronomer. Volume 1 of *Rekisho Shinsho* described observed behaviors of solar and planetary motions based on a heliocentric point of view. Volume 2 introduced new technical words into Japanese, such as dynamic force (dō-ryoku 動力), acceleration (kasokudo加速度), centrifugal force (enshin-ryoku 遠心力), gravitational force (ju-ryoku 重力), and others, and described planetary motions based on Newton's theory. In Volume 3, he introduced Western mathematical concepts, such as Newton's differential calculus, trigonometry, Napier's logarithm, the trigonometric logarithm, and others. Keill's original book was composed of six volumes, among which Shizuki translated for about a half of the whole volume in Rekisho Shinsho, for a part limited around the half of the whole volumes. Shizuki also translated some other parts under different titles, such as *Dogaku Shinan* (動学指南, *Lecture on Dynamics*), *Nisshoku Sousan* (日食総算, *General Calculation of Solar Eclipses*), and others. Furthermore, at the end of Volume 2 of *Rekisho Shinsho*, Shizuki put forth an original hypothesis on the origin of the Solar System, which somewhat resembled the hypotheses of Pierre-Simon de Laplace and Immanuel Kant.

Although Shizuki was not a scientist, the main purpose of these books was to introduce Western science and the new cosmology based on Newtonian mechanics and Kepler's laws of planetary motions. However, the mechanical sciences presented in *Rekisho Shinsho* attracted no interest from professional astronomers of the time. In the middle and late Tokugawa period, astronomy was not sufficiently advanced to recognize the need for Newtonian mechanics.

1.3.5 *Yamagata Banto and Criticism of Traditional Cosmology*

Yamagata Banto (山片蟠桃, 1748–1821) was a wealthy merchant and a famous scholar in Osaka known by his radical criticism of traditional Japanese and Chinese thought (Fig. 1.2).

Yamagata was born in Kamizume Mura (神爪村, present-day Takasago City, Hyogo Prefecture) the son of a wealthy farmer and moved to Osaka as an adopted son of his uncle at age 13 and began to work as a clerk in Masuya (升屋), a money exchange merchant. While there, he studied Confucianism and science from Asada Goryu at the Kaitokudo School (懷徳堂), which had been established by wealthy merchants in Osaka out of their own cultural interest. Despite his poor eyesight, he devoted 18 years late in his life to write a book, *Yume no Shiro* (夢の代、*Instead of Dream*) (Yamagata 1820), which was a series of essays in 12 volumes concerning all aspects of human life, including astronomy, geography, history, political systems, and atheism. On cosmology, he adopted the heliocentric theory and criticized all traditional Chinese and Japanese cosmologies based on his atheism. He accepted the Newtonian theory and attractive forces as the origin of the motion of heavenly bodies. In *Yume no Shiro*, Yamagata also wrote on the solar calendar, in which 1 year is composed of 365 days and one leap year is inserted every 4 years. Yamagata was one of the first proponents of a solar calendar in the Tokugawa period (Araki 1940).

His cosmology was not confined to the Solar System (Arisaka 1982). He believed that the Sun was one among the many stars that were distributed widely outside the Solar System and that the Milky Way might be a collection of faint stars (Fig. 1.3). He also argued that each star might have a planetary system and that some stars might even have a habitable planet like Earth. His views were quite advanced for the time.

Fig. 1.2 Statue of Yamagata Banto

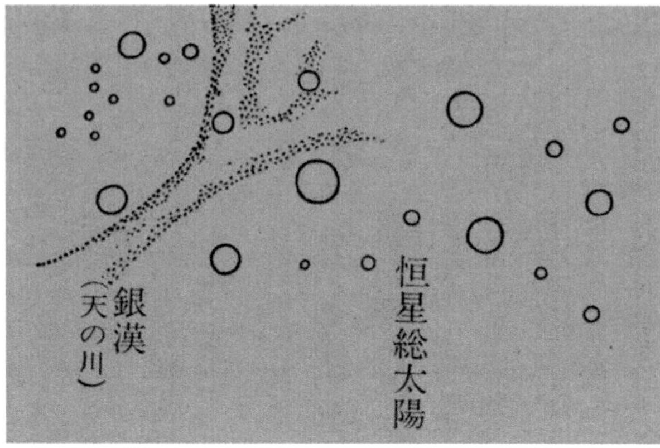

Fig. 1.3 Stellar system and the Milky Way. The left letter 銀漢(天の川) denotes the Milky Way and right letter 恒星総太陽indicates that every star is a sun (*Yume no Shiro* 1820)

1.3.6 Kawamoto Komin and Modern Physics

In the early nineteenth century, Dutch learning had already spread among scholars in many areas of the country. Its focus was mainly science, technology, and medicine. It was Kawamoto Komin (川本幸民,1810–1871) who first introduced modern physics and chemistry to Japan in the mid-nineteenth century (Kita 2008).

Kawamoto was born in Sanda (三田), Hyogo Prefecture, as the youngest son of Kawamoto Shu'an (川本周安), medical doctor of the Sanda clan. Komin studied Chinese and Dutch medicine in Osaka and then moved to Edo for a broader Dutch education. He remained in Edo for most of his life, working as a physician and as a teacher of the Dutch language at his private school Seishu Do (静修堂). His interest was gradually widened to the basic science of physics and chemistry.

In 1835, Kawamoto married Aochi Hideko (青地秀子), a daughter of Aochi Rinso (青地林宗, 1775–1833), who was also a Dutch scholar with a deep interest in the basic sciences. In 1827, he wrote a book entitled *Kikai Kanran* (気海観瀾、*General View on Air and Water*) (Aochi 1827), which presented a concise physical interpretation of meteorological phenomena of air, water, light, sound, clouds, thunder, and others in a total of 98 pages (Fig. 1.4).

Kawamoto was extremely impressed on this Aochi's book. He then started the study of physics and wrote a series of books called *Kikai Kanran Kogi* (気海観瀾廣義, *General View of Air and Water, Widely Extended*) (Kawamoto 1851), a great expanded of *Kikai Kanran* into 15 volumes. His books were published from 1851 to 1856. This series was based on Buijs's book (1828) on *Natural Science in School* as well as several other Dutch books and arranged according to his own scientific views. The series contained a general overview of physics, including astronomy, meteorology, and optical instrumentation.

On astronomy, Kawamoto provided a quantitative description of the Solar System, as shown in Fig. 1.5. The Sun is located at a distance of 25 years by the velocity of bullets from the Earth (one bullet velocity is about 200 m/s). Seven planets revolve around the Sun under Bode's law: Mercury, Venus, Earth, Mars, Jupiter,

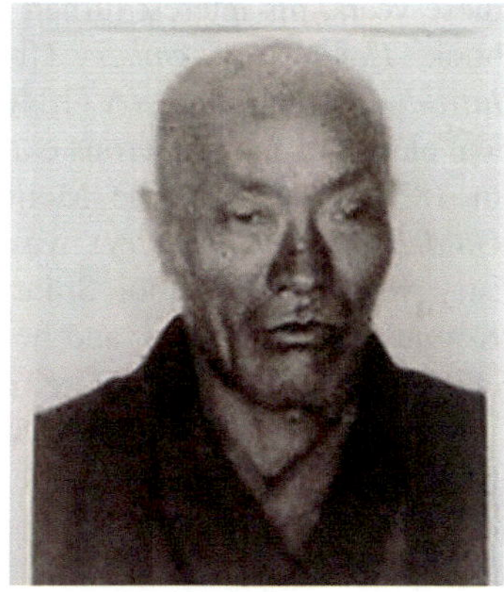

Fig. 1.4 Portrait of Kawamoto Komin (ca. 1861). (Japan Academy of Sciences, Library)

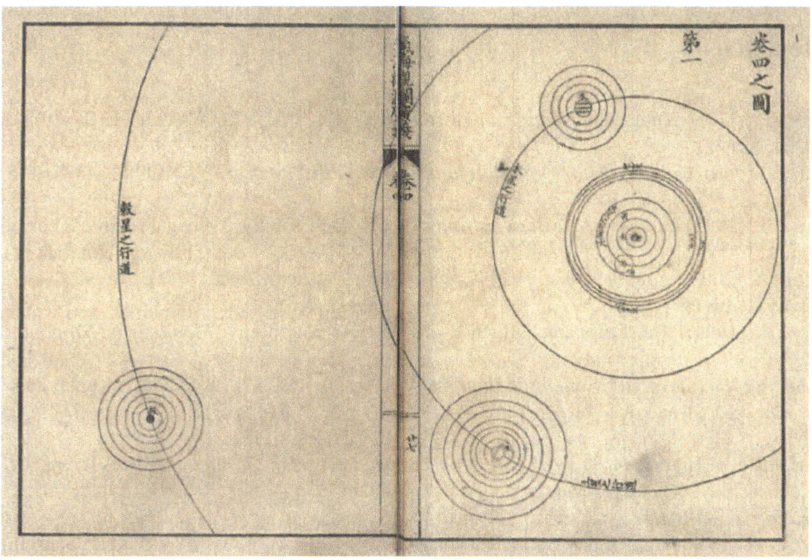

Fig. 1.5 Solar System in Kawamoto's *Kikai Kanran Kogi*, Volume 4 (Kawamoto 1856)

Saturn, and Kaku-sei (殻星, Uranus). Between Mars and Jupiter are four small planets, which according to Kawamoto were the remnants of an ancient explosion of a large planet moving in this orbit.

Kawamoto also wrote that stars were distant suns distributed far beyond our Solar System. The distance to nearby stars from the Sun may be around five million years by the velocity of bullets. On the origin of stars' variability, he suspected a combination of the rotation period and the variation in the size and position of star spots, by analogy with sunspot phenomena. In this way, he presented the contemporary state of modern physics and astronomy to Japan.

In 1859, Kawamoto was appointed professor of Dutch at Bansho Shirabe Dokoro (蛮書調所, Institute of Foreign Book Study) of the Tokugawa government. During these years, his interest turned from physics to chemistry. In 1861, he published a book, *The New Chemistry* (化学新書), a translation of J. A. Stickhardt's book *Introduction to Chemistry* (1846). It contained the properties of more than 50 chemical elements and numerous examples of experiments involving chemical reactions. In 1874, already in the Meiji era, a new book, *Kagaku Tokuhon* (化学読本, *Guidebook for Chemistry*), was published; it included a new part on organic chemistry written by Tsuboi Shinryo (坪井信良). These old and new books made Komin's name famous, and he came to be called the father of chemistry in Japan. He died in 1871 at age 70 in Tokyo.

Later on, Bansho Shirabe Dokoro was renamed Kaisei Gakko (Kaisei School, 開成学校) and served as the base of the Imperial University in Tokyo in the Meiji era (Chap. 2).

References

Aochi, R. (1827). *General view on air and water* (青地林宗: "気海観瀾") (Meizan-Kaku, Edo, 名山閣, 江戸).

Araki, T. (1940). *General history of the Japanese Calendar*, Koseisha (荒木俊馬、日本暦学史概説、恒星社).

Arisaka, T. (1982). *Studies of Western learning IV*, 181–204, On the Yamagata Banto's cosmology (日本洋学史の研究IV, 有坂隆道: 山片蟠桃の大宇宙論について) Sogensha (創元社).

Gordon, A. (2003). *A modern history of Japan – From Tokugawa times to the present*. New York: Oxford University Press.

Hirose, H. (1964). The European influence on Japanese astronomy. *Monumenta Nipponica, 19*, 295–314.

Hirose, H. (1971a). *Modern scientific thought,* Vol. II, pp. 465–470, Kobayashi Kentei and Nigi-Ryakusetsu, Iwanami Shoten (近世科学思想 下, 広瀬秀雄: 小林謙貞と二儀略説、岩波書店).

Hirose, H. (1971b) *Nihon Shiso Taikei,* Vol. 63, *Modern scientific thought,* Vol. II, pp. 10–107, Kobayashi Kentei, Nigi-Ryakusetsu, Iwanami Shoten ("日本思想大系 63, 近世科学思想 下", 広瀬秀雄現代語訳: 小林謙貞、 二儀略説、岩波書店).

Hockey, T. et al. (Eds.). (2007). *Biographical Encyclopedia*, Vol, 3, 1533, Mukai Gensho; Vol. 4, 2003, Shizuki Tadao, Springer.

Kawamoto, K. (1851–1856). *Extended general view on air and water*, Seishu Do, Edo (川本幸民, 気海観瀾廣義, 静修堂, 江戸).

Kita, Y. (2008). *Kawamoto Komin, a Dutch scholar* (北 康利: "蘭学者 川本幸民"), PHP Kenkyu Sho (PHP 研究所), Tokyo.

Matsuo, R. (2007). *Dutch learning in Nagasaki – Shizuki Tadao and his age* (松尾龍之介: 長崎蘭学の巨人— 志筑忠雄とその時代), Gen Shobo (弦書房), Fukuoka.

Nakamura, T. (2012). Astronomy in Edo – Shibukawa Harumi and Astronomy in the Edo period (中村士: 江戸の天文学).

Nakayama, S. (1969). Chapter 3: The period of recognition of western supremacy: From the mid eighteenth century to the late nineteenth century. In *A history of Japanese astronomy – Chinese background and Western impact* (pp. 163–224). Cambridge, MA: Harvard University Press.

Nakayama, S. (1971). *Nihon Shiso Taikei,* Vol. 63, pp. 497–511, *The tradition of Chinese astronomical calendar and Shibukawa Harumi*, Iwanami Shoten (中山茂: 中国系天文暦学の伝統と渋川春海, 日本思想大系 63、、岩波書店).

Nakayama, S. (1972). Takahashi Yoshitoki and a private view of Lalande's astronomy, Iwanami Shoten (中山茂: 高橋至時とラランデ暦書管見, 日本思想大系65、岩波書店). *Nihon Shiso Taikei, 65*, 473–478.

Ohmori, M. (1971). Hosei Rekigaku (法政暦学), 25, 29–37, *Rekisho Shinsho* and a history of the study of Shizuki Tadao (大森実: 『暦象新書』および志筑忠雄の研究史).

Takahashi, Y. (1803). *A private review of Lalande's astronomy, selected* (高橋至時: ラランデ暦書管見抄), translated to modern Japanese by Nakayama, S. 1972, *Nihon Shiso Taikei* (日本思想大系)、Vol. 65, pp. 167–206, Iwanami Shoten (岩波書店)。.

Uchida, M. (1986). *Encyclopedia of calendar and time*, Yuzankaku (内田正男: 暦と時の事典、雄山閣).

Yamagata, B. (1820). *Yume no Shiro* (山片蟠桃: 夢の代, *Instead of dream*) partly reproduced in *Nihon no Meicho*, Vol. 23, 1971, Chuo Koronsha ("日本の名著", 中央公論社).

Chapter 2
Astronomy from Meiji to Taisho Period 1868–1926

Abstract In 1868, the Meiji government was established by the Meiji Emperor. The new government adopted a policy of modernization in political, economic, and cultural affairs. The first step was to hire foreign instructors in a wide range of affairs. The next step was to send Japanese researchers to Western countries. Astronomy started with positional astronomy in the adoption of the Gregorian calendar and the construction of a new observatory. The study of physics was launched in the fields of geophysics, spectroscopy, and atomic physics. In the early twentieth century, Shinjo Shinzo of Kyoto University and Ichinoe Naozo of the Tokyo Astronomical Observatory (TAO) were greatly impressed by astrophysics in their foreign studies. This was the initial phase of astrophysics in Japan.

2.1 Meiji-Taisho Period and Background of Astronomy

In 1868, the Tokugawa Shogun restored the reins of government to Emperor Meiji (1852–1912). This is known as the Meiji Revolution. The Meiji government moved the capital from Kyoto to Tokyo (formerly Edo), and it adopted a policy of modernization in Japan, implementing a Western system in political and military affairs, as well as social customs and culture. The Meiji period lasted 45 years, from 1868 to 1912, and the Taisho period 15 years, from 1912 to 1926. In these periods, the groundwork for modern astronomy was laid by the government (Bartholomev 1989; Nakayama 1969).

2.1.1 Calendar Reform

In 1870, the Meiji government abandoned the Asakusa Observatory in Tokyo and established Daigaku (大学, the Institute of Learning) in Kyoto instead of On'yo Ryo. The Tsuchimikado family (土御門家) was put in charge of Tenmon-Rekido (天文暦道, astrology and calendar-making) under the jurisdiction of Daigaku and

T. Kogure, *The History of Modern Astronomy in Japan*, Historical & Cultural Astronomy, https://doi.org/10.1007/978-3-030-57061-3_2

engaged in the promulgation of a yearly calendar edited based on the Tenpo calendar (天保暦), a lunisolar calendar.

In the fall of the same year, 1870, the main office of astronomy was moved to Tokyo and renamed the Seigaku Kyoku (星学局, Bureau of Astronomy) for the purpose of calendar reform. The Meiji government expedited reform to showcase to the world Japan's modernity. The Gregorian calendar was abruptly promulgated by changing the day from December 3, 1872 to January 1, 1873. This sudden change impacted the year's regular events, which elicited strong reactions.

2.1.2 *Construction of a New Observatory*

The Bureau of Astronomy in the Ministry of Education planned to build a new astronomical observatory with three functions: (i) yearly calendar making, (ii) keeping time and time information, and (iii) determination of latitude and longitude. During the same time, the Ministry of Interior Affairs and the Ministry of the Navy had different plans for the new observatory and saw it serving different purposes, so that confusion surrounding it lasted several years (Tokyo University 1987).

Finally, after negotiation within the government, the TAO was established in 1887. It was an attached institute of the College of Science, Imperial University, which managed all three of the aforementioned functions. The observatory was first located on the campus of the Imperial University in Hongo, Bunkyo-ku, and then moved to Azabu, Minato-ku, both in Tokyo, in 1893.

The main instruments were a Repsold transit instrument (aperture 13.5 cm), Merz and Repsold meridian circle (aperture 12.3 cm, diameter of circle 56.4 cm), Troughton and Simms equator (aperture 20 cm), and Merz equator (aperture 16.2 cm) (Fig. 2.1).

With the development of astronomical observations, the Azabu campus underwent intensive construction of additional astronomical facilities and, in addition, the night sky became too bright due to Tokyo's city lights. The TAO moved

Fig. 2.1 Main campus of Astronomical Observatory at Azabu, Tokyo (Tokyo University 1987)

from Azabu to Mitaka Village (now Mitaka City), around 20 km west of Azabu, in 1924, as will be seen in Chap. 3.

2.1.3 Higher Education in Early Meiji Era

In the late Edo period, there were three principal schools for higher education established by the Tokugawa government: Kaisei Gakko (開成学校, Kaisei School), which primarily taught foreign languages and science, Igakko (医学校, Medical school), which taught Chinese medicine (漢方医学), and Shoheiko (昌平黌, Shohei School), where Japanese and Chinese classics were the main focus.

The Meiji government planned for the first time to establish a comprehensive university in the Western style. However, this was far from simple. For this purpose, numerous foreign instructors were hired to conduct a variety of specialized courses.

After some preparation, Kaisei and medical schools were unified into the Imperial University on April 12, 1877. The Kaisei School became the core of the new university with its Colleges of Law, Science, and Literature. The medical school later joined the university as the College of Medicine. Shoheiko was abandoned due to internal conflicts. At first, the faculties offered 8-year courses divided into 4 years of preparatory work at the preparatory school and 4 years of specialization at the university.

In 1897, a new university system was enacted, and the university was renamed the Tokyo Imperial University, and the Kyoto Imperial University was newly established. As a preparatory course for the university, a high school system with 3-year courses was established in the same year. This high school system is referred to as high school (old) throughout this book, in order to distinguish from the high school in the postwar time (Chap. 8).

The third university was established in Sendai, Miyagi Prefecture, as Tohoku Imperial University in 1907, and seven Imperial Universities in total were established by 1939. Of these, astronomy was promoted at Tokyo, Kyoto, and Tohoku Imperial Universities. The TAO was attached to Tokyo Imperial University.

2.1.4 Taisho Period (1912–1926)

Emperor Meiji died in July 1912, and Emperor Taisho (1879–1926) acceded to the throne. The Taisho era, lasting 15 years, may be characterized as a transformative stage from a clanship state policy to the age of party cabinet. Since the emperor had been sickly and mentally unstable, society was not in a suitable state. Nevertheless, modernization of lifestyles penetrated entirely from the upper to the middle classes. This period is often called the age of *Taisho democracy* as the Japanese people enjoyed a climate of political and social liberation (Gordon 2003).

In astronomy, Japan still focused on foreign studies as in the late Meiji period. In this chapter, a brief history of positional astronomy and the early stage of astrophysics in the Meiji and Taisho eras are presented (Nakamura 2008).

2.2 Positional Astronomy

2.2.1 E. Lépissier and H. M. Paul, First Foreign Instructors

The Meiji government took up the modernization of astronomy at Kaisei Gakko (Nakamura 2008). Emile Lépissier (1824–1874), who was a former Deputy Astronomer at the Paris Observatory, was invited to be the first professor of astronomy at Kaisei Gakko in 1870. He taught positional astronomy to about 15 students until 1873. Besides teaching, he was consulted on plans for the new observatory, but he retired in 1874 due to illness. In 1877, Kaisei School was reorganized as Imperial University, and the Departments of Physics and Astronomy were installed in the College of Science the following year. In parallel, an astronomical-meteorological observatory was built on campus. The first professor of astronomy in the Department of Science at Imperial University was Henry M. Paul (1851–1931), who had been working at the Naval Observatory in Washington, D.C., in the fields of positional astronomy and geodetic surveying. In Japan, he served as a professor of astronomy at Imperial University from 1880 to 1883 (Watanabe 1975). Although his lectures were limited to spherical astronomy and geodesy, he cultivated several able students, including Tanakadate Aikitsu (Sect. 2.3.2). In 1883, Paul returned to the Naval Observatory, his home institute.

2.2.2 Terao Hisashi and Position Astronomy

Terao was the first Japanese professor of astronomy at Tokyo Imperial University, and he contributed to the progress of positional astronomy in Japan (Hirayama et al. 1923).

Terao Hisashi (寺尾寿, 1855–1923) was born in the city of Fukuoka (at present) in Kyushu as the eldest son of Terao Kiheita (寺尾喜平太), a samurai of the Fukuoka clan in the Edo era. Terao entered Kaisei School in 1874, just after the retirement of Lépissier. After graduation, Terao studied positional astronomy at the Paris Observatory from 1879 to 1883, in a government study-abroad program. In Paris, he learned astronomy at the Paris Observatory under Francoix Felix Tisserand (1845–1896), known as the author of the *Treatise on Celestial Mechanics*.

In 1882, he participated in a French expedition to Martinique Island in the Caribbean Sea for the purpose of observing the transit of Venus on the solar disk. Thereafter he visited several observatories in the USA and returned to Japan in 1883,

Fig. 2.2 Portrait of Terao Hisashi, painted by Kuroda Kiyoteru (黒田清輝) in around 1913 (Nakamura 2008))

His first work was to establish the Tokyo School of Science (now Tokyo University of Science, 東京理科大学) and served as the first rector of the school. The following year, however, Terao was appointed professor of astronomy at Tokyo Imperial University as Henry Paul's successor. In 1888, he was also appointed Director of Tokyo Astronomical Observatory (TAO), serving until 1919. Among his students, Hirayama Kiyotsugu (平山清次), Kimura Hisashi (木村栄), Sotome Kiyofusa (早乙女清房), and Hirayama Shin (平山信) played important roles in the history of modern astronomy in Japan. Terao measured latitude by making use of a new meridian circle at Sendai in 1883. He also organized the first Japanese expedition to observe a solar eclipse in India in 1898 (Fig. 2.2).

Following retirement from TAO, he spent his old age at Ito, Shizuoka Prefecture, enjoying reading books. He died in 1923 at age 67.

2.2.3 Kimura Hisashi and Latitude Variations

Kimura Hisashi (木村栄, 1870–1943) was born in Izumino (泉野, now in the city of Kanazawa, Ishikawa Prefecture). He learned astronomy from Terao and geophysics from Tanakadate Aikitsu at the Tokyo Imperial University. He devoted most of his career to the study and measurement of latitude variations at the International Latitude Observatory (緯度観測所), Mizusawa, Iwate Prefecture (Hashimoto 1943; Yokoyama 2008) (Fig. 2.3).

In 1891, Seth Carlo Chandler (1846–1913) at the Longitude Department of the U.S. Coast Survey discovered irregular variations in latitude, which came to be known as Chandler's wobble (Chandler 1891). The Coast Survey organized international stations to study wobble at six stations in the U.S., Turkestan, Japan, and Italy, located along the same latitude of 39 degrees north. Mizusawa was one of the

Fig. 2.3 Photograph of Kimura Hisashi with the Order of Culture (文化勲章) in 1937 (Hashimoto 1943)

stations. Twelve pairs of stars were selected and observed continuously at these stations. The results of the variation in latitude were plotted on the two-dimensional (*x*, *y*) plane of the position of the Earth's pole. In 1902, Kimura found that the variations should be traced in three-dimensional space and added a new *z*-term as follows:

$$\Delta\varphi = x \cos \lambda + y \sin \lambda + z \tag{2.1}$$

where $\Delta\varphi$ indicates the latitude variation; x, y, and z denote the position of the Earth's pole; and λ denotes the west longitude of each observatory (Kimura 1902). The path of the (x, y) position and the mean curve of the z-term are illustrated in Fig. 2.4. The amplitude of x, y variations was around 0.17 arcsec, which corresponds to around 5 m of pole displacement. The existence of *the* z-term indicates that the Earth could not be a solid body. It is now thought that the z-term is due to the presence of a liquid core deep in the Earth's interior, though there remain problems to be studied to understand its real origin.

For this discovery he won the Gold Medal of the Royal Astronomical Society in 1936, and the Order of Culture (文化勲章) from the Japanese government in 1937. He died in Tokyo at age 73.

2.2.4 Hirayama Shin and Astronomy

Hirayama Shin (平山信, 1868–1945) worked on positional astronomy in the Meiji era and on astrophysics from Taisho to the early Showa era (1918–1932) (Hagihara 1947).

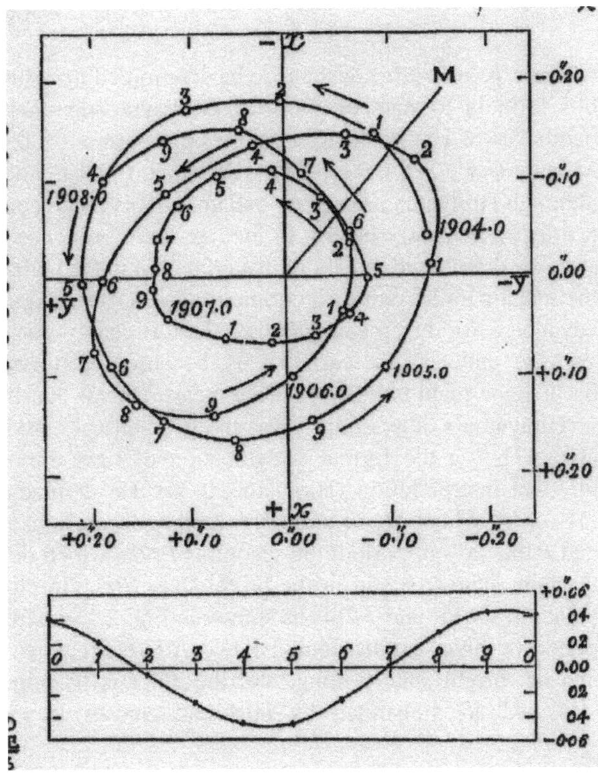

Fig. 2.4 Variation of the North Pole. Upper panel: displacement of pole from 1904 through 1908, each year is divided into ten parts, M indicates the direction to Mizusawa. Lower panel: average curve of z-term for 8 years (Kimura 1908)

Hirayama Shin was born in Tokyo as a son of a samurai family. He learned astronomy from Terao Hisashi in the Department of Astronomy, Tokyo Imperial University. In 1888, he graduated from university and the next year was sent to Europe to study astronomy. He first stayed at Greenwich Observatory to study positional astronomy and then moved to Potsdam Observatory in 1890. In Potsdam, he learned astrophysics from Herman C. Vogel (1841–1907), who was working on the revision of the spectral classification of stars at that time. After returning to Japan, Hirayama was appointed professor of astronomy at the Tokyo Imperial University, where he taught positional astronomy, celestial mechanics, geodetic surveying, and, later, astrophysics. In the Meiji era, his research interests were restricted to positional astronomy, surveying asteroids, and solar eclipse observations.

From 1918, Hirayama's interest turned to astrophysics, particularly statistical astronomy. His first work in the field was the measurement of mean parallax of stars, for every spectral type from B to K and M stars.

The mean parallax was defined by Jacobus C. Kapteyn (1851–1922) of Groningen University, the Netherlands, as a tool to estimate the distance of stars lying beyond the limit of trigonometric parallax. For stars with magnitude m and proper motion μ (arcsecond per year), the mean parallax $<\pi>$ is defined by

$$\log \langle \pi \rangle = a + bm + c \log \mu \tag{2.2}$$

where a, b, and c are constants to be determined from observation. Hirayama applied this formula to stars of every spectral type separately and found that the mean parallaxes of F, G, and K stars are larger ($\langle\pi\rangle \backsim 0.05$) than those of B, A, and M stars ($\langle\pi\rangle \backsim 0.02$–$0.01$) (Hirayama 1922b). The reason for this difference, although Hirayama offered no explanation, is made comprehensible by the fact that nearby stars mostly belong to the late main-sequence stars, whereas B, A, and M stars are distributed in a more distant region in the Galaxy. In addition, he compared the mean parallax with trigonometric and spectroscopic parallaxes and claimed an advantage for his mean parallax. Unfortunately for him, the role of the *mean parallax* ended in the early 1920s, because spectroscopic and Cepheid parallaxes became the main tools of distance measurements starting at that time.

Hirayama's other subject was the geometrical classification of long-period variables. This is the formal classification of light curves without entering into its physical interpretation (Hirayama 1924). He defined two parameters: the first is $(M-m)/P$, where $M-m$ indicates the days from light maximum to light minimum and P the days of period; the second is Ft/Br, which denotes the ratio of the days in the faint state (Ft) and in the bright state (Br). He plotted observed points on the plane $(M-m)/P$ and Ft/Br, as shown in Fig. 2.5, where it is apparent that the long-period variables are distributed in two different regions: one is that $(M-m)/P$ is large and the bright state is long, and the other is its opposite. The two regions thus classified are illustrated by dots and crosses in Fig. 2.5. Philip (1918) also

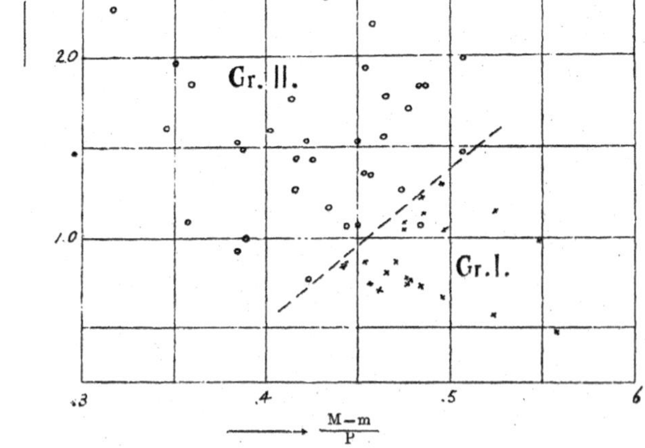

Fig. 2.5 The geometrical classifications of long-period variables by Hirayama (dotted and crossed points) and by Philip (dashed line and Gr. I and II) are illustrated (Hirayama 1924; Philip 1918)

geometrically classified long-period stars into two groups Gr.I and Gr. II separated by a dotted line in Fig. 2.5.

At the TAO, Hirayama was appointed Second Director in 1919, succeeding Terao, and served until his retirement in 1928. As Director of TAO he was elected member of the executive committee of the International Astronomical Union (IAU), twice in 1925 and 1928, as will be seen in Chap. 3.

In 1945 he died in Tokyo at age 77.

2.2.5 Hirayama Kiyotsugu and Asteroids

Hirayama Kiyotsugu (平山清次, 1874–1943) contributed to celestial mechanics, observations of asteroids, and study of the history of astronomy (Hagihara 1943). He was born in Sendai and learned astronomy at Tokyo Imperial University. Upon completion of his study he was appointed associate professor (1906) and professor (1919) of astronomy at the same university. Among his disciples, Hagihara Yusuke (萩原雄祐), Kaburaki Masaki (鏑木政岐), and Hirose Hideo (広瀬秀雄), all served as professors of astronomy at Tokyo Imperial University. In 1921, he moved to the TAO as an engineer of the calendar and time until his retirement in 1935. He died in 1943 in Tokyo at age 68 (Yoshida 2019) (Fig. 2.6).

In 1916, Hirayama was sent to the United States by the Japanese government and worked on nautical almanacs at the Naval Observatory in Washington, D.C. He then moved to Yale University to develop lunar theory. During his stay at Yale, Hirayama (1918) wrote a paper, "Groups of Asteroids Probably Having Common Origins." He found three families of asteroids, each consisting of 16–22 asteroids that share similar orbital elements, such as a semimajor axis, eccentricity, and orbital inclination (Hirayama 1918). Hirayama suggested that the asteroids

Fig. 2.6 Photograph of Hirayama Kiyotsugu (Hagihara 1943)

belonging to these families had the same origin, namely, that they were created from one or two large bodies by unknown cause (Hirayama 1922a). After returning to Japan, he continued his asteroid searches and discovered most of the major families known today.

Hirayama worked as an engineer at the TAO from 1921 on calendar-making and time systems. In parallel with these routine works he was also interested in the history of astronomy, and he wrote a number of essays on the calendar and time systems in the world. Some of them are collected in his essays *Calendar and Time* (Hirayama 1938). In1920, Kanda Shigeru (神田茂, 1894–1974) entered the TAO, as chief editor of the annual *Chronological Science Tables* (理科年表). In parallel, Kanda worked on the history of astronomy in Japan and collected astronomical data (Chap. 3, Sect. 3.9).

2.3 Dawn of Astrophysics and Geophysics

In the Meiji period, astrophysics was still not a distinct subject from physics and geophysics. Takamine Toshio (see below) wrote in his reminiscences:

> "In the history of physics in Japan, the first developed field was geophysics, which was promoted by Tanakadate Aikitsu and Nagaoka Hantaro, and next was spectroscopy, also promoted by Nagaoka" (Fujioka 1964)

This was affected by the first foreign instructor, Thomas C. Mendenhall, who was a specialist of geophysics and spectroscopy. Among the first generation that studied under Mendenhall were Tanakadate Aikitsu and Nagaoka Hantaro. Both worked extensively in physics and geophysics and trained the next generation, such as Takamine Toshio, Terada Torahiko, Ishiwara Jun, and others. However, they worked exclusively in the field of physics, without a direct connection to astronomy.

The dawn of astrophysics among astronomers began with scientists associated with foreign studies, such as Hirayama Shin, Shinjo Shinzo, and Ichinohe Naozo. Of these, Hirayama worked on positional astronomy in the Meiji era under Terao, Director of the TAO, while Shinjo and Ichinohe were deeply involved in the development of astrophysics in Western countries and devoted efforts to promoting astrophysics in Kyoto and Tokyo, respectively.

In the discussion that follows, the lives and works of these pioneers are successively presented.

2.3.1 T. C. Mendenhall, First Professor of Physics

In 1878, Thomas Corwin Mendenhall (1841–1924), a professor of physics and mechanics at Ohio State University, was recruited to help modernize Meiji Era Japan as one of the hired foreign instructors, and he served as a professor of physics

Fig. 2.7 Photograph of T.C. Mendenhall (center) and students of Department of Physics in 1881. His interpreter is sitting to his left (Nakamura 2008)

at Imperial University in Tokyo (Crew 1934). Because his main field of study was geophysics, he established a meteorological observatory on campus to conduct systematic observations of weather and solar phenomena during his stay in Japan. He also measured surface gravity at sea level and at the summit of Mt. Fuji, 3776m high from the sea level, with his student Tanakadate Aikitsu. He estimated the density of the Earth, based on the difference in gravity at different heights of observed points from the sea level. Mendenhall also made a series of elaborate measurements of the wavelengths of the solar spectrum employing the best grating spectrograph of the time. The results of his measurements of Fraunhofer lines were published in the *Memoirs of Imperial University* in 1881. He also became interested in earthquakes while in Japan, and he was one of the founders of the Seismological Society of Japan.

In 1881, Mendenhall was recalled to Ohio State University as the director of the newly established Ohio State Meteorological Service. He then successively served as President of the Rose Polytechnic Institute in Terre Haute, Indiana, Superintendent of the U.S. Coast and Geodetic Survey in Washington, D.C., and President of the Worcester Polytechnic Institute. He died in Ohio in 1924 at age 83 (Fig. 2.7).

2.3.2 Tanakadate Aikitsu and Geophysics

Tanakadate Aikitsu (田中館愛橘, 1856–1952) was a physicist with diverse interests who made a significant impact in Japan (Hagihara and Hashimoto 1952; Yoshida 1999). He was born in Ninohe (二戸), Iwate Prefecture. He had originally intended to study the traditional samurai duties of governing the country. However, he entered Imperial University in Tokyo in 1879, where he majored in physics under Mendenhall and studied geophysics under another foreign professor, Cargill G. Knott (1856–1922, Scottish physicist and geophysicist), focusing in particular on geomagnetism and seismology. Having graduated in 1882, Tanakadate

Fig. 2.8 Portrait of Tanakadate Aikitsu (Science Museum of Tanakadate Aikitsu Memorials, 田中館愛橘記念科学館、二戸市)

was appointed lecturer and then associate professor in 1883 at the university, and he started a geomagnetic survey of Japan as an assistant of Knott. Together they delineated the first earthquake hazard map of Japan.

Tanakadate was sent to Europe by the government from 1888 through 1891. He first studied under William T. Kelvin at Glasgow University, focusing on electricity and magnetism. He then moved to Berlin University for 1 year before returning to Japan (Fig. 2.8).

From 1893 to 1896, Tanakadate carried out a survey of gravity and geomagnetism in Japan for geophysical research. He founded the Institute of Seismology at Tokyo Imperial University. The International Latitude Observatory at Mizusawa was also established in 1899 prompted by his proposal. Kimura Hisashi was one of his students. Tanakadate was also an early proponent of aviation. In the Russo-Japanese war (1904–1905), he served as an advisor to the Japanese army and recommended the use of air balloons for military reconnaissance purposes. This led to the establishment of the Aviation Laboratory at Tokyo Imperial University. Tanakadate participated in the Paris Conference in 1906 on the organization of the metric system of measures and weights, which led to a bill on the metric system in Japan.

Tanakadate's contributions to science and technology, teaching, and cultural understanding were recognized by the Japanese government, which awarded him the Order of Culture (文化勲章) in 1944. He died in 1952 at age 86.

2.3.3 Nagaoka Hantaro and Physics

Nagaoka Hantaro (長岡半太郎, 1865–1950) is one of the pioneers in physics and geophysics in Japan during the Meiji period (Itakura 1976; Fujioka 1963).

He was born in Nagasaki, Kyushu, and learned physics at Tokyo Imperial University. Upon graduating in 1887, Nagaoka began to work on magnetism, with a particular interest in magnetostriction (deformation of a metallic solid or liquid upon the onset of magnetization) in nickel and iron.

While a graduate student, Nagaoka conducted an experiment on Herz's radio wave experiment. In 1888, Heinrich R. Hertz (1857–1894) conducted the first experiment on the transmission and reception of radio waves in the laboratory. Upon hearing this news, Nagaoka immediately constructed a new device and similarly confirmed the transmission and reception of radio waves in 1889. He presented his experiment at a meeting of the Physical Society of Japan, and his presentation made a deep impression on the several hundred attendants (Fig. 2.9).

In 1893, Nagaoka was sent to Germany for 3 years to study physics, mostly at the University of Berlin, where he continued his study of magnetostriction along with atomic and molecular structures, based on the mathematical theory of molecules developed by James Clerk Maxwell (1831–1879). In 1894, he moved to Munich University for 1 year to study the kinetic theory of gases under Ludwig Boltzmann (1844–1906). After returning to Berlin University, he studied hydrodynamics and electromagnetic theory under Max Planck (1856–1947).

After visiting Vienna and London, he returned to Japan in 1896 and was appointed professor of physics at Tokyo Imperial University, where he served until his retirement in 1925.

In the early twentieth century, physicists began to investigate the structure of atoms. In 1904, Joseph John Thomson (1856–1940) at Trinity College of Cambridge University proposed the so-called plum-pudding model, where an atom was a uniform sphere with a positive charge, with electrons scattered freely inside the sphere, like plums in a pudding. Nagaoka could not accept Thomson's model, since he thought that opposite charges repel each other. In contrast, Nagaoka proposed an alternative Saturnian model of the atom in which a positively charged central core is

Fig. 2.9 Portrait of Nagaoka Hantaro (Nagaoka Hantaro Memorial Museum, Yokosuka, 長岡半太郎記念館、横須賀市)

surrounded by revolving electrons, in a manner like Saturn and its rings (Nagaoka 1904).

In addition to the study of atomic structure, Nagaoka undertook the spectroscopic study of Zeeman effects for some metals. A Zeeman effect appears as the splitting of a spectral line into several components in the presence of a magnetic field. Nagaoka was highly impressed by the discovery of George Ellery Hale (1868–1938) in 1908 of the existence of strong magnetic fields in sunspots, which were derived using the Zeeman effect behaviors of many Fraunhofer lines. Hale found the maximum strength of the field ranged from 2900 to 4500 Gauss in different spots (Hale 1908a, b).

In the early twentieth century, spectroscopic observations were not yet initiated in Japan. Impressed by the role of spectroscopy and the Zeeman effect, Nagaoka wrote in the Astronomical Herald (Nagaoka 1909):

> Recently, the application of optics, especially spectroscopy, to the study of the physical state of celestial objects has been gaining more and more importance. Spectroscopy is the oldest and the most useful application of optics.

In the article, Nagaoka explained the mechanism of Zeeman effects in spectroscopy and its importance in astrophysics.

From a physical point of view, he was also interested in astrophysics, and some of his results were presented in the *Astrophysical Journal*. One example is "Vacuum Arc for Obtaining Spectra Extending from Visible Light to Soft X-Rays" (Nagaoka and Sugiura 1923).

Geophysics was also one of Nagaoka's research subjects. He studied geomagnetism under C. S. Knott at Tokyo Imperial University. When Knott and Tanakadate planned to conduct geomagnetic surveys all over Japan in 1887, Nagaoka accompanied Knott to survey eastern Japan. When a great earthquake occurred in Aichi and Gifu Prefectures (Nobi earthquake, 濃尾地震), causing severe and widespread damage in 1891, Tanakadate and Nagaoka analyzed the variation of geomagnetism before and after the earthquake and discovered a relationship between them. The following year, the Meiji government established the Investigation Council for the Prevention of Earthquake (震災予防調査会) to which Tanakadate and Nagaoka were both appointed as members.

When Nagaoka returned from Europe in 1896, he brought a Repsold gravimeter that was of greater precision than that used by Tanakadate in the early 1880s. He used this gravimeter to measure the gravity at Kyoto, Kanazawa, Tokyo, and Mizusawa in collaboration with his student Shinjo Shinzo. Nagaoka also made comparative measurements of the gravitational force at Potsdam and Tokyo, from which the study of the large-scale distribution of gravity in Europe and Japan became possible. Subsequently the work of Nagaoka was taken over by Shinjo (next section).

After retirement from Tokyo University in 1925, Nagaoka was appointed head scientist of Riken (理研, Institute of Physical and Chemical Research) and then served as the first president of Osaka Imperial University, from 1931 to 1934. In 1925, he was awarded an honorary doctorate together with W. W. Campbell,

B. Baillaud, W. de Sitter, and F. Schlesinger, from the IAU at the General Assembly held at Cambridge, England. For his achievements in science, he was awarded the Order of Culture (文化勲章) by the Japanese government in 1937. He died in 1950 in Tokyo at age 85.

2.3.4 Takamine Toshio and Spectroscopy

Takamine Toshio (高嶺俊夫, 1885–1959) followed up on the works of Nagaoka in spectroscopy, particularly the Stark effect (Fujioka 1964; De Vorkin 2002).

Takamine was born in Tokyo and was the second son of Takamine Hideo (高嶺秀夫), a samurai of the Matsudaira clan at Aizu-Wakamatsu, Fukushima Prefecture.

In 1906, Takamine entered Tokyo Imperial University and studied physics under Nagaoka Hantaro. One day, during a lecture on spectroscopy, Nagaoka showed one of his solar spectra, which he himself had observed, and murmured, "Why doesn't anyone study the solar spectrum?" This comment excited Takamine's mind, and he decided to study spectroscopy at the graduate level under Nagaoka.

In 1915, Takamine was appointed associate professor at Kyoto Imperial University and worked on the Stark effect for hydrogen, helium, and other elements. The Stark effect, which was discovered in 1913 by Johannes Stark in Germany, is the shifting and splitting of spectral lines of atoms due to the presence of an external electric field. In 1918, Takamine traveled to the Optical Laboratory of the Mt. Wilson Observatory at Pasadena, California, and measured the Stark effect of metallic lines as a laboratory astrophysicist. This work was published in the *Astrophysical Journal* (Takamine 1919). By then he was developing an interest in stellar physics and started collaborating with astrophysicists in Japan (Fig. 2.10).

In 1921, Takamine returned to Tokyo and began working at Riken (理研 = 理化学研究所, Institute of Physical and Chemical Research), which had been established in 1917 but was not fully equipped with laboratory systems and buildings. This was still true in 1921. Nagaoka served as the head of the physics section, where three laboratories were staffed with three chief researchers, Nagaoka, Takamine, and Nishikawa Shoji (西川正治). In his laboratory, Takamine promoted the study of spectroscopy. In particular, he built a vacuum spectrograph, by which he measured the line and continuum spectra in a wavelength range of 3000 to 900 Å. He thus identified the lines of Lyman series, ionized neon, oxygen, silicon, and others (Takamine 1941). He then became close friends with Hisose Hideo (広瀬秀雄) of the TAO and Fujita Yosio (藤田良雄) of Tokyo University to deepen the relationship between physicists and astronomers through spectroscopy. He was a physicist as well as a laboratory astrophysicist and a pioneer of astrophysical spectroscopy in Japan in the 1920s and 1930s.

He died in Tokyo in 1959 at age 74.

Fig. 2.10 Portrait of Takamine Toshio (Fujioka 1964)

2.4 Shinjo Shinzo and Astrophysics

2.4.1 Life and Works

Astrophysics at Kyoto University was initiated and promoted by Shinjo Shinzo (新城新蔵 1873–1938) (Kabumoto 1996; Noda 1979; Yabuuchi 1992).

Shinjo was born in the city of Aizu-Wakamatsu, Fukushima Prefecture, in 1873, the sixth son of Shinjo Hyoemon (新城平右衛門), a sake brewer, and his new son was named Shinzo (*new cellar* in Japanese) in celebration of the construction of a new sake cellar that year.

Shinzo was a boy genius and went through primary and secondary school at an accelerated pace. He entered Dai-Ni High School (old) in Sendai at the age of 15. After 3 years he enrolled in the Science College of the Tokyo Imperial University and studied physics and geophysics under Tanakadate Aikitsu. Shinjo concentrated in his studies on geodetic surveying. After graduation from the university, he was appointed lecturer in physics at the Military Academy of the Army, and he also served as a geodetic surveyor in the Ministry of Education (Fig. 2.11).

He actively undertook the survey of geomagnetism and surface gravity in many island locations in Japan and even in China. On his survey activity, Nagaoka Hantaro, Shinjo's senior and collaborator, said the following in his memorial address for Shinjo:

> He [Shinjo] has been working as a lecturer of physics at the Military Academy of the Army. After returning from foreign study, in around 1897, I felt the need to obtain gravity measurements in Tokyo and other important places. In carrying out this work with him, we had daily discussions and then, together with Ohtani Ryokichi (大谷亮吉), we measured the absolute gravity in Tokyo, Kyoto, Mizusawa, and other places. Later on, we measured

Fig. 2.11 Photograph of Shinjo Shinzo in China, December 1932. (Shinjo Bunko, No. 649, Kyoto University)

> the relative gravity between Potsdam and Tokyo, and this was applied in many points in Japan. In those days, since we could not use radio time signals, we had to measure the time using a transit instrument at every point of gravity measurement. He demonstrated an excellent ability in this work, and he established the competent for his future as an astronomer. (Nagaoka 1939)

In 1898 Shinjo married Watanabe Waka (渡辺わか), the third daughter of Watanabe Nozomu, the descendant of a samurai. Shinjo moved to Kyoto to serve as an associate professor of physics at the Science College of Kyoto Imperial University. While measurements of gravity remained a research interest, he gradually moved to the study of astronomy.

Shinjo obtained a private fellowship from Fujihara Chu'ichiro (藤原忠一郎), a wealthy merchant in Kyoto, and traveled to Europe in January 1905. He mainly stayed at Göttingen University for the study of physics under Woldemar Voigt (1850–1919), who was working on crystal physics and electro-optics at the time. In Göttingen, Shinjo also attended the lectures of Karl Schwarzschild (1873–1916) on astrophysics. Karl was the director of the Göttingen University Observatory and worked on the photometry and theory of solar and stellar atmospheres in this period. Shinjo was deeply impressed by his lectures on new developments in astrophysics. Shinjo's eyes were opened to astrophysics. He came to believe strongly that the future of astronomy should consist entirely of astrophysics. In his day, the study of astrophysics had not yet been launched in Kyoto. In Europe, he spent much of his time visiting many observatories and universities in order to widen his knowledge on astrophysics as well as on geodesy.

After returning to Kyoto, he began delivering lectures on astrophysics. In 1907, he was promoted to professor of physics in the Department of Physics, College of Science, Kyoto Imperial University. While he continued with his measurements of gravity in many parts of the Japanese islands, he began to develop his own theoretical model on the evolution of the Universe.

In 1921, under the leadership of Shinjo, a new Department of Astronomy was established. The Ministry of Education initially opposed the establishment of a new Department on astronomy, since the ministry thought that one astronomy department in Tokyo was enough in Japan. To persuade the ministry, a new idea was needed. Shinjo asserted the importance of astrophysics as a new science that would develop in a way that was different from classical astronomy. Thus was established the Department of Astrophysics with two chairs. As the first chair, Shinjo taught astrophysics and the history of oriental astronomy, and Yamamoto Issei (山本一清, Chap. 4), an associate professor, taught observational astronomy. In the next year, the second chair was created, in which Ueta Joe (上田穣), who had been invited from the International Latitude Observatory, Mizusawa, taught positional astronomy.

The idea of building a new astronomical observatory was pushed by Shinjo. In 1925, he established the Kyoto University Observatory equipped with a Sartorius 18-cm refractor in a corner of the university campus. In those days, the city of Kyoto launched a plan to consolidate the street and transportation system of the city. On just the west side of the observatory, a new boulevard was constructed, and city trams started to run. The observational environment deteriorated considerably. The issue of moving the observatory to a better site became an urgent one. Under Shinjo's efforts, a new observatory was constructed in 1929 at the summit of Mt. Kwasan equipped with a Cooke 30-cm refractor and Sartorius 18-cm refractor as the main telescopes; it was named the Kwasan Observatory, and its first director was Yamamoto Issei.

In 1929, Shinjo was elected eighth president of Kyoto Imperial University for a four-year term. During his tenure, Japan became increasingly militaristic, *A History of 100 Years of Kyoto University*, published in 1998, describes these years as an age of national distress. Japan broke out war against China and occupied some of its territory. Shinjo, in cooperation with former presidents, fought for academic freedom and autonomy at the university. After his term as president, he was appointed Director of the Shanghai Institute of Science, which had been newly established in Shanghai. The most important issue he was assigned to address was the preservation of scientific data and documents in China and preventing their destruction in the war. However, he died suddenly in 1938 at a hospital in Shanghai during his tenure at the age of 65.

In a memorial addresses by his supporters, Araki Toshima (荒木俊馬) wrote that "Shinjo devoted his life to science with genuine devotion to the truth," and Noda Churyo (能田忠亮) wrote that "Shinjo was a man of deep and tender affection and was kind to all his friends. He was also an indulgent parent to his children at his home."

2.4.2 *The Meteoroid Theory of Stellar Evolution*

In the 1910s, Shinjo started contemplating an idea on the evolution of stars based on his own observations of meteors. His initial aim was to explain the mechanism of Cepheid variables, but gradually, Shinjo extended this idea to the evolution of stars in general. His basic assumption on was the condensation, rotation, and conservation of the angular momentum of meteoroid clouds.

2.4.2.1 Universal Existence of Meteoroids

Shinjo first observed the inflow of meteors from space into the terrestrial atmosphere. He estimated the total mass of meteors falling into the atmosphere through an excess of rotational velocity of the upper atmosphere affected by the falling meteorites. He also assumed that meteors are ranging widely in size, from 1 μm up to several 10 km. Furthermore, he regarded all meteor showers, comets, zodiacal light, and Saturn's rings as being composed of meteoroids. Outside the Solar System, dark clouds may be a conglomeration of meteoroids (Shinjo 1914).

In this way, he claimed that meteoroids were ubiquitously distributed in space and that stars were formed in gravitationally condensed parts of the Universe. Condensation of meteoroid cloud takes place in usual accompanying some overall rotation. The combination of condensation and rotation in the meteoroid clouds is the basic factor for the formation of stars and their variabilities. Whether the primordial matter is gas or meteoroid could be distinguished by the existence of average angular momentum in the primary conglomeration. The overall rotation observable in most celestial objects yields reliable evidence for the meteoroid origin of star formation.

2.4.2.2 Theory of Stellar Variability

Shinjo proposed an eccentric-nucleus theory of Cepheid variables, against the pulsation theory of Harlow Shapley (1914). Shinjo supposed that the primordial state of stellar evolution was a kind of meteoric swarm of immense dimension, and the subsequent evolution consisted of progressive condensation as a result of mutual attraction. Each condensation episode may develop in several different ways, based on the initial values of mass and angular momentum. Systems with large angular momenta would naturally condense into two or more separate nuclei and, hence, develop into binary or multiple systems, while those with comparatively small values of angular momentum may condense into a single star. In the latter case, however, Shinjo supposed that the condensing nuclei might not coincide with the centers of mass of the whole system. This notion of eccentric condensation of nuclei was the starting point of Shinjo's theory on the nature of Cepheid variables (Shinjo 1922a, b).

As a simple case, he considered a system of meteoric swarm as shown in Fig. 2.12, where the system is composed of a central sphere (center O_1) and an eccentric nucleus (center O, radius R, mass m, mean density ρ). The central sphere is

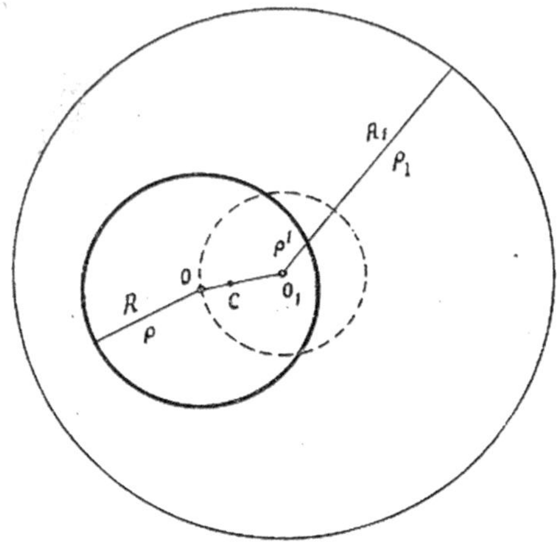

Fig. 2.12 Shinjo's eccentric-nucleus model (Shinjo 1922a)

divided into an outer part (radius R_1, mass m_1, mean density ρ_1) and an inner core (radius $R\,'$, mass m', mean density ρ'). Point C indicates the center of mass of the whole system. By setting $m_1/m = \alpha$ and P the period of revolution of the eccentric nucleus around C, Shinjo (1922a) derived the relation

$$P^2 = \frac{1}{\frac{4}{3}\pi\rho'} \quad \frac{\alpha}{1+\alpha}. \tag{2.3}$$

On the Cepheid variables, Harlow Shapley (1914) had already developed the pulsation model, according to which the pulsation period is related to the mean density of stars in the form

$$P^2 = \frac{1}{\frac{4}{3}\,\pi\,\rho}. \tag{2.4}$$

In his theory, Shinjo obtained a proportional relation similar to Shapley's, and he claimed the priority of his model in two points.

First, the period of a Cepheid variable may be regarded as the fundamental mode of oscillation in the pulsation theory. If the Cepheids are radial oscillators, some of them would show some higher order of oscillation. However, there is no evidence of such higher-order oscillations among the Cepheids. On the other hand, the Cepheids should always be of a single period in his eccentric-nucleus model, which is in good agreement with observations. (For higher-order oscillations, see Chap. 7, Sect. 7.3.1).

Second, the eccentric-nucleus model is applicable to all types of variable stars except Algols. Furthermore, this model describes the basic process of stellar evolution, as shown in the next subsection.

2.4.2.3 Stellar Evolution

The basic assumption of Shinjo's theory of stellar evolution is to regard the Sun as a variable star (Shinjo 1922b, 1931). He claimed that sunspot phenomena and the equatorial acceleration of solar rotation were the results of falling meteors on the surface. The high-speed impact of meteors gave rise to acceleration in equatorial gas flow, where sunspots were formed by the effects of its turbulent motion.

Shinjo compared solar phenomena with other types of variable stars such as Cepheids and long-period variables, which were, according to his view, in an earlier stage of stellar evolution. By extending his meteoroid hypothesis on variable stars, he described the evolutionary path of stars as shown in Fig. 2.13. This figure indicates that a star is born as a long-period variable star, which is very huge, cool, and slowly rotating. The star slowly shrinks as the surface temperature increases and the variation period gets shorter in order from the left ascending branch of M, K, and G up to the B type. Thereafter, the star turns to the right descending branch, where the present Sun exists and continues contracting toward a red dwarf.

Figure 2.13 clearly resembles the so-called Lockyer's arch by Norman Lockyer (see Appendix, Fig. A.3). Lockyer' arch represents the arrangement of his unique chemical classification of stars. Lockyer argued that stellar spectra could be divided into two, ascending- and descending-temperature sequences, like the pattern in Fig. 2.13, and thus proposed the two-way evolution of stars. In contrast, Shinjo arranged the ascending branch in the order of the period of variable stars and the descending branch in the order of contraction toward red dwarfs. He placed the position of the Sun in the middle of the descending branch with a question mark. Though Shinjo did not directly refer to Lockyer's arch, we can guess that he might have seen Lockyer's arch anywhere during his stay in Europe.

Shinjo also considered the nature of planets, satellites, and comets based on the same meteoroid hypothesis and thus built a theory of unified cosmic evolution.

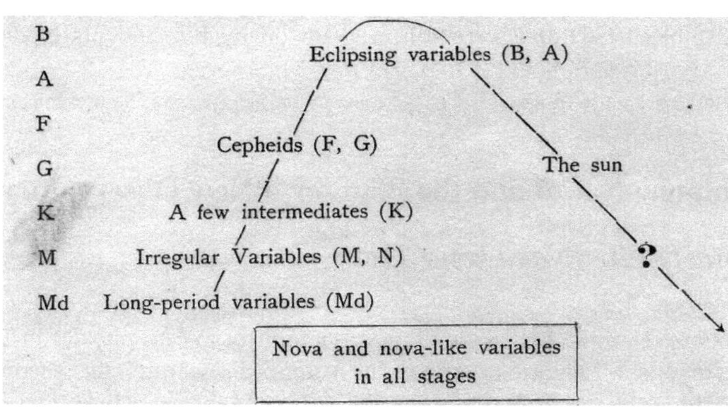

Fig. 2.13 Sequence of variable stars and the path of stellar evolution (Shinjo 1922b)

Although his theory was not advanced thereafter, his study was the first step in astrophysics in Kyoto.

2.4.3 History of Oriental Astronomy

Shinjo also contributed to the study of the history of oriental (mostly Chinese) astronomy. Shinjo's work began when a question was posed in around 1905 by Kano Naoki (狩野直喜 or Kano Kunzan 狩野君山), professor of literature and a friend of Shinjo, on the astronomical chronology of ancient China. Shinjo had been reading Chinese classics from his youth, so he could jump into this study immediately. His method was to find the position of the main constellations from classic literature at specified times and on certain days, such as the evening of the solstice or day of the equinox each year. The position of constellations slowly advances due to the precession of the Earth. With this knowledge, Shinjo could estimate the chronology of some dynasties in ancient China. Shinjo also studied the history of calendar making. In China, the lunisolar calendar had already been adopted in early Zhou (周初, ~1040 B.C.) and was gradually improved. An improvement was made especially with the adoption of the leap month, where an additional month is inserted once every 2 or 3 years to reconcile the seasons with the calendar. As for the origin of leap-month system, Iijma Tadao (飯島忠雄, Gakushuin University in Tokyo) asserted its Greek origin, since the Metonic cycle was discovered by Meton of Athens in 432 B.C. This cycle indicates that the length of 19 solar years corresponds to 235 synodic months. This suggests that seven leap months should be inserted in the calendar every 19 years (Iijima 1939). In contrast, Shinjo insisted on the leap month's Chinese origins, based on the examinations of classical references on the lengths of the solar year and synodic months from the early Zhou to Han Dynasty and showed the gradual adoption of a leap-month system independently of Meton's cycle (Shinjo 1938). While the debate lasted a long time, Yabuuchi Kiyoshi finally settled the problem in favor of Shinjo's view (Yabuuchi 1992).

Shinjo's works were collected in the *History of Astronomy in East Asia* (東洋天文学史) (Shinjo 1938) and were followed up on by his students Noda Churyo, Yabuuchi Kiyoshi, and others (Chap. 4).

2.5 Ichinohe Naozo and the Plan for a New Observatory

2.5.1 Early Life (Nakayama 1989)

Ichinohe Naozo (一戸直蔵, 1878–1920⟩was a pioneer in astrophysics in Tokyo. He was born in Nishi-Tsugaru (西津軽) in Aomori Prefecture, the second son of Ichinohe Tomosaku (一戸友作), a farmer. In the Meiji era, Nishi-Tsugaru was a remote village, located near the coast of the Japan Sea. His father Tomosaku was an influential person, sometimes elected to the village assembly. Tomosaku, however, lacked an understanding of the need to educate his five children. He thought that primary school was adequate for farmers. In opposition to his father, Naozo left the village secretly with his mother's support and went to a middle school in the city of Hirosaki.

In 1896, he went to Dai-Ni High School (old) in Sendai, where he became acquainted with Shinjo Shinzo, four years Naozo's senior, and formed a friendship that lasted the rest of their lives. In 1900 (at age 22), Ichinohe enrolled in the Science College of Tokyo Imperial University and studied positional astronomy under Terao Hisashi and Hirayama Shin. After graduation, he was hired as a technician in the TAO in 1903. His duties at the observatory involved routine work on time keeping and yearly calendar making. Freedom of scientific research was quite limited under the bureaucratic system, to which Ichinohe was unaccustomed.

In his own work, he was strongly attentive to the progress of astronomical observations in the USA.

The Harvard College Observatory changed from a traditional to a modern astronomical observatory when E. C. Pickering was appointed its fourth director in 1876. Pickering took up two big survey projects: photographic photometry and spectral classification of stars in the entire sky.

The Lick Observatory was constructed in 1888, through a bequest from James Lick, a carpenter and piano maker. This observatory was the first permanent mountain-top observatory at Mt. Hamilton near San Francisco. It was equipped with a 91-cm (36-inch) refractor, the largest one in the world at the time. The first discovery made with this telescope was a new satellite of Jupiter. Then spectroscopic observations were promoted by W. W. Campbell (1862–1938), particularly on the radial velocities of stars. The observatory's first director was Edward S. Holden (1846–1914).

The Yerkes Observatory was founded in 1897 by George Ellery Hale (1868–1938) with funding from industrialist Charles T. Yerkes. The observatory, located in Williams Bay, Wisconsin, was operated by the University of Chicago. It houses a 102-cm (40-inch) refractor, the largest ever successfully used for astronomy.

Hale was a pioneer of solar astrophysics in the USA, as well as a capable administrator. He also succeeded at securing the construction funds for the Mount Wilson Observatory. He was subsequently appointed first director of the observatory in 1904.

With the development of these new modern observatories, the center of astrophysical observation essentially moved from Europe to the USA by the end of the nineteenth century.

Ichinohe was conscious of the trend in astronomical research and made up his mind to study astronomy in the USA. In the Meiji period, because governmental grants for foreign study were limited to Western Europe, he saved his money for travel expenses by writing several popular books. In 1904, he married Kikuchi Ine (菊池イ子), and they had a daughter the following year. Leaving his family back home, Ichinohe set out on his own for San Francisco (Fig. 2.14).

Fig. 2.14 Photograph of Ichinohe Naozo (Archive News No. 15, NAOJ)

2.5.2 Study of Astrophysics at Yerkes Observatory

In September 1905, Ichinohe arrived at Yerkes Observatory. Edwin Brant Frost (1868–1935), its second director after G. E. Hale, welcomed the first young astronomer from Japan. He recognized that Ichinohe had no training in astrophysical observations, so Frost set up two research programs for Ichinohe to pursue at the Yerkes Observatory.

The first was photometric observations of the long-period variable star Omicron Ceti (o Ceti) (named Mira, Latin for *wonderful* or *astonishing*, by Johannes Hevelius in his 1662 work *Historiola Mirae Stellae*). He observed Mira using the finder of a 30-cm refractor, opera glass, and the naked eye. Standard stars were selected from the Bonn Photographic Catalogue and the Photometric Catalogue of the Harvard College Observatory. The result of his observations is shown in Fig. 2.15, where epoch of maximum light in December 1906 and somewhat irregular light curve are indicated (Ichinohe 1907a). From 1906 to 1907, he observed seven other variable stars and derived their light curves.

The second research project assigned to Ichinohe was the spectroscopic analysis of close binaries (Ichinohe 1907b). He analyzed spectrograms obtained with a 101-cm refractor for three stars in the constellations Sagittarius, Cancer, and Virgo, all showing marked absorption lines in hydrogen, helium, and ionized calcium. Among them, Fig. 2.16 illustrates the radial velocity curve of μ Sagittae. Based on this curve, Ichinohe derived the orbital elements: orbital period 180.2 days, orbital eccentricity 0.441, and total mass of primary and its companion 13.5 $M_{\odot}/\sin^3 i$, where $M_{\odot}$ and i denote the solar mass and the angle of orbital inclination, respectively.

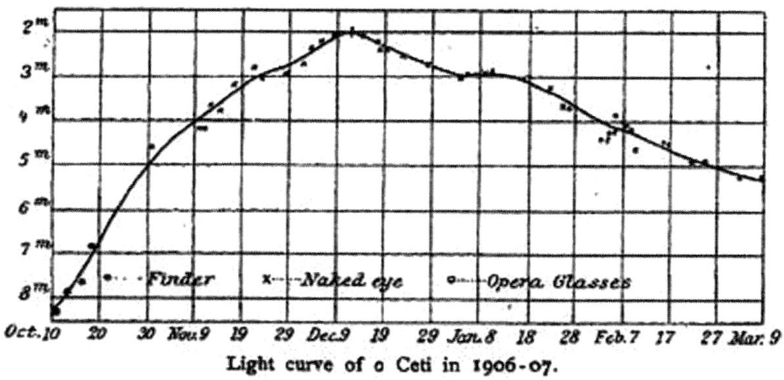

Fig. 2.15 Light curve of o Ceti, observed by Ichinohe (Ichinohe 1907a)

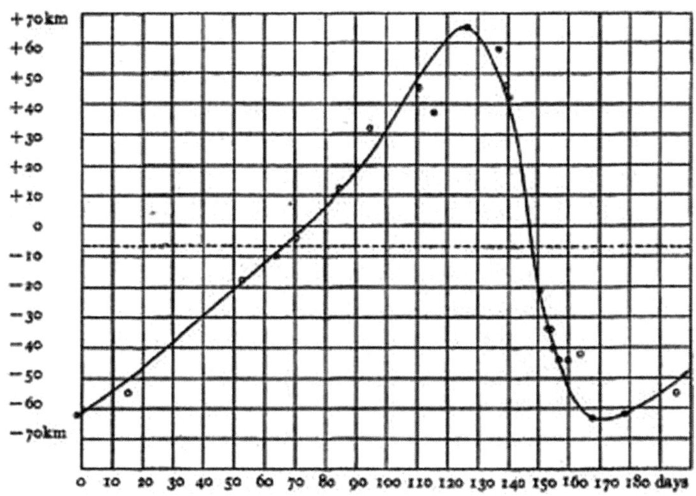

Fig. 2.16 Radial velocity curve of μ Sgr (Ichinohe 1907b)

Since spectroscopic observations require a large telescope and a spectrograph of adequate resolution, Yerkes Observatory made a strong impression on Ichinohe, not only for its instrumentation, but also in its free conception for observational projects. Ichinohe thus became convinced that the future of astronomy in Japan should follow the American style in instrumentation as well as in its operations.

Ichinohe was a talented writer. Before setting off for the Uunited States, he wrote a book and produced a translation. Both were published during his stay in America. The first one was *Koto Tenmon Gaku* (高等天文学, *Advanced Astronomy*), a mathematical treatise on spherical astronomy and the structure of the Solar System. The second one was *Seishin Tenmon Gaku – Uchu Kenkyu* (星辰天文学•宇宙研究, *The Stars: A Study of the Universe*), which is a Japanese translation of Simon

Newcomb's book, published in 1901. Both books represented the first introduction of astrophysics written for high-level readers.

2.5.3 Plan of a New Astronomical Observatory in Taiwan

After returning home in 1907, Ichinohe continued his photometric observations of variable stars, which were the first astrophysical observations of stars in Japan. For spectroscopic observations, a sufficiently large telescope was indispensable, so he began to design the ideal observatory. The basic concepts behind it were as follows:

(i) It should be a modern observatory equipped with a large telescope comparable to that in the USA.
(ii) It should be located on a high mountain.
(iii) It should be located as far south as possible in Japanese territory.
(iv) It should be a private observatory to ensure freedom of research (Fig. 2.17).

As for potential observatory sites, Ichinohe selected the summit of Mt. Niitaka (新高山、now Yushan 玉山, 3997-meters above sea level) in Taiwan. At this juncture of history, Taiwan was under Japanese rule. Ichinohe planned to construct an observatory equipped with two main telescopes, a 90-cm reflector and an 84-cm refractor. Under the financial support of Goto Shinpei (後藤新平), who was the Head of Civilian Affairs of Taiwan and switched jobs in 1907 to become First Director of the South Manchuria Railway, Ichinohe made two exploratory investigations into the selection of the site near the top of Mt. Niitaka. He decided on two suitable sites for domes and a main building. In the end, unfortunately, he lost Goto's support for the project because of its enormous construction costs. He returned to Tokyo expecting to meet with investors who would offer financial support for his

Fig. 2.17 Image of main building designed for Mt. Niitaka Observatory (Archive News No. 35, NAOJ)

plan for the next 20 years. Despite his hopes, he was forced to retire from the TAO in 1911. It was said that Director Terao at that time considered Ichinohe's plan crazy and unrealistic and said it was made without any consideration for costs.

Ichinohe was lonely at the TAO. No one was interested in astrophysics, which was an unallied field for practical astronomers. Moreover, even if Mt. Niitaka Observatory could be brought into existence, it would be almost impossible to maintain and perform modern astrophysical observations. At long last, Ichinohe's plan was implemented more than a half century later in Japan.

Three months after Ichinohe returned home, in January 1908, the Astronomical Society of Japan was founded. He was an outspoken proponent of the society. A meeting of promoters was soon held, and the society was launched under its first president, Terao Hisashi, director of the TAO, and editor in chief of the newly created monthly journal *Tenmon Geppo* (*Astronomical Herald*) was Ichinohe, who wrote many popular articles for amateur astronomers.

2.5.4 Later Life

His first effort after leaving the TAO was to publish a science journal *Gendai no Kagaku* (「現代乃科学」, *Modern Science*) that aimed at high-level popularization and the exchange of scientific information, like the British journal *Nature* and American journal *Science*. The first volume was published in 1913. For this journal, many academic scientists submitted papers and reports, including Terada Torahiko (寺田寅彦), Honda Kotaro (本多光太郎), Kusakabe Shirota (日下部四郎太), and Shnjo Shinzo (新城新蔵). The publication of this journal was beset with financial difficulties, and it was made possible only through Ichinohe's strenuous efforts. In parallel with the publication of *Gendai no Kagaku*, Ichinohe also wrote several popular books, among which *Tsuki* (*The Moon*, 1909) and *Hoshi* (*The Stars*, 1910) were written in a unique style that mixed astrophysics with poetry (Nakayama 1989).

During this period, however, Ichinohe was in an advanced stage of tuberculosis, and her died in November 1920, age 42. Ichinohe's supervisor, Hirayama Kiyotsugu, wrote in his memorial address:

> "I recognized your iron will in everything. At the same time, I was hoping that your strong will would work in such a way as to not injure your health and cause you to come into conflict with others. However, both hopes were dashed. Your life as an astronomer was truly short, like a comet" (Hirayama 1920).

His friend Shinjo Shinzo wrote:

> "An unruly horse sank down without having the chance to give full scope to its genius" (Nakayama 1989).

References

Bartholomev, J. R. (1989). Chapter 3: Formation of Meiji scientific community, and Chapter 4: Laying the institutional foundation of Science. In *The formation of science in Japan*. New Haven/London: Yale University Press.

Chandler, S. C. (1891). On the variation of latitude. *Astronomical Journal, 11*, 59–61.

Crew, H. (1934). *Thomas C. Mendenhall, 1841–1924, biographical memoires*. National Academy of Sciences of the USA, Washington D. C.

De Vorkin, D. (2002). Toshio Takamine's contact with western astrophysics. *Astrophysics and Space Science Library, 275*, 145–157.

Fujioka, Y. (1963). T. Takamine and H. Nagaoka, Pioneers of spectroscopy in Japan (分光研究、藤岡由夫; 高嶺敏夫と長岡半太郎. 日本における分光学の開拓者たち). *Research of Spectroscopy, 12*, 4–7.

Fujioka, Y. (1964). *Takamine Toshio and spectroscopy* ((藤岡由夫: 高嶺敏夫と分光学), Ohyokagaku Kenkyusho (応用光学研究所刊).

Gordon, A. (2003). Chapter 9: Economy and society. In *A modern history of Japan* (pp. 139–160). Oxford University Press.

Hagihara, Y. (1943). Reminiscence of Professor Hirayama Kiyotsugu, and bibliographical notes (天文月報, 萩原雄佑: 平山清次先生の思い出). *Astronomical Herald, 36*, 65–68.

Hagihara, Y. (1947). Obituary of Professor Hirayama Shin (天文月報、萩原雄佑: 平山信先生を偲ぶ). *Astronomical Herald, 40*, 18–20.

Hagihara, Y., & Hashimoto, M. (1952). Obituary of Professor Tanakadate Aikitsu. (天文月報,萩原雄佑及び橋元昌矣: 田中館愛橘博士を悼む). *Astronomical Herald, 45*(7), 99–101.

Hale, G. E. (1908a). On the possible existence of a magnetic field in sunspots. *Astrophysical Journal, 28*, 315–342.

Hale, G. E. (1908b). The Zeeman effect in the Sun. *Publications of the Astronomical Society of the Pacific, 20*, 287–288.

Hashimoto, M. (1943). Reminiscence of Dr. Kimura Hisashi (天文月報, 橋元昌矣: 木村栄博士の思い出). *Astronomical Herald, 36*, 117.

Hirayama, K. (1918). Groups of asteroids probably of common origin. *Astronomical Journal, 31*, 185–188.

Hirayama, K. (1920). On the reminiscence of Ichinohe Naozo (天文月報. 平山清次: 一戸直蔵の思い出). *Astronomical Herald, 13*, 185.

Hirayama, K. (1922a). Families of asteroids. *Japanese Journal of Astronomy and Geophysics, 1*, 55–95.

Hirayama, S. (1922b). The mean parallax of stars of different spectra. *Annales de I'Observatoire astronomique de Tokyo, Apendice, 10*, 1–6.

Hirayama, S. (1924). Note on the geometrical classification of long-period variables. *Japanese Journal of Astronomy and Geophysics, 2*, 143–146.

Hirayama, K. (1938). *Calendar and time*, Kouseisha (平山清次: 暦法及時法、恒星社).

Hirayama, S., Nakamura, K., et al. (1923). Obituary of Professor Terao Hisashi (天文月報、平山信, 中村恭平、他:寺尾寿先生追悼文集). *Astronomical Herald, 16*(9), 131–143.

Ichinohe, N. (1907a). Maximum of Ceti in 1906. *Astronomische Nachrichten, 176*, 311–314.

Ichinohe, N. (1907b). Orbit of the spectroscopic binary μ Sagittarii. *Astrophysical Journal, 26*, 157–163.

Iijima, D. (1939). *Ancient history and astronomy in China*, (飯島忠雄: 支那古代史と天文学), Kouseisha-Kobunnkaku (恒星社厚生閣).

Itakura, S. (1976). *Nagaoka Hantaro*, Asahi Shinbun Sha (板倉聖宣: *長岡半太郎*、朝日新聞社).

Kabumoto, K. (1996). The role of Shinjo Shinzo in the Japanese modern astronomy (科学史研究, 株本訓久: 日本現代天文学史における新城新蔵の役割). *Study of Science History, 35*, 260–270.

Kimura, H. (1902). A new annual term in the variation of latitude, independent of the components of the poles motion. *Astronomical Journal, 22*, 107–108.

Kimura, H. (1908). Latitude variation in 1907 (天文月報、木村栄: 1907 年における緯度変化). *Astronomical Herald, 1*(5), 50.

Nagaoka, H. (1904). On a dynamical system illustrating the spectral lines and the phenomena of radio activity. *Nature, 69*(1791), 392–393.

Nagaoka, H. (1909). On Zeeman effect (天文月報, 長岡半太郎: ゼーマン効果につき). *Astronomical Herald, 2*, 13–16.

Nagaoka, H. (1939). Obituary of late Shinjo Shinzo (長岡半太郎: 故新城新蔵君を憶う). *Shizen (Nature), 8*, 82–85. (Shizen, 自然, is a Japanese magazine, published by the Shanghai Institute of Science).

Nagaoka, H., & Sugiura, Y. (1923). Vacuum arc for obtaining spectra extending from visible light to soft X-rays. *Astrophysical Journal, 57*, 86–91.

Nakamura, T. (2008). Chapter 1: Dawn of Astronomy in Japan (中村士、日本天文学の黎明). In *One hundred years of the astronomical society of Japan* (日本天文学会編、日本の天文学の百年)、Hyoronsha (評論社), (pp. 3–12)

Nakayama, S. (1969). Chapter 16: The transition to modern astronomy. In *A history of Japanese astronomy – Chinese background and Western impact* (pp. 218–225). Cambridge, MA: Harvard University Press.

Nakayama, S. (1989). *Ichinihe Naozo – A resigned man of great ambitious—*, (中山茂:一戸直蔵一野におりた志の人–), Reproport Sha (リプロポート社).

Noda, C. (1979). Hoshi-no Techo (星の手帖), Biography of Shinjo Shinzo (能田忠亮: 新城新蔵伝), pp. 93–95

Philip, T. E. R. (1918). Private communication.

Shapley, H. (1914). On the nature and cause of Cepheid variation. *Astrophysical Journal, 40*, 448–465.

Shinjo, S. (1914). *Meteorinfalle als Ursache des vermuteten Zurückhalten der obersten Atmösphäre* (Vol. 1, pp. 1–20). Memoirs of College of Science, Kyoto Imperial University.

Shinjo, S. (1922a). On the physical nature of Cepheid variation. *Japanese Journal of Astronomy and Geophysics, 1*, 7–21.

Shinjo, S. (1922b). General consideration of the variable stars from the standpoint of stellar evolution. *Japanese Journal of Astronomy and Geophysics, 1*, 183–190.

Shinjo, S. (1931). Iwanami course "Physics and chemistry", Part II-C, Cosmic evolution (岩波講座、物理学および化学、新城新蔵; 宇宙進化論, 岩波書店).

Shinjo, S. (1938). *History of astronomy in the far East*, (新城新蔵: *東洋天文学史*), Kobundo⟨弘文堂⟩.

Takamine, T. (1919). The stark effect for metals. *Astrophysical Journal, 50*, 23–41.

Takamine, T. (1941). Absorption spectra of gases in the extreme ultraviolet. *Astrophysical Journal, 93*, 386–390.

Tokyo University. (1987). *One hundred years of the Tokyo astronomical observatory*. Tokyo University. (東京大学百年史—部局史—東京天文台).

Watanabe, M. (1975). *M. Paul – American astronomer and teacher in Japan*, National Diet Library (国会図書館刊、渡辺正雄 : M. ポール -- 日本におけるアメリカ人の天文学者、教育者).

Yabuuchi, K. (1992). In Egami, N. (Ed.), *Heritage of oriental study*. Shinjo Shinzo (*東洋学の伝承*, 江上波夫編; 藪内清 :新城新蔵).

Yokoyama, K. (2008). Part I, Chapter 5: The 90 years of the history of international latitude observatory (横山紘一: 緯度観測90年の歴史). In *One hundred years of the astronomical society of Japan* (日本天文学会編、日本の天文学の百年)、Hyoronsha (評論社), pp. 43–51.

Yoshida, H. (吉田晴代) (1999). *Aikitsu Tanakadate and the beginning of physical researches in Japan*, PhD thesis of Hokkaido University.

Yoshida, S. (2019). The life and researches of Kiyotsugu Hirayama (1874 – 1943) – Various aspects of the discovery of Hirayama families⟨天文月報、吉田省子: 平山清次の生涯と研究 – 小惑星の族発見をめぐる諸相.). *Astronomical Herald*, *112*, 601–612.

Chapter 3
Astronomy in Early Showa. I. Tokyo 1926–1945

Abstract In early Showa, Japan was under military rule and ultimately rushed headlong into World War II.

Japanese astronomers in this era were removed from the age of foreign studies. Research in astronomy and astrophysics was mostly carried out at the Tokyo Astronomical Observatory (TAO) and three Imperial Universities (Tokyo, Kyoto, and Tohoku), where some notable results were obtained. The astronomical activities in these areas will be presented separately in the next three chapters through the works of some leading astronomers.

This chapter is devoted to the Tokyo area: Tokyo Imperial University and the TAO. The astrophysical works of Hagihara Yusuke, Hatanaka Takeo, Fujita Yoshio, Kaburaki Masaki,

Osawa Kiyoteru, and Hirose Hideo are successively discussed (Unno, 2008). Studies of the history of oriental astronomy in the postwar period are also presented.

3.1 The TAO and the IAU

In international relations, the period from Taisho to early Showa was when Japan was first welcomed into the International Astronomical Union (IAU) and thereafter rejected because of Japan's extreme militarism (Blaauw 1994).

The IAU was established and the first general assembly (GA) was held in Rome in 1922. From the early phase, Japan was warmly welcomed into the IAU. The successive GA and Japanese delegates who were selected from the TAO are as given in Table 3.1 (Fig. 3.1).

No Japanese delegate attended the sixth GA in 1938. Japan had already entered the Sino-Japanese war, and the Japanese army occupied some Chinese territory.

In 1939, World War II broke out between the Axis Powers (Germany, Italy, and Japan) and the Allied Powers (England, France, Netherlands, USA, and others), and the IAU experienced prolonged interruption in its activities mainly due to the

T. Kogure, *The History of Modern Astronomy in Japan*, Historical & Cultural Astronomy, https://doi.org/10.1007/978-3-030-57061-3_3

Table 3.1 Early IAU and General Assembly (Blaauw 1994)

GA	Year	Site	New president	Japanese delegate
1st	1922	Rome	W. W. Campbell	–
2nd	1925	London	W. de Sitter	Hirayama Shin (V-P)
3rd	1928	Leiden	F. Dyson	Hirayama Shin (E-C)
4th	1932	Cambridge (Mass.USA)	F. Schlesinger	Hirayama Kiyotsugu
5th	1935	Paris	E. Esclangon	Tanakadate Aikitsu, Sotome Kiyofusa
6th	1938	Stockholm	A.S. Eddington	–

Note: *V-P* vice president, *E-C* executive committee member

Fig. 3.1 Hirayama Shin in 1928 at Leiden with IAU executive committee members, 1925–1928. Front row, left to right: H. Deslandres, W. de Sitter (President), S. Hirayama (Vice-President); back row: F. Schlesinger, F. J. M. Stratton (General Secretary), and A. S. Eddington (Blaauw 1994)

difficulty of communication and a poor financial state. The war ended with the defeat of the Axis Powers, and the executive committee of the IAU met at the Copenhagen Conference in November 1946, calling for widespread participation of the most important astronomers in the conference. This was the first international meeting after World War II. At this meeting, aside from general reviews of certain specific fields of astronomy, several resolutions to be submitted to the coming next General Assembly were adopted. One resolution was concerned with the organization's attitude toward Japan and Germany.

General Secretary J. H. Oort wrote a letter to the president, in which he said, "Also with respect to the Japanese, ... I am against expelling them from the Union..." Vice President Andre Danjon replied to Oort: "Personally, I am of the opinion that, for the moment, German and Japanese astronomers should not be admitted to the Union...."

Danjon's opinion was put forth as the formal attitude of the IAU. The next GA was held in 1948 in Türich, where General Secretary B. Strömgren and predecessor Oort supported the readmission of Japan into the IAU, so that President Spencer

Jones sent an invitation to Hagihara Yusuke (Director of the Tokyo Astronomical Observatory), but he was unable to attend the meeting the time.

In February 1949, Strömgren, with the consent of the executive committee, suggested to Hagihara that he might take the first step toward the readmission of Japan. Indeed, by the time of the GA of 1952 in Rome, membership had been reestablished. Subsequently, Hagihara served twice as vice-president (1961–1964, 1964–1967) of the IAU.

3.2 Tokyo Astronomical Observatory and Astronomy

The Tokyo Astronomical Observatory (TAO), currently the National Astronomical Observatory of Japan) (NAOJ), has been the center of astronomy in Japan since its foundation.

While its early phase was discussed in Chap. 2, a brief history of the TAO up to the early Showa era is summarized here (Tokyo University 1987).

The TAO was established in 1888 on the campus of Tokyo Imperial University in Hongo (本郷) and moved to Azabu (麻布) in 1893, both in Tokyo. The number of staff was first limited to 8 and gradually increased up to 12 by 1922. The main routine activities were time keeping, calendar making, and observations on positional works and solar observations.

In 1924, the TAO moved to Mitaka, west of Tokyo, and the staff increased to 15 members, and new instruments were constructed gradually from two transit instruments to a 20 cm equatorial refractor, a 20 cm comet seeker, and the Tower Telescope equipped with a 60 cm coelostat and high-dispersion spectrograph (1928). A Zeiss 65 cm equatorial refractor (1929) and others were successively constructed, and astrophysical observations also started along with positional astronomical observations.

By 1944, the number of staff increased up to 32, and observational activity expanded, including the position of stars with two transit instruments, photographic observations of asteroids and comets with an astrograph, solar spectroscopy with 65 cm coelostat, and others. Expeditions of solar eclipse and some theoretical works were also promoted (Fig. 3.2).

In February 1945, the main building was destroyed by fire, and a large amount of observed data, measuring machines, and clocks were lost in the fire. Also, in the same year, a plan was put in place to evacuate works related to calendar making and time keeping to Nagano Prefecture to avoid air attacks of US armed forces, but the war ended before the evacuation. Up to this time, the TAO had a classical system, similar to most European observatories in the nineteenth century. The observatory adopted an engineer-technician system for carrying out routine activities, nominally without research staff.

Table 3.2 shows past directors, along with their specific fields. The director was mostly a positional astronomer, reflecting the primary function of the observatory.

Fig. 3.2 Main building of TAO established in 1924 at Mitaka (Tokyo University 1987)

Table 3.2 Past Directors of TAO (I)

	Name	Term	Specific fields
1.	Terao Hisashi	1888–1919	Positional astronomy
2.	Hirayama Shin	1919–1928	Positional astronomy, statistical astronomy
3.	Sotome Kiyofusa	1928–1936	Positional astronomy, Solar System
4.	Sekiguchi Koikichi	1936–1945	Solar physics, meteorology
5.	Hukumi Naobumi (Acting Director)	1945–1946	Positional astronomy

Directors in postwar period will be given in Table 6.2 in Chap. 6.

Research activities were promoted as joint works of the TAO and the Department of Astronomy at Tokyo Imperial University.

The modernization of the observatory, particularly its transformation into a research institute, took place in the postwar era and will be considered in Chap. 6.

3.3 Hagihara Yusuke and Celestial Mechanics

3.3.1 Life and Works

In Tokyo, astrophysics was promoted by Hagihara Yusuke (萩原雄祐, 1897–1979), who devoted his professional life to celestial mechanics and, additionally, physical processes in planetary nebulae (Suemoto et al. 1979).

Hagihara was born in Osaka in 1897. His early years were full of difficulties, his parents were divorced, and his mother deserted him soon after he was born. His father's factory faced financial difficulties during Hagihara's school years. In middle school, in Osaka, he needed help from his teacher Origuchi Shinobu (折口信夫, 1887–1959). He was a poet, known by his pen name Shaku Choku (釋迢空), and he was later appointed professor of Japanese literature at Kokugakuin University in

Tokyo. With the support and encouragement of Origuchi, Hagihara was able to enter Tokyo Imperial University and finished his education in astronomy in 1921. In 1923, Hagihara was appointed associate professor at the university and was sent to England by the government for 2 years. He stayed mainly at Cambridge University, where he studied celestial mechanics under H. F. Baker, a mathematician, and astrophysics and general relativity under A. S. Eddington. After returning to Tokyo in 1925, Hagihara started to give lectures at Tokyo Imperial University. From 1928 through 1929 he had another opportunity to stay at Harvard University, where he studied celestial mechanics under George Burkoff. In 1930, he completed his doctoral dissertation on the stability of satellite systems.

In 1935, Hagihara was promoted to a full professor of astronomy. In the fall of 1946, Hagihara was appointed Director of the TAO at Mitaka, when the TAO was almost completely destroyed during the war. He had to start with the reconstruction of the main buildings, observation rooms, and basic observational instruments. After several years, the buildings and facilities had mostly been recovered. In 1949, a coronagraph with a 10 cm aperture was installed on the summit of Mt. Norikura, in Nagano Prefecture, as the first remote observational facility outside Mitaka. A 200-MHz solar radio telescope was also installed at the Mitaka campus. In 1951, a new dome equipped with a 30 cm Cooke refractor was completed. In 1957, construction began on a large telescope.

After retirement in 1957, Hagihara moved to Sendai as a professor of astronomy at Tohoku University and then to Utsunomiya as President of Utsunomiya University from 1961 to 1967. In January 1979, he died suddenly of illness at age 82 (Fig. 3.3).

Fig. 3.3 Hagihara Yusuke, Vice President of IAU, at General Assembly in Hamburg in 1964 (Blaauw 1994)

3.3.2 Celestial Mechanics

Hagihara's works on celestial mechanics cover three main fields: (1) two-body problems in the general-relativistic Schwarzschild field (1930), (2) stability of planetary and satellite systems and the theory of secular perturbations (1927–1944), and (3) extension of the fundamental theory of celestial mechanics (1933–1944). These works were later collected and rearranged in his five-volume *Celestial Mechanics*, published in the period 1970–1976.

Hagihara's first work was on general relativity. The exact equation of the gravitational field around a mass point was derived by Karl Schwarzschild in 1916 and is called the Schwarzschild field. Hagihara traced the trajectories of a test particle in a strong Schwarzschild gravitational field under various initial conditions. As a result, he derived various types of trajectories classified as, for example, pseudocircular, quasi-elliptic, quasi-parabolic, quasi-hyperbolic, quasi-hyperbolic spiral, and others. A common feature of these types is the remarkable secular variation, so, he used the term *quasi* or *pseudo* for the type of trajectory (Hagihara 1930). Some examples are shown in Fig. 3.4.

Figure 3.4a illustrates a quasi-elliptic trajectory that corresponds to the orbit of ordinary Keplerian planetary motion and is accompanied by secular motions in advance of perihelion and changing orbital forms. Figure 3.4b shows a case of quasi-circular motion, where a moving particle cannot cross over the outer circle. Figure 3.4c shows a type of quasi-hyperbolic spiral in which the trajectory is opened

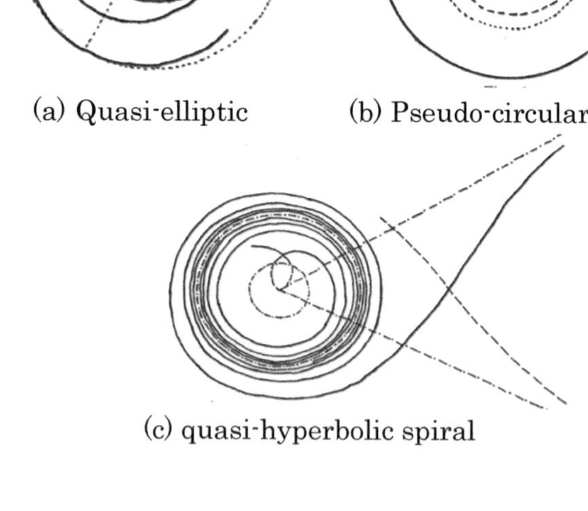

Fig. 3.4 Trajectories of a test particle in Schwarzschild gravitational field (Hagihara 1930); (**a**) quasi-elliptic; (**b**) pseudo-circular; (**c**) quasi-hyperbolic spiral

to the outer space. This trajectory may be applicable to the motion of a particle falling onto a black hole or a white dwarf.

3.3.3 *Physics of Planetary Nebulae*

From 1937 through 1944, Hagihara worked on the physics of planetary nebulae in two fields: the physical state of nebulae in radiative equilibrium and velocity distribution of electrons in nebulae.

3.3.3.1 Planetary Nebulae in Radiative Equilibrium

Planetary nebulae are tenuous gaseous objects surrounding central hot stars, which emit strong ultraviolet radiation and ionize hydrogen in nebulae. In 1937, Hagihara constructed a nebular model, under the assumption that planetary nebulae are composed of pure hydrogen in a state of isothermal and radiative equilibrium (Hagihara 1937). Hagihara's work started almost the same year as J. G. Baker and D. H. Menzel in the United States. He solved the equations of radiative transfer inside nebulae for every radiation of Lyman, Balmer, and Paschen series, including both their continuum and line series. Starting from the transfer of the Lyman continuum as the first approximation, he dealt with the Lyman, Balmer, and Paschen lines as the second approximation. Some of the results of his calculations were as follows.

(a) Nebular hydrogen atoms are ionized almost from the ground energy levels by the Lyman continuum, and the ionization from the second and third energy levels is negligible.
(b) In the Lyman series, Lyman alpha ($L\alpha$) is exceedingly intense, as compared to $L\beta$ and higher members, which are several orders of magnitude weaker.
(c) In the Balmer and Paschen lines, in contrast to the Lyman series, the emergent fluxes of the series lines gradually decrease toward higher members. He computed the relative intensities of Balmer lines (so-called Balmer decrement) in several cases of electron temperature T_e and optical depths τ_{Lc} for the diffuse Lyman continuum. The decrement in the case of $T_e = 20{,}000$ K and $\tau_{Lc} = 0.2$ is shown in Table 3.3, as well as the decrement of Baker and Menzel for $T_e = 20{,}000$ K in Case B (optically thick for Lyman line radiations), cited as BM (Case B) (Baker and Menzel 1938). Both decrements present roughly good coincidence.

Table 3.3 Balmer decrements in planetary nebulae (Hagihara 1937)

Balmer lines	Hα	Hβ	Hγ	Hō	Hε	H8
Hagihara (1937)	3.03	1.0	0.51	0.25	0.19	0.11
BM(case B)(1938)	2.50	1.0	0.50	0.30	0.19	0.13

(d) A large part of the radiative moments transferred to the nebular gas is due to the Lyman radiation generated inside nebulae, and the moments increase rapidly from the inner to the outer boundary. This is simply a result opposite to that of J. H. Jeans' theory, in which radiation pressure becomes zero at the outer boundary (Jeans 1923). Hagihara argued that the amount of radiative moments depends upon the temperature of the central stars and the optical depth for the Lyman continuum. In some conditions, nebulae became unstable for expansion (Hagihara and Hatanaka 1946).

3.3.3.2 Velocity Distribution of Electrons in Planetary Nebulae

The velocity distribution of free electrons in planetary nebulae is usually expressed by the Maxwell distribution for an ideal gas in thermal equilibrium. The distribution function $f(v)$ takes the form

$$f(v) = \left(\frac{m}{2\pi kT}\right)^{3/2} \exp\left[-\frac{m}{2kT}\left(v_x^2 + v_y^2 + v_z^2\right)\right] \tag{3.1}$$

where v ($v_{x,}$ v_y v_z) denotes the velocity of the gas, T the kinetic temperature of the gas, and m and k are the particle mass and the Boltzmann constant, respectively.

Hagihara examined the velocity distribution of electrons and whether it might deviate from a Maxwell distribution when nebulae are exposed to strong ultraviolet radiation from hot central stars (Hagihara 1939, 1940).

In addressing this problem, Hagihara derived the approximate solution of the Boltzmann equation, which involves interaction terms among the constituting electrons, protons, and hydrogen atoms. He made use of a series-expansion method and adopted up to the second term of the series. Under the condition of planetary nebulae, he obtained a new velocity-distribution function $F(v)$ in the form

$$F(v) = \beta f(v) \tag{3.2}$$

where β denotes the deviation factor from the Maxwell distribution $f(v)$ given in Eq. (3.1). The significance of Eq. (3.2) for the effect of a non-Maxwell distribution lies in the deviation of electron temperature, without changing the functional form of the velocity distribution.

The deviation factor can be obtained through an approximate numerical solution of the Boltzmann eq. A revised electron temperature is given based on Eq. (3.2).

The electron temperatures thus derived are shown for some cases in Table 3.4. He thus demonstrated that the electron temperature of a nebula decreases around 30% when the non-Maxwell distribution is taken into consideration, as compared to the case of Maxwell velocity distributions.

Hagihara applied his theory of non-Maxwell distribution to many spectroscopic problems, such as the intensity distribution of continuous radiation, relative

Table 3.4 Effect of non-Maxwell distribution on values of electron temperature (Hagihara 1940)

Stellar T_s	Maxwell T_e	Non-Maxwell T_e	Rate of decrease
5000 K	4500 K	3200 K	0.29
10,000	9000	6300	0.30
20,000	17,500	12,300	0.30
50,000	41,000	28,000	0.28

intensities of Balmer and Paschen lines, conditions of radiative equilibrium, and so on. However, Hagihara's theory was not accepted subsequently, for two main reasons: first, the effects of a non-Maxwell distribution on the spectroscopic behaviors of planetary nebulae are relatively small compared to the variety of observed behaviors, and, second, the numerical computation of the effects is complex compared to a simple Maxwell distribution.

Hagihara's theoretical works on planetary nebulae were taken up by his students Hatanaka Takeo (next section) and Unno Wasaburo (Chap. 7), both of whom were promoted to professors of astronomy at the University of Tokyo.

3.3.4 Construction of the Okayama Astrophysical Observatory

In the early 1950s, Hagihara began to consider the construction of a large optical telescope in Japan. He argued for the necessity of a large telescope by citing an analogy with one of the three legs of a tripod, that is, Japan, which is located around 120° in longitude from Europe and America.

Due to his patient persuasion, the telescope project was endorsed in 1954 and construction started. The reflecting telescope with 188 cm apertures was ordered from Grubb-Parson in England, and the observatory was sited atop Mt. Chikurinji in Okayama Prefecture.

In 1961, the Okayama Astrophysical Observatory (OAO) was established and Hagihara's dream of a tripod was finally realized, though he had already retired from the TAO by 1957 (see Chap. 6).

3.4 Hatanaka Takeo and Astrophysics

3.4.1 Life and Works

Hatanaka Takeo (畑中武夫, 1914–1963) is known as a pioneer in radio astronomy in Japan. He also worked on the theoretical study of planetary nebulae and of stellar evolution (Hitotuyanagi et al. 1964).

Fig. 3.5 Hatanaka Takeo (Physics Today, 17, No.3, 1964)

Hatanaka was born in the city of Shingu (新宮市), Wakayama Prefecture, and was adopted at age 3 into the Hatanaka family. In 1934, he entered Tokyo Imperial University and studied astrophysics under Hagihara. Upon graduation in 1937, he began to work at the university as a research assistant and was eventually promoted to professor of astronomy in 1953. He remained there for all his later life. He worked mainly in three fields: the physics of planetary nebulae (1942–1947), radio astronomy (1948–1962), and stellar evolution in the early Universe (1956–1964).

His contribution to the popularization of astronomy was also evident in the writing of popular books, frequent appearances on public media (radio and TV), popular lectures, and other factors.

In November 1963, he suddenly died of apoplexy at age 49, before his pioneering efforts bore full fruit (Fig. 3.5).

3.4.2 Physics of Planetary Nebulae

The spectra of planetary nebulae show strong emission lines and weak continuum. The emission lines are generally classified into three groups: (1) permitted lines of HI, HeI, HeII, etc., arising from the recombination of free electrons with atoms or ions, (2) forbidden lines of OIII, NeIII, and others, excited by electron collisions, and (3) permitted lines of OIII and NIII, known as Bowen fluorescence lines (Bowen 1934). From the late 1930s to the early 1940s, studies of planetary nebulae were promoted worldwide as a useful application of quantum physics, which was formulated in the late 1920s. In Japan, the physics of planetary nebulae was studied by Hagihara from 1937 to 1944 for recombination spectra (Sect. 3.1). Hatanaka took over Hagihara's work on the study of planetary nebulae in the period 1942–1947.

Hatanaka's first area was the analysis of forbidden lines of ionized oxygen (OII and OIII) in planetary nebulae (Hatanaka 1942). He solved the radiative equations related to two metastable states and derived the relative populations of OII and OIII, *N* (OII)/*N* (OIII), as a function of stellar temperature *T*s, and compared them with observations for several planetary nebulae. The relative population gradually decreases with increasing stellar temperature, in good agreement with observations.

Hatanaka's next work was on the analysis of fluorescence lines, observed in OIII and NIII ions. It was shown by Ira S. Bowen that fluorescence lines are produced by the close coincidence of the wavelengths of the two lines of different ions HeII and OIII, or OIII and NIII, as shown in Table 3.5, where the level transitions 1→2 and 1→3 are approximately shown in the energy level diagram in Fig. 3.6 (Bowen 1935).

Hatanaka mainly considered optical interactions between OIII and HeII ions. As seen in Table 3.5, both HeII and OIII ions can be excited from level 1 to level 2 by ultraviolet radiation from a central star. In addition, OIII ions can be excited by the emission line of HeII from level 2 to level 1 due to the high abundance of HeII

Table 3.5 Coincidence of wavelengths between ions (Bowen 1935)

Ion	HeII	OIII	NIII
Transition	1→2	1→2	
Wavelength (Å)	303.799	303.744	
Transition		1→3	1→3
Wavelength (Å)		374.436	374.434, 374.442

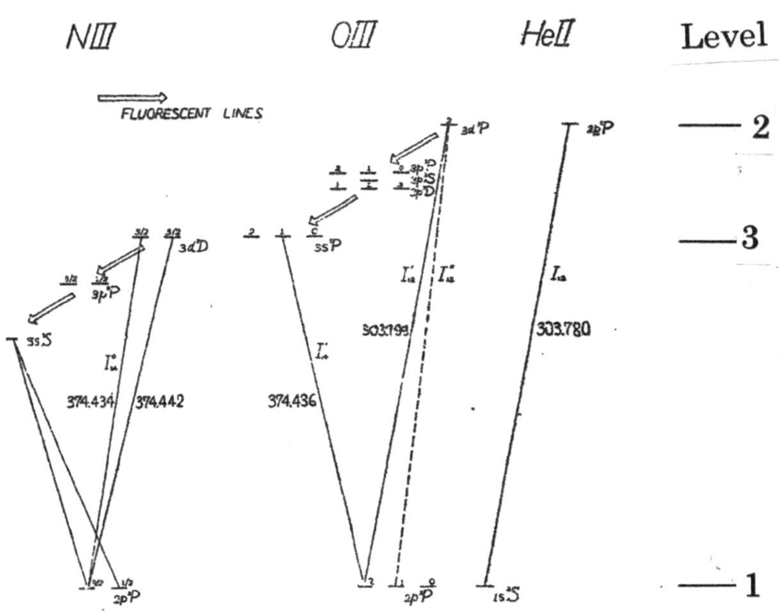

Fig. 3.6 Energy diagram related to fluorescence lines (Bowen 1935; Hatanaka 1946)

compared with that of OIII. By this coincidence, the level 2 of OIII becomes overpopulated, and this causes additional cascade transitions from this level to low-lying levels, forming strong emission lines compared to the cascade lines in other transition series (Hatanaka 1946).

Hatanaka first derived the basic equations of the interaction process and obtained formal solutions. Numerical calculations were performed on the overpopulation of the level 2 of OIII. Numerical calculations were performed on the intensity of emission lines due to the overpopulation of the level 2 of OIII. He calculated the line λ3444Å (energy-level transition 2 to 3 in Fig. 3.6). He also calculated the line strength of HeII λ3203Å (the same energy-level transition). The relative intensities of these lines were in range from 1.1 to 1.4 under suitable boundary conditions. Hatanaka compared his theoretical values with the relative intensities of three planetary nebulae observed by Aller as follows (Aller 1941):

Planetary nebula	NGC 7009	NGC 7027	NGC 7662
I(λ3444)/I(λ3203)	1.54	1.24	1.21

He concluded that these results were acceptable.

Hatanaka also claimed that the optical interaction between the lines of OIII and NIII was quite analogous to the case of HeII and OIII, though no numerical calculations were carried out.

3.4.3 Radio Astronomy

Inspired by Hagihara, Hatanaka turned his research subjects to solar radio astronomy in 1948 (Tokyo University 1987). Together with Moriyama Fumio (守山史生) and Suzuki Shigemasa (鈴木重正), Hatanaka constructed a 5 × 2.5 m equatorially mounted array at 200 MHz for solar radio observations at Mitaka. Radio noise was received from the whole surface of the quiet Sun with some long-term variations (Hatanaka and Moriyama 1953).

In the 1950s, Hatanaka and his group continued their observations of the mapping and polarization properties of active solar regions (Hatanaka et al. 1955) and some types of solar radio bursts (Hatanaka 1957). An example of radio spots observed on the occasion of a partial solar eclipse on June 20, 1955, is illustrated in Fig. 3.7, where the radio spot showed remarkable agreement with the sunspot and calcium plages (Hatanaka et al. 1955).

In the late 1950s, a project on the construction of a 24 m radio telescope to observe galactic objects such as HII regions and the galactic center was launched. Hatanaka, however, could not participate in this project because of his sudden death in 1963 at age 49.

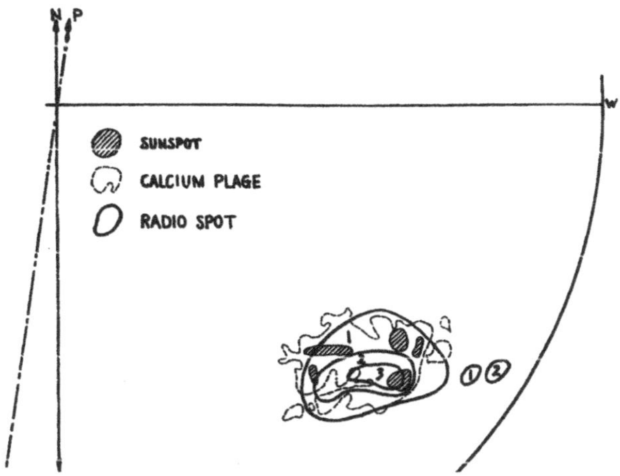

Fig. 3.7 Brightness distribution of a solar radio spot appearing in active solar region observed on June 20, 1955 (Hatanaka et al. 1955)

3.4.4 Evolution of Galaxies

Around 1956, Hatanaka became interested in the study of galaxies and joined a study group on galactic evolution composed of Taketani Mitsuo (武谷三男), Hayakawa Sachio (早川幸男), Obi Shin'ya (小尾信也), and others. They were widely interested in the origin of stellar populations and in the early evolution of galaxies.

3.4.4.1 Origin of Stellar Population

In galaxies, stars are globally separated into two populations: Population I: the groups of young stars distributed in the disk, and Population II: the group of old stars distributed in galactic halo and globular clusters.

Maartin Schwarzschild and Lyman Spitzer (1953) considered the existence of white dwarfs and metal-poor stars in Population II stars. Both of these stars were the remnants of primordial galaxies. White dwarfs were originally massive stars. They were the cores of massive stars that remained after their supernova explosions. The ejected gas, containing metal components, was mixed with primordial interstellar matter. Population I stars were formed from such metal-rich interstellar matter (Schwarzschild and Spitzer 1953).

In 1956, Taketani and Hatanaka's group extended Schwarzschild and Spitzer's scenario, based on the recalculations of nuclear reaction rates converted from helium to carbon. Their arguments are as follows (Taketani and Hatanaka 1956):

(a) Protogalaxies are composed of primordial elements (mostly hydrogen and helium) in the form of turbulent gas and dust.
(b) Stars of the first generation were formed from their dense interstellar clouds.
(c) Low-mass stars, less than about one solar mass, remained Population II stars due to their long evolutionary time as a result of their slow nuclear burning rate.

(d) Massive stars evolved much faster than in Schwarzschild and Spitzer's case, accompanying the production of heavy elements during their later stage of evolution, and those massive stars ended up as white dwarfs.
(e) The enrichment of heavy metals in the interstellar medium is the result of heavy metals ejected from massive stars mixed with primordial interstellar matter.
(f) Population I stars were formed from the mixed interstellar matter.
(g) As a consequence, a galaxy is composed of four types of objects: interstellar matter, Population I and II stars, and white dwarfs.

3.4.4.2 Evolution of Galaxies

Galaxies are generally classified by Hubble type as elliptical (E) to spiral (S), barred spiral (SB), and irregular (Ir).

Hatanaka and his group considered the formation of galaxies in the Hubble framework by dividing it into four groups, as summarized in Table 3.6 (Hatanaka et al. 1964). Based on the Hubble type, they distinguished dwarf E (Es) from normal E type by their mass (Hatanaka et al. 1964).

Hatanaka's group depicted the scenario as follows:

(1) The primordial galaxy is spherical and assumed to be a mass of gas clouds, whereas present galaxies are roughly separated into spherical (E) and flat (S) systems, as seen in Table 3.6. The evolution of primordial galaxies may be explained in terms of the role of two parameters: total angular momentum and random velocity of cloud motion. In spherical systems, both parameters take small values, whereas in flat systems, both take larger values.
(2) In spherical systems, cloud-cloud collisions take place with small heating effects due to the small collision velocity. As a result, star formation is favorable to forming low-mass stars. Over a long evolutionary time, interstellar matter is almost dispersed to outer space. This suggests that E and S0 galaxies are composed of mostly red dwarfs with a small amount of interstellar matter.
(3) In flat systems, large random motion of clouds gives rise to strong shock waves at cloud collisions, producing hot and dense regions, where it becomes possible to form massive stars. Interaction between clouds and massive stars sustains the rich gas content throughout the evolution of galaxies. Due to high angular momentum, a protogalaxy evolves into a flat disk system, often producing spiral arms or central bar systems.

Table 3.6 Main features of galaxies (Hatanaka et al. 1964)

Shape	Hubble type	Mass in $M_\odot$	Gas content	Composite spectral type
Spherical	E, S0	$\sim 8 \times 10^{11}$	Poor	Dwarf star (K)
Flat	S, SB	$\sim 3 \times 10^{10}$	Rich	Early star (A – K)
Irregular	Ir	—	Rich	Early star (A – F)
Dwarf spherical	Es	3×10^{9}	Poor	No bright star (K)

(4) Nuclei of galaxies. Almost independent of morphological types, galaxies retain compact nuclei, roughly of radius 10 pc and mass 10^7 $M_\odot$. The mass of a nucleus may have been supplied from the whole region of the galaxy through cloud collisions. In collision, parts of gas fall into the central region of the galaxy, where accumulated gases produce massive stars, which evolve fast and give rise to the explosion of supernovae. Repeated supernova explosions are the origin of active nuclear phenomena.

Theoretical models of Taketani and Hatanaka's group belong to the early phase of galactic astronomy in postwar development.

3.5 Fujita Yoshio and Cool Stars

3.5.1 Life and Works

Fujita is a pioneer in the spectroscopy of cool stars in Japan. During the war he theoretically derived the dissociation equilibrium of diatomic molecules, and in the postwar period he promoted spectroscopic observations in the USA and in Japan. He contributed to the construction of a 188 cm telescope at the OAO and devoted his later life to the observations of cool stars (Fujita 1986; Osaki 2008; DeVorkin 1997; Yamashita et al. 2013).

Fujuta Yoshio (藤田良雄, 1908–2012) was born in the city of Fukui in 1908, the eldest son of Fujita Teizo (藤田貞造), chief editor of the Fukui newspaper. Fukui Prefecture is in Japan's snow country in northern part of the country. In the wintertime, the region is buried in heavy snow, and most of the time it is cloudy or snowy. In summer, in contrast, the night sky often offers a display of stunning constellations. According to his reminiscences, his interest in astronomy was sparked by such beautiful constellations.

In 1928, he entered Tokyo Imperial University and studied theoretical astronomy under Hagihara and observational astronomy under Hirayama Kiyotsugu. Under Hirayama's guidance Fujita conducted spectroscopic observations of the asteroid Eros with an objective prism and revealed the existence of a thin gaseous envelope around the solid body of Eros.

After graduating in 1931, he worked at the TAO as a technician. The TAO's director was Sotome Kiyofusa, and Fujita was assigned two duties there. One was to install the newly imported telescope, the Zeiss tower telescope, which consisted of a heliostat and a grating spectrograph. Upon completion of the telescope at the solar tower, Fujita began to observe the Sun with this grating spectrograph, but this lasted for a short period of time only, because he was more drawn toward the second assignment, laboratory spectroscopy of molecules. He assisted Tanaka Tsutomu (田中務) in the Department of Physics, Tokyo Imperial University, who visited the TAO every week to carry out spectroscopic experiments with a vacuum spectrograph for diatomic molecules, particularly for heavy water, D_2O. Tanaka's intention

Fig. 3.8 Fujita Yoshio (Yamashita et al. 2013)

was to elucidate the quantum mechanical structure of this molecule. Fujita was simply fascinated by the beauty of molecular band spectra and made up his mind to study stellar molecular spectra (Fig. 3.8).

In 1937, Fujita moved to Tokyo Imperial University as a lecturer and worked on the theoretical study of molecular bands. In those days Takamine Toshio (Chap. 2) held a joint seminar of physicists (Takamine, Kotani Masao (小谷正雄) and others) and astrophysicists (Hagihara, Fujita and Hatanaka) every week at Takamine's laboratory in the Riken (Institute for Physical and Chemical Research). This seminar continued until about 1941, when the Japanese government declared war against the Allied countries and basic research became difficult. In early 1945, the staff and students of the Department of Astronomy evacuated to the mountain site of Nagano Prefecture, while Tokyo, including Tokyo University, was almost destroyed by US air attacks. In August of 1945, the war ended with the defeat of Japan, and the reconstruction of research facilities became urgent, but it faced many difficult problems. Under the leadership of Hagihara, Fujita and the university staff devoted every effort to the reconstruction.

In 1950 Fujita had the opportunity to visit Lick Observatory for 3 months on a US fellowship, where he carried out spectroscopic observations of carbon stars with a prism spectrograph attached to a 36-inch (91-cm) refractor. He was then able to move to Yerkes Observatory for an additional 9 months, where he continued observations of U Cygni and other carbon stars with a 40 inch (100 cm) refractor.

In 1951, Fujita returned home deeply impressed by the enormous power of large telescopes and the urgent need for the construction of a large telescope in Japan. In the same year he was promoted to full professor of astronomy at Tokyo University and oversaw graduate students in their studies of late-type stars; the first group of students included Yamashita Yasumasa (山下泰正), Tsuji Takashi (辻隆), and Utsumi Kazuhiko (内海和彦).

In 1954, a project to construct a large telescope was undertaken in the TAO. Fujita served as chairman of the construction committee. He paid particular attention to the design of a high-dispersion Coudé spectrograph. After the establishment of the OAO, he began to work on spectroscopic observations of carbon stars, work that endured for his long professional life.

After retirement in 1969, Fujita moved to Tokai University as a professor of astronomy and physics. In 1965 he was nominated as a member of the Japan Academy (日本学士院), and later elected President of the Academy (1994–2006).

Fujita enjoyed his later years. In 2009, he was the oldest member, at age 100, of a group that observed a solar eclipse from a ship in the central Pacific Ocean. He died in January 2013 at age 104.

Fujita is also known as a poet of *waka* poetry, which is a type of poem in classical Japanese literature:

Praise the Subaru Telescope
Calmly awaking to observe the stars
In the blue sky of Mauna Kea
(Aosora no Hosi wo kiwamu to Mauna Kea Ugoki somenishi Suharu tata emu)

This poem was presented at the Utakai Hajime (First Poetry Reading), convened by the Emperor on New Year's Day 1999.

3.5.2 Theoretical Spectroscopy

Fujita's work on cool stars can be divided into early theoretical and later observational studies. In the1930s and 1940s, his main subject was the theoretical spectroscopy of diatomic molecules, and after the 1950s, his study became observational in the US and in Japan. All his work was collected and published as a private issue (Fujita 1997).

Fujita's theoretical works were on the dissociation equilibrium of diatomic molecules, such as CH, C_2, and Z_rO, observable in late-type stars (Fujita 1938).

When a molecule AB is produced by a combination of atom A and atom B, the dissociation equilibrium can be expressed by the formula

$$\frac{p_A p_B}{p_{AB}} = K_{AB} \tag{3.3}$$

where p_A, p_B, p_{AB} are the partial pressure of the respective atoms and molecule, and K_{AB} is a constant including the atomic and molecular constants concerned. Fujita calculated the relative strengths of numerous spectral bands of diatomic molecules under conditions of dissociation equilibrium in stellar atmospheres. He paid particular attention to molecules combined with the elements O, C, and N, such as CO, CN, and C_2. He calculated the relative intensities of these bands as a function of stellar temperature and surface gravity and applied them to the atmospheres of giant

Table 3.7 Branches in late-type stars (Fujita 1939)

Notation	Abundance	Band intensity (Shane 1928)			Spectra type
		C_2	CN	TiO, ZrO	
A	O > C and N > C	w	w	s	K – M
B	O > C > N	w	s	s	S
C	N > C > O	int	s	w	S′
D	C > O and C > N	s	s	w	R – N

Band intensity: *w* weak, *s* strong, *int* intermediate

and dwarf stars. As a result, Fujita found that molecular abundance was higher in giant than in dwarf atmospheres (Fujita 1935).

Fujita also applied his calculations of the relative intensities of molecules to the interpretation of spectral sequence in the Harvard classification (Appendix, Sect. 1). This sequence is widely accepted as a steadily decreasing temperature sequence, based on Saha's ionization theory. In this theory, however, it is not clear why spectral sequence separate into three branches at G type.

Fujita's idea was to examine the variety of relative abundance of O, C, and N from the relative intensities of molecular bands in stellar spectra. By comparing these with observed intensities prepared by C. D. Shane (Shane 1928), Fujita discovered the existence of four branches: K – M, R – N, S, and S′, instead of three branches, as shown in Table 3.7. In addition, Fujita pointed out that there exists an upper boundary of stellar temperature above which the spectral branch vanishes due to weak intensities of the bands, which are related to the separation of the branch. For the temperature of G-type stars it was found to be around 5600 K for giants and around 6300 K for dwarfs; these are the temperatures of G-type stars (Fujita 1939, a, b).

3.5.3 Spectroscopic Observations at Lick and Yerkes Observatories

In the fall of 1950, Fujita arrived at the Lick Observatory where the director was C. Donald Shane (1895–1983). His thesis in the 1920s was the spectroscopic observations of late-type stars, particularly carbon stars. Fujita employed Shane's data for his theoretical studies of spectral sequence as shown above.

Thus, under the support of Shane, Fujita began to observe the carbon star χ Cygni. When he obtained the first spectrum, it made a strong impression on him. In his memoirs, he wrote, "When I saw the first spectrum of χ Cygni, which I had been seeking for a long time, I could not escape from my deep impression. . . ."

Figure 3.9 shows the spectrum of χ Cygni as obtained by Fujita at Lick Observatory. The strong emission lines in the Balmer series and numerous molecular bands are remarkable.

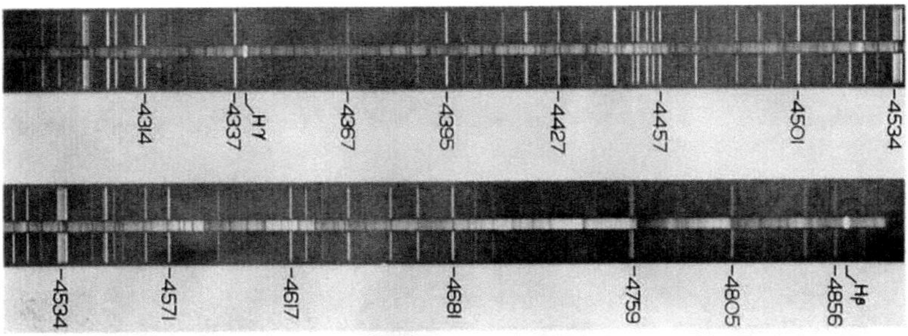

Fig. 3.9 Spectrum of carbon star χ Cygni. The comparison spectrum on both sides of the stellar spectrum denotes the Titan emission lines (Fujita 1951)

At the end of 1950, Fujita moved to the Yerkes Observatory and continued the spectroscopic observations of several carbon stars using the 40 inch (100 cm) refractor. At this time, he worked on the spectral analysis to determine the effective temperature, abundance of elements, and structure of stellar atmospheres of carbon stars.

After a 9 month stay at Yerkes, he returned home in 1951, with great anticipation for making observations with a large telescope.

3.5.4 Okayama Astrophysical Observatory and Observations of Carbon Stars

The OAO was established in 1961, and Fujita began to work on spectroscopic observations of late-type stars. His observations aimed at the infrared spectral region from λ 7800 through λ 8900 Å of carbon stars with a high-dispersion spectrograph. He selected this region because of the lack of previous spectroscopic observations for carbon stars.

The spectra of carbon stars in this region are mainly composed of CN, $C^{13}N^{14}$, and the telluric bands of water vapor, as shown in Fig. 3.10. From the relative intensities of these lines Fujita and Utsumi derived the relative isotope abundance of C^{12} and C^{13} for some subtypes of carbon stars (Fujita and Utsumi 1963).

Fujita continued his infrared spectral observations of carbon stars and other late-type stars for a long time at the OAO. Based on his theoretical considerations of the dissociation equilibrium of diatomic molecules, he refined the classification scheme of carbon stars and the branching system of late-type stars. His works were taken up by Yamashita Yasumasa and others.

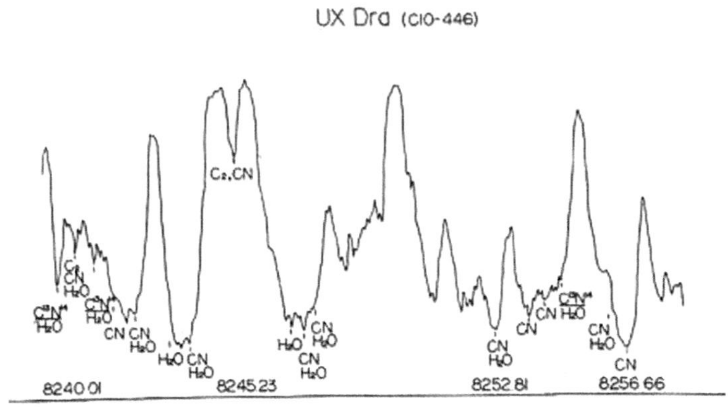

Fig. 3.10 Spectral tracing of the carbon star UX Dra in the photographic infrared region. The H_2O lines denote terrestrial water vapor (Fujita and Utsumi 1963)

3.6 Kaburaki Masaki and Stellar Astronomy

3.6.1 Life and Works

Kaburaki worked on stellar astronomy and popularization of astronomy (Kaburaki 1963; Takase 1988).

Kaburaki Masaki (鏑木政岐, 1902–1987) was born in the city of Kanazawa, Ishikawa Prefecture, the son of a Shinto priest. He studied spherical astronomy under Hirayama Kiyotsugu and stellar astronomy under Hirayama Shin in the Department of Astronomy, Tokyo Imperial University. After graduation in 1926, he began to work as a technician at the Meridian Circle Laboratory at the TAO.

His main duties were meridian observations and time keeping. In 1935, he moved to the Department of Astronomy, Tokyo Imperial University, as an assistant professor and promoted to professor in 1946 (Fig. 3.11).

In 1961, Kaburaki of Tokyo University and Shimizu Tsutomu (清水彊), a professor at Kyoto University, organized the first interuniversity meeting on stellar astronomy. This annual meeting was called the Stellar Astronomy Meeting (SAM). In its 10-years span, SAM contributed greatly to the collaborative study of galactic astronomy in Japan and was also involved in the initial planning stage of a project that became the large Schmidt telescope with an aperture of 105/150 cm, which was realized after long discussions in the SAM (Chap. 6).

After retirement in 1963, Kaburaki was invited to serve as a professor at Tokai University in Tokyo. Along with his research activities, he also diligently continued his work on education and the popularization of astronomy. He wrote about 20 popular books in his later years. In addition, he served as Director of the Goto Planetarium in Tokyo until 1986 and died a year later at age 85.

Kaburaki's stellar astronomy was taken up by his graduate students Takase Bunshiro (高瀬文志郎) and Ishida Keiichi (石田蕙一) at the TAO.

Fig. 3.11 Kaburaki Masashi (left) and Shimizu Tsutomu of Kyoto University at the SAM meeting in 1964 (By author)

3.6.2 *Local System of Galaxy*

Kaburaki worked on two areas of stellar astronomy: moving clusters and velocity ellipsoids of nearby stars around the Sun in the Galaxy.

3.6.2.1 Moving Clusters

A moving cluster is a group of nearby stars with a common motion with respect to the local standard of rest. Rasmuson (1921) classified eight moving clusters into two types, one composed of low-velocity stars and the other of high-velocity stars, as given in Table 3.8 (Rasmuson 1921). All moving clusters are distributed within 30° in galactic latitude, and the direction of motion is opposite to that of solar motion. A notable difference in the composition of stars in the two types of clusters is the spectral type, that is, stars of low-velocity clusters are almost early-type (B and A stars), whereas high-velocity clusters are dominated by stars of the G and K types.

Kaburaki considered the statistical properties of moving clusters through the study of radial velocities and proper motions by taking into account the effects of solar motion and galactic rotation for low-velocity and high-velocity clusters, successively (Kaburaki 1933).

(a) Low-velocity moving clusters. These clusters consist of early-type stars and are distributed in low galactic latitudes. They are located at distances comparatively far from the Sun. Kaburaki calculated the radial velocity (V_r) and proper motion in longitude (μ_l) of the cluster members as a function of galactic longitude. His calculated mean curve of proper motion is illustrated in Fig. 3.12, where one can see good agreement for three clusters, Perseus, Sco-Cen, and Pleiades, against a

Table 3.8 Moving clusters (Rasmuson 1921)

Type of moving cluster	Cluster	Number of stars	Convergent		Space velocity	Distance
			l	b	km s^{-1}	pc
Low velocity	Perseus	42	175°	+1°	6	95
	Sco-Cen	147	314	+10	6	83
	Pleiades	12	323	−5	5	72
	Orion	17	80	−9	6	540
High velocity	Taurus	39	149	+11	28	35
	Ursa Major	22	0	−8	30	50
	Praesepe	8	158	+28	28	112
	61 Cygni	57	173	+5	80	38

Note: Galactic coordinates of convergent (l, b) denote the old system. The present longitude L is given by $L = l + 33$

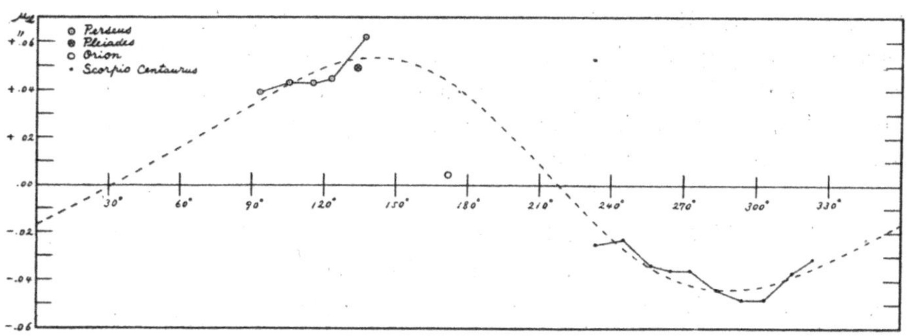

Fig. 3.12 Relation between observed and computed values of proper motions in longitude of low-velocity moving clusters (Kaburaki 1933)

remarkable deviation for the Orion cluster. This deviation may be explained by the large distance from the Orion cluster as seen in Table 3.8. Kaburaki supposed that the Orion cluster may belong to a different stellar system in the Galaxy.

(b) High-velocity moving clusters. These clusters occupy a position nearer to the Sun than the low-velocity group. Kaburaki found that the convergent points of these clusters are separated into two directions, which approximately coincides with the directions of the two-star streams proposed by Jacobus C. Kapteyn. The asymmetry of a star's motion was first noticed by Kapteyn by subtracting the solar motion with respect to nearby stars (Kapteyn 1904、 see appendix). To express the asymmetry, he introduced the two-stream hypothesis: stars in the vicinity of the Sun are a mixture of two stellar populations having different mean motions, and two streams move toward the Orion and Sagittarius constellations, respectively (Kapteyn 1904; Kapteyn and Weersma 1912). Kaburaki noticed high-velocity clusters also moving in these two directions: Taurus, Praesepe, and 61 Cygni clusters toward Orion and the Ursa Major cluster toward Sagittarius.

3.6.2.2 Velocity Ellipsoids of Nearby Stars

On the asymmetry of stellar motion, Karl Schwarzschild showed that an asymmetric distribution could be represented by an ellipsoidal distribution, instead of the two-stream hypothesis, that is, by a normal frequency function with unequal dispersions in different directions. The center of the ellipsoid indicates the mean motion of the stars relative to the Sun (Schwarzschild 1907).

Kaburagi adopted the ellipsoidal hypothesis of Schwarzschild and derived the velocity ellipsoids of nearby stars for different radial velocity steps. He derived the velocity ellipsoids in six steps for low-velocity stars (0–15, 15–30, ..., 65–70 km s^{-1}) and four steps for high-velocity stars (70–100, 100–150, ..., 200–250 km s^{-1}), and the projections of ellipsoids on the galactic plane are delineated in Fig. 3.13, where the horizontal direction corresponds to the galactic longitude $l = 0°$ and 180°, and vertical direction to $l = 90°$ and 270° (in the old system of galactic coordinates).

He noticed that the velocity ellipsoids represented different forms for different velocity steps. The major and intermediate axes of the ellipsoid lie nearly in the galactic plane, and the minor axis is perpendicular to it. The directions of the major

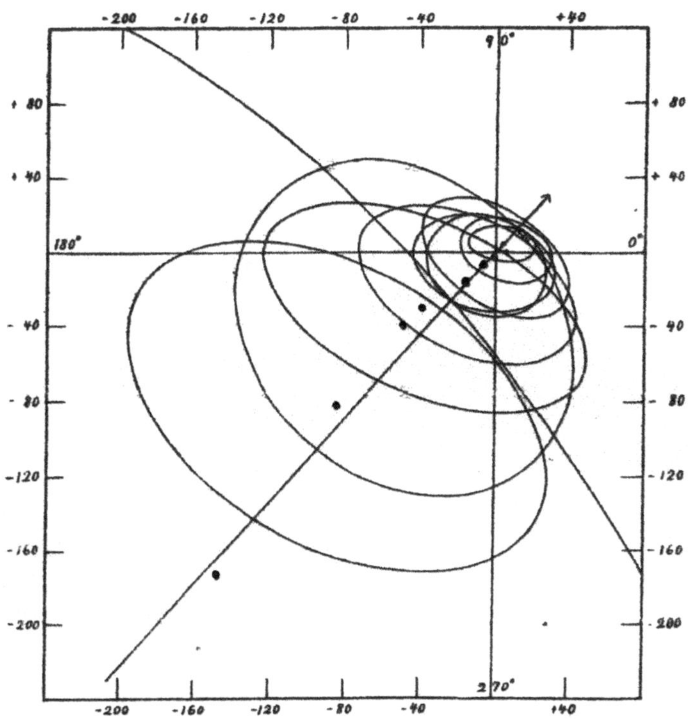

Fig. 3.13 Projection on galactic plane of velocity ellipsoids in relation to stars in solar neighborhood. The dots represent the centers of the ellipsoids (Kaburaki 1933)

axes are nearly parallel to each other extending from the second to fourth quadrants, as seen in Fig. 3.13. Higher-velocity stars indicate a larger-velocity ellipsoid with a higher mean motion with respect to the Sun, and the ellipsoids extend toward the third quadrant.

Kaburagi thus demonstrated that the velocity distribution of stars in the solar vicinity had remarkable asymmetries, which may be related to the structure of the Galaxy.

3.7 Osawa Kiyoteru and Stellar Physics

3.7.1 Life and Works

Osawa Kiyoteru (大沢清輝, 1917–2005) was a pioneer in astrophysical observations in Japan, particularly in photoelectric photometry (Yamashita 2006; Nariai 2006).

He was born in Tokyo and studied astronomy at Tokyo Imperial University. Upon graduation, in 1941, he was hired by the TAO as a technician. In this period, he learned modern electronics in the Faculty of Technology of Tokyo Imperial University. Afterward he was promoted to engineer at the Observatory, where his duty was the maintenance of observational instruments. In this period, Osawa set up a photoelectric photometer and started photometric observations of stars with a 30 cm telescope at the Mitaka campus of the TAO (Fig. 3.14).

In 1953–1956, he stayed at Yerkes Observatory as a guest observer and carried out two works: the spectral classification of B and A stars and calculation of model atmospheres of A-type stars. After returning home, he was promoted to professor of astronomy at the University of Tokyo working at the TAO in 1956.

Fig. 3.14 Osawa Kiyoteru (Yamashita 2006)

The project of constructing a 188-cm telescope and a new observatory was under way at the time. Osawa, as a member of the construction committee, contributed to resolving technical problems, including the site search.

In 1960, the OAO was established and Osawa was appointed first director of this observatory.

He then served as director of the TAO from 1963 up to his retirement in 1977. Thereafter, he served as a professor of astronomy at Chiba University for 5 years.

Osawa died in 2005 at age 88.

3.7.2 *Astrophysical Works*

3.7.2.1 Theoretical Study of A-Type Star Atmospheres

During his stay at Yerkes Observatory, Osawa calculated model atmospheres of A-type stars having effective temperatures of *Te* = 8900 K and 7500 K and surface gravity log g = 3.5, 4.0 and 4.5. In this calculation he took into consideration the continuous absorption by hydrogen, negative ion of hydrogen, and ionized hydrogen molecule. The spectral energy distributions in two stellar types of A3V and A9V are illustrated in Fig. 3.15. The Balmer discontinuity at λ 3647 Å is remarkable in A3 stars. Based on this model, he calculated the profiles of the Balmer H-gamma line and compared with observed line profiles. He found fair agreement between model and observations compared to existing models. He showed that this better agreement

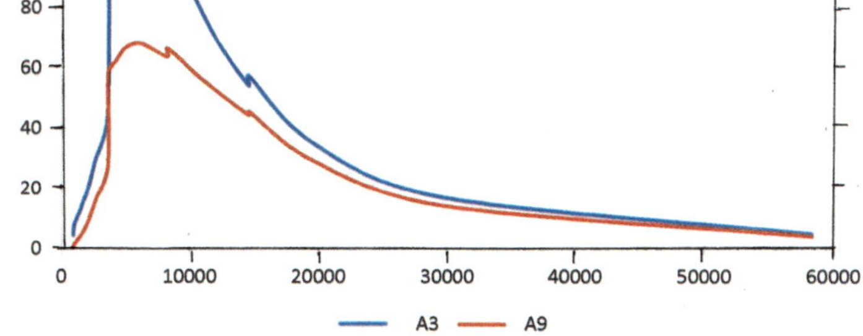

Fig. 3.15 Spectral energy distribution of model atmosphere of A-type main-sequence stars (A3 star, blue line: *Te* = 8900 K, log g = 4.5, and A9 star, red line: *Te* = 7500 K, log g = 4.5). The abscissa is the wavelength (Å), and the ordinate is the energy flux ($F_\nu \times 10^6$) (erg Hz^{-1} cm^{-2} s^{-1}) (Osawa 1956)

was obtained by the inclusion of continuous absorption by hydrogen molecules (Osawa 1956).

3.7.2.2 Photometric Observations of Metallic-Line Stars

Osawa installed a new photoelectric photometer on the 30 cm reflector of the TAO and carried out three color (UBV) observations of metallic-line stars and found that these stars indicated considerable ultraviolet deficiency compared to main-sequence stars (Osawa 1958). The distributions of metallic-line stars on the two-color diagram (U – B, B – V) are illustrated in Fig. 3.16, where the continuous curve shows the approximate mean colors of normal main-sequence stars. The U – B color indicates strong ultraviolet deficiency toward a lower direction in this figure.

3.7.2.3 Spectral Classification

At Yerkes Observatory, Osawa took up the MK spectral classification of early A-type stars. The MK Spectral Atlas by W. W. Morgan et al. yields the two-dimensional fine spectra for around 200 normal and peculiar stars over all spectral types (Morgan et al. 1943). The extension of the spectral classification to a sufficiently large number of stars was an urgent subject in the 1950s. In particular, Osawa carried out the MK classification of stars in spectral classes B8 to A2. Spectral classification near A0 stars is especially difficult because of faint absorption lines, except Balmer series. Osawa overcame this difficulty with sufficiently deep observations and determined the MK spectral class for 533 stars in this spectral region (Osawa 1959).

In 1963 and 1966, Osawa took up the MK classification of faint A-type stars at the OAO and compared it with the HD classification. He found that MK classifications made by Slettebak for bright stars showed a nearly parallel relation with HD

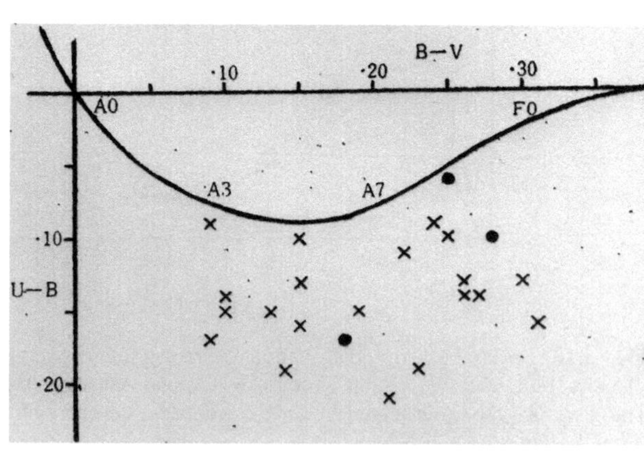

Fig. 3.16 Colors of metallic-line stars. Crosses and dots represent Osawa's measurements and Johnson and Morgan' observations, respectively (Osawa 1958)

classification (Slettebak 1954), whereas the MK types classified by Osawa for faint stars markedly deviated toward the late type, as shown in Fig. 3.17.

3.7.2.4 Observations at OAO

Osawa and his group mainly used a 91 cm reflector for photoelectric observations. He observed several types of variable objects, as shown in Table 3.9, where the characteristics for some stellar types are briefly given as follows:

Hydrogen-deficient stars show little or no hydrogen in their atmosphere and generally belong to early-type stars in a wide luminosity class.

A-type peculiar stars belong to the type of chemically peculiar stars, including metallic-line stars (Am stars) and stars of strong magnetic field.

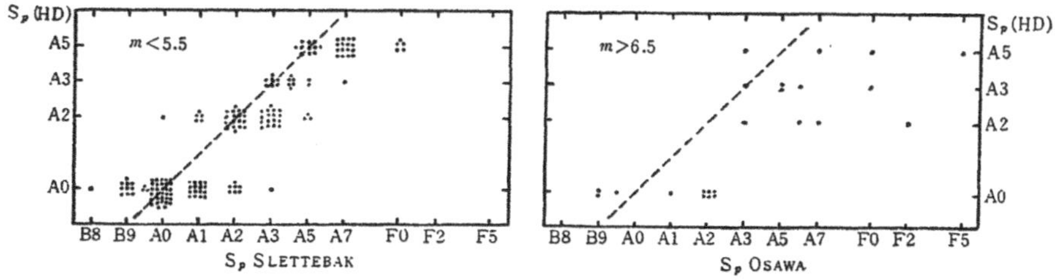

Fig. 3.17 Comparison of HD spectral types with MK types for A-type stars. Left panel: classification by Slettebak for bright stars ($m < 5.5$); right panel: classification by Osawa for faint stars ($m > 6.5$) (Slettebak 1954; Osawa 1963, 1966)

Table 3.9 List of stars observed by Osawa at OAO

Stellar type	Star name	Observation year
Hydrogen-deficient star	HD 30353	1963
A-type peculiar star	HD 221568	1963
Flare star	UV Ceti	1968–1973
	YZ CMi	1968–1973
	EV Lac	1969–1971
	AD Leo	1970–1974
Novae	N Cygi	1975
	N Sagittarius	1975
Nova-like star	In Ophiuckus	1976
Comets	Kohoutek	1973
	Heck-Same	1973
	Gehrels 2	1973
	Kobayashi et al	1975
	Mori-Sato et al.	1975
	Sato	1975

Flare stars are variable stars that undergo an unpredictable dramatic increase of brightness for a few minutes. It is believed that the flares are analogous to solar flares. The brightening occurs across the spectrum from X-ray to radio waves. Most flare stars are dim red dwarfs.

Novae and comets are well-known objects. Osawa's works given in this table were largely taken up by Nishimura Shiro, Nariai Kyoji, and others at the OAO.

In addition, Osawa also carried out the optical identification of X-ray source Sco X-1 with photometric and spectroscopic observations at the OAO in 1966.

3.8 Hirose Hideo and Astronomy

3.8.1 *Life and Works*

Hirose Hideo (広瀬秀雄, 1909–1981) showed deep scholarship in a wide array of fields in astronomy, including observations of small bodies of the Solar System, the design of optical systems and the construction of large telescopes at the Dodaira Station of the TAO, and the history of oriental astronomy (Kawaguchi et al. 1982).

Hirose was born in the city of Himeji, Hyogo Prefecture, entered Tokyo Imperial University, and studied astronomy under Hirayama Kiyotsugu on celestial mechanics, positional astronomy, and the history of astronomy. Hirose's interests in astronomy were highly affected by his teacher throughout his life.

After graduating in 1932, Hirose began to work at the TAO.

His first study was on celestial mechanics. After completing his thesis on the position of moving celestial bodies in 1949, he was appointed professor of astronomy at the University of Tokyo in 1951. Besides the teaching of astronomy, he pursued the observation of asteroids. In 1963, he moved to the TAO and served as director until his retirement in 1970. After retirement, he served as professor successively at Saitama University and Senshu University. He died in 1981 in Tokyo at age 72 (Fig. 3.18).

3.8.2 *Observations of Minor Objects*

One of Hirose's major works was the observation of small celestial bodies in the Solar System. His interest was focused on asteroids in 1949–1956 and then comets in 1965–1970. For both objects, observations had an astrometric purpose, including identifications, accurate positions, orbital elements, ephemerides, and photographic photometry. The results of his observations were reported to the Minor Planet Circulars of the IAU for asteroids and to the IAU Circulars for comets.

In 1960s, Hirose's group, which included Nagasawa Ko and Tomita Koichiro, organized photometric and spectroscopic observations of meteors in several meteor streams from three observation sites (Hirose et al. 1968). For this purpose, they

Fig. 3.18 Hirose Hideo (Kawaguchi et al. 1982)

constructed special wide-field cameras with sky field 60° × 80°, and the sites were selected at the Dodaira Station of the TAO, the Mitaka campus, and the Tatebayashi Station. Tatebayashi is located about 40 km northeast of Dodaira, the Mitaka campus about 40 km southeast of Dodaira, and Tetebayashi about 60 km north of the Mitaka campus. Simultaneous observations from these sites were carried out during the active periods of the major streamers, such as Perseids, Leonids, and Geminids, for the purpose of dynamical and physical studies of meteors. Observations were made with automatic cameras at three sites, whereas observations with an objective grating were made at Dodaira. This combination allowed for the measurement of the position, height, velocity, and spectrum for each meteor. Figure 3.19 illustrates an example of the spectra of a Leonid meteor taken on November 16, 1965. The 0th, first, and second order spectra are shown in the same photographic films. The heights of this meteor are indicated by arrows. This meteor appeared at a height of around 100 km and disappeared at 85 km. The number of emission lines appeared at around 90 km. Other members of the Leonid stream also showed emission lines with different chemical species.

In these observations they found many emission lines in the meteor spectra. The numbers of emission lines were different for different meteor streams, as given in Table 3.10. Hirose's group thus found the emission lines in meteors to be very different from stream to stream. They also showed that the formation of emission lines depended on the velocity and height of meteors. They noticed that the spectral features were different for the fast and slow meteor streams. In the fast meteors, such as Leonids and Perseids, CaII H and K lines were generally the strongest, as well as in some other lines such as Si II, Na I, Fe I, and others, depending on the brightness of the meteors. The forbidden line of O I in bright meteors showed quite peculiar aspects, appearing long before the other lines at heights over 100 km and disappearing first. In the case of slow meteors, such as Geminids, nearly all the lines are identified with Fe I except for the D lines of Na I. They supposed that the difference in the spectra of the fast and slow meteors might have been due to

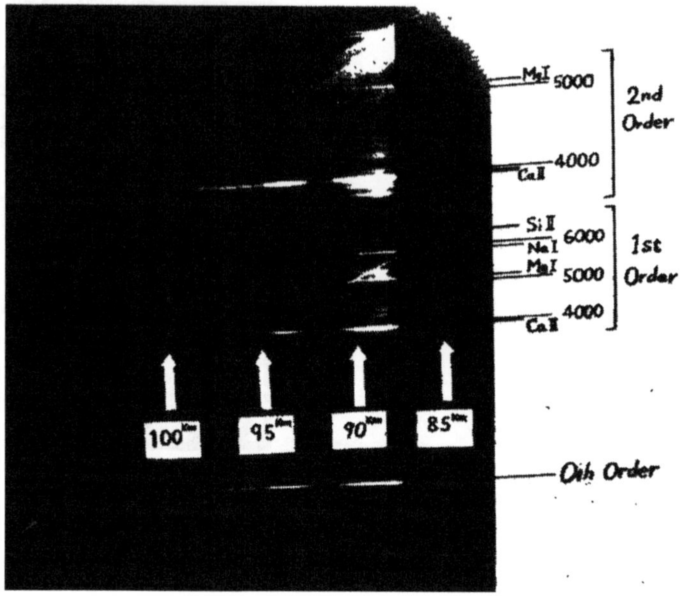

Fig. 3.19 Spectra of a Leonid meteor taken November 16, 1965. The 0th, first, and second order spectra are shown separately. Heights of various points are indicated by arrows (Hirose et al. 1968)

Table 3.10 Number of emission lines detected in meteors (Hirose et al. 1968)

	Number of lines			
Stream	More than 20	10–19	1–9	Total number of meteors
Perseids	–	2	12	14
Taurids	–	–	2	2
Leonids	7	6	25	38
Geminids	3	1	11	15
Sporadics	–	–	10	10
Total	10	9	60	79

the velocity difference between these two groups or to their original chemical composition, though the evidence was not conclusive.

3.8.3 *History of Oriental Astronomy*

Hirose's works on the history of oriental astronomy were produced in the years around his retirement.

Hirose's first work were the studies of Juji Reki (授時暦, Shou-Shi Calendar, see Chap. 1) in two problems. The first one (Hirose 1969) was on the influence of the mathematical method of the Shou-Shi calendar on Japanese mathematics. The

Shou-Shi calendar was produced by Kuo Shoujing (郭守敬) in 1281 in the Yun Dynasty (元) and was introduced into Japan around 1670. Kuo Shoujing applied Arabian astronomy based on spherical trigonometry to the Chinese calendar. In Japan, Seki Takakazu (関孝和,1642–1708) introduced Kuo's mathematics into Japanese mathematics.

Hirose's second topic (1970) was the determinations of the days of the winter solstice in the Shou-Shi calendar and other old calendars in China. He found that the days of the solstice in old calendars are different when compared to the values calculated from the modern calendar.

In the 1970s, Hirose turned his attention to the translation of some astronomical classics from the Tokugawa period into modern Japanese with some critical notes. These translations were presented in the *Series on Japanese Thought*, Volumes 63 and 65 (日本思想大系) in 1971–1972.

3.9 History of Astronomy in Postwar Period

Studies of the history of astronomy in Tokyo area started with Kanda Shigeru and Hirose Hideo. This section discusses the works of Kanda and of the generations following him (Miyajima 2008).

Kanda Shigeru (神田茂, 1894–1974) of the TAO collected records of astronomical phenomena, such as solar and lunar eclipses, arrangements of planets, occultation, comets, guest stars, and meteorological phenomena, observed in the years up to 1600 (Kanda 1935). This collection was extended by Osaki Shoji (大崎正次) to the modern age between 1600 and 1867 (Osaki 1994). Osaki also collected the names of historical constellations in China (Osaki 1986). In the early Zhou period (around 1000 BC), the numbers of constellations are limited to the ecliptic, north polar region, and regions of bright stars like Orion. Over time, the constellations were expanded to the whole sky and gradually introduced to Japan. Osaki identified their names and compared them with Western constellations. Uchida Masao (内田正男), who was working in the same section with Kanda in the TAO, edited an encyclopedia on the calendar and time in Japanese history (Uchida 1986).

Saito Kuniji (斎藤国治, 1913–2003) worked on solar physics at the TAO and, following his retirement in 1974, started studying the history of ancient astronomy. He was interested in various appearances of unusual stars and their effects on social events. He summarized astronomical phenomena in the Asuka period (seventh century, age of Prince Shotoku) (Saito 1982) and in ancient China (Saito and Ozawa 1992). Saito also proposed the "ancient astronomy" as a new science field. This field treats the studies on the reliability of ancient records and on the accurate determination of the ages of historical events (Saito 1982). This new field yields close connection of humanistic science (literature and history) and natural science (astronomy). For example, Fig. 3.20 illustrates the paths of a total eclipse (July 3, 1742) and a total ring eclipse (January 30, 1786). These eclipses were observed in Hirosaki (弘前), Sendai (仙台), Kyoto (京都), and other locations,

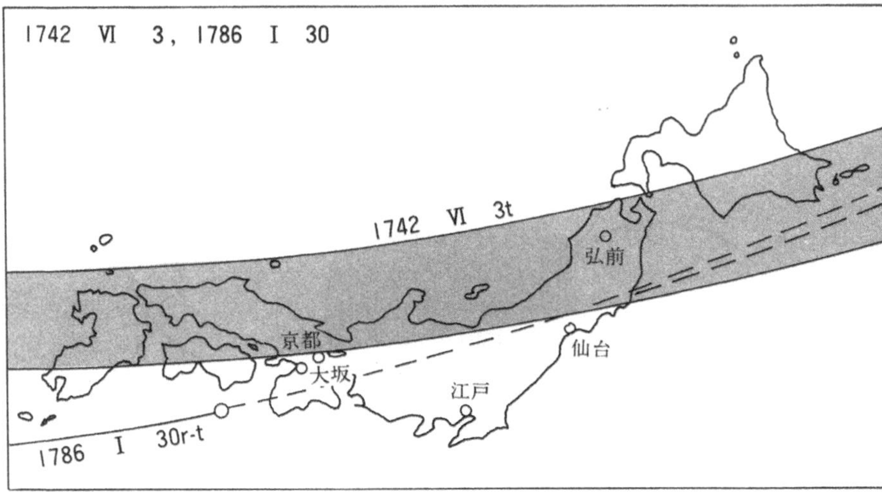

Fig. 3.20 Total eclipses observed in eighteenth century. Total eclipse (July 3, 1742) and total ring eclipse (January 30, 1786). The eclipse paths were calculated by Saito (1990)

among which the observations of eclipse and social reactions were recorded at Hirosaki in detail under the collaboration of Clan peoples (Saito 1990).

Nakayama Shigeru (中山茂, 1928–2014) studied the history of astronomy at Harvard University in the 1960s and his thesis constituted a comprehensive history of Japan up to the early Meiji era. Astronomy in Japan has been advanced through the impacts from China and Western countries. In the late nineteenth century, Western supremacy was finally recognized, and modern astronomy started in Japan (Nakayama 1969). Nakamura Tsuko (中村士) of the TAO collected bibliographical data on Japanese astronomy in the pre-Meiji period (Nakamura and Ito 2006). He also promoted international cooperation on the history of astronomy in Asian countries, particularly on the rise of astrophysics in Asia (Nakamura and Orchiston 2017).

References

Aller, L. H. (1941). Physical processes in gaseous nebulae. XIV. *Astrophysical Journal, 93*, 236–243.

Baker, J. G., & Menzel, D. H. (1938). Physical processes in gaseous nebulae. III. The Balmer decrement. *Astrophysical Journal, 88*, 52–64.

Blaauw, A. (1994). *History of the IAU*, Kluwer Academic Publishers.

Bowen, I. S. (1934). The excitation of the permitted OIII nebular lines. *Publications of the Astronomical Society of of the Pacific, 46*, 146–148.

Bowen, I. S. (1935). The spectrum and composition of the gaseous nebulae. *Astrophysical Journal, 81*, 1–16.

DeVorkin, D. (1997). *Oral history interview, Yoshio Fujita*, American Institute, Physics Publishing.

Fujita, Y. (1935). Dissociation of diatomic molecules in the stars abundant of hydrogen. *Japanese Journal of Astronomy and Geophysics, 13*, 21–42.
Fujita, Y. (1938). Dissociation of molecules in the carbon stars. *Proceedings of Physical and Mathematical Society of Japan, 20*, 149–159.
Fujita, Y. (1939). An interpretation of the spectral sequence for the late-type stars. *Japanese Journal of Astronomy and Geophysics, 17*, 17–57.
Fujita, Y. (1941a). On the M-S differentiation of the late type stars. *Japanese Journal of Astronomy and Geophysics, 18*, 45–49.
Fujita, Y. (1941b). On the M-S differentiation of the late type stars (Second paper). *Japanese Journal of Astronomy and Geophysics, 18*, 177–183.
Fujita, Y. (1951). Absorption lines and bands in the spectrum of Chi Cygni. *Astrophysical Journal, 113*, 620–629.
Fujita, Y. (1986). *My half century spent with stars* (藤田良雄、星とともに半世紀), Private issue.
Fujita, Y. (1997). *Collected papers on the spectroscopic behavior of cool stars*, Private issue.
Fujita, Y., & Utsumi, K. (1963). Spectral features in the infrared region of some carbon stars. II. *Proceedings of Japan Academy, 39*, 358–363.
Hagihara, Y. (1930). Theory of the relativistic trajectoris in a gravitational field of Schwarzschild. *Japanese Journal of Astronomy and Geophysics, 8*, 67–167.
Hagihara, Y. (1937). Radiative equilibrium of a planetarium. *Japanese Journal of Astronomy and Geophysics, 15*, 1–136.
Hagihara, Y. (1939). Electron velocity distribution in a planetary nebula. *Japanese Journal of Astronomy and Geophysics, 17*, 199–264.
Hagihara, Y. (1940). The electron velocity distribution in the planetary nebulae. *Monthly Notices of the Royal Astronomical Society, 100*, 631–634.
Hagihara, Y., & Hatanaka, T. (1946). On the radiative transfer in an expanding planetary nebula. *Japanese Journal of Astronomy and Geophysics, 21*, 45.
Hatanaka, T. (1942). Intensity of forbidden lines and abundances of OII and OIII atoms in planetary nebulae. *Japanese Journal of Astronomy and Geophysics, 20*, 19–35.
Hatanaka, T. (1946). Theory of optical interaction among HeII, OIII and NII atoms in planetary nebulae. *Japanese Journal of Astronomy and Geophysics, 21*, 1–53.
Hatanaka, T. (1957). Polarization of solar radio bursts. *IAU Symposium*, (4), 358–362.
Hatanaka, T., & Moriyama, F. (1953). A note on the long-period variation in the radio-frequency radiation from the quiet Sun. *Publications of the Astronomical Society of Japan, 4*, 145–151.
Hatanaka, T., Akabane, K., et al. (1955). Solar radio spot and calcium plage. *Publications of the Astronomical Society of Japan, 7*, 161–162.
Hatanaka, T., Hayakawa, S., et al. (1964). Part I. Evolution of galaxies – Development of its research in Japan. *Progress in Theoretical Physics, 31*, 2–34.
Hirose, H. (1969). Studies of Zyuzi-Reki or Shou-Shih-li (1)(広瀬秀雄. 東京天文台報, 授時暦の研究.1). *Tokyo Astronomical Observatory Report, 14*, 499–511.
Hirose, H. (1970). Studies of Zyuzi-Reki or Shou-Shi-Li (2) (広瀬秀雄, 東京天文台報, 授時暦の研究 2) *Tokyo Astronomical Observatory Report, 14*(2), 376–386.
Hirose, H., Nagasawa, K., & Tomita, K. (1968). Spectral studies of meteors at the Tokyo astronomical observatory. *IAU Symposium*, (33), 105–118.
Hitotuyanagi, Z., et al. (1964). Astronomivsl Herald, 57, 33–36, *Obituary of Hatanaka Takeo, collection of Hitotuyanagi, Z., Miyamoto, S., Hayakawa, S., Unno, W., and Moriyama, F.* (天文月報、一柳寿一、他: 畑中武夫追悼文集).
Jeans, J. H. (1923). The mechanism and structure of planetary nebulae. *Monthly Notices of the Royal Astronomical Society, 83*, 481–491.
Kaburaki, M. (1933). On the motion of the local system as viewed from that of moving clusters, nearer stars, and high velocity stars. *Japanese Journal of Astronomy and Geophysics, 10*, 313–367.
Kaburaki, M. (1963). Reminiscences of 40 years (天文月報、鏑木政岐: 40年を回顧して). *Astronomical Herald, 56*, 57–59.

Kanda, S. (1935). *Collection of old astronomical data in Japan* (Ed.) Koseisha (神田茂編、日本天文史料、恒星社刊).

Kapteyn, J. C. (1904). Statistical method in stellar astronomy. *International Congress of Arts and Sciences, St. Louis, 4*, 417–422.

Kapteyn, J. C., & Weersma, H. A. (1912). Stars. Motion in space, Second note: The position on the derivation of the constants for the two star stream. *Monthly Notices of the Royal Astronomical Society, 72*, 745–756.

Kawaguchi, I., et al. (1982). Obituaries of Professor Hirose Hideo (天文月報、川口市郎他: 広瀬秀雄教授追悼文集). *Astronomical Herald, 75*(3).

Miyajima, K. (2008). *One-hundred years of the Astronomical Society of Japan* (日本天文学会編、日本の天文学の百年), Hyoronsha (評論社), 153–158, History of astronomy (宮島一彦, 天文学史).

Morgan, W. W., Keenan, P. C., & Kellman, E. (1943). *An atlas of stellar spectra with an outline of spectral classification*. The University of Chicago Press.

Nakamura, T. & Ito, S. (2006). *General catalog of Japanese astronomical bibliography and materials in the pre-Meiji period* (中山士及び伊藤節子: 明治期前日本天文暦学・測量の書目事典、第一書房).

Nakamura, T., & Orchiston, W. (Eds.). (2017). *The emergence of astrophysics in Asia*. Springer.

Nakayama, S. (1969). *A history of Japanese astronomy – Chinese background and western impact*. Harvard University Press.

Nariai, K. (2006). Obituary of Professor Osawa Kiyoteru (天文月報、成相恭二: 大沢清輝先生の思い出). *Astronomical Herald, 99*, 279–280.

Osaki, S. (1986) *A history of Chinese constellations*, Yuzankaku (大崎正次: 中国の星座の歴史、雄山閣)、.

Osaki, S. (1994). *A historical data of the modern Japanese astronomy,* Hara Shobo (大崎正次: 近世日本天文史料、原書房).

Osaki, Y. (2008). Interview to Yoshio Fujita. In *One-hundred years of astronomy in Japan,* Astronomical Society of Japan, 263–270, Kouseisha-Kouseikaku (日本の天文学の100年, 尾崎洋二: 藤田良雄先生へのインタビュー. 恒星社厚生閣).

Osawa, K. (1956). Model atmosphers for A-type stars. *Astrophysical Journal, 123*, 513–520.

Osawa, K. (1958). On colors of metallic-line stars. *Publications of the Astronomical Society of Japan, 10*, 102–103.

Osawa, K. (1959). Spectral classification of 533 B8 – A2 and the mean absolute magnitude of A0 V stars. *Astrophysical Journal, 130*, 159–177.

Osawa, K. (1963). Spectral classification and three-color photometry of A-type stars. *Publications of the Astronomical Society of Japan, 15*, 274–276.

Osawa, K. (1966). Spectral classification and three-color photometry of A-type peculiar stars. *IAU Symposium, 24*, 15–20.

Rasmuson, N. M. (1921). A search on moving clusters. *Meddelanden fran Lunds Astronomiska Observatorium, Series II, 26*, 3–74.

Saito, K. (1982). *Astronomy in Asuka period*, Kawade Shobo-Shinnsha (斎藤国治: 飛鳥時代の天文学、河出書房新社).

Saito, K. (1990). *The road to ancient astronomy*, Hara Shobo (斎藤国治: 古天文学の道、原書房).

Saito, K., & Ozawa, K. (1992). *Inspection of astronomical records in the ancient China*, Yuzannkaku (斎藤国治、小沢賢二: 中国古代の天文記録の検証、雄山閣出版).

Schwarzschild, K. (1907). Über die Eigenbewegung der Fixsterne. *Nachrichten von der Gesellschaft der Wissenschaften zu Göttingen, 615*, 614–632.

Schwarzschild, M., & Spitzer, L. (1953). On the evolution of stars and chemical elements in the early phases of a galaxy. *Observatory, 73*, 77–79.

Shane, C. D. (1928). The spectra of carbon stars. *Lick Observatory Bulletin*, (396), 123–129.

Slettebak, A. (1954). The spectra and rotational velocities of the bright stars of Draper types B8 – A2. *Astrophysical Journal, 116*, 146–166.

Suemoto, Z., et al. (1979). Reminiscences of Professor Hagihara Yusuke (天文月報、末元善三郎他: 萩原雄佑先生追悼集). *Astronomical Herald, 72*, 117–124.

Takase, B. (1988). Obituary of Professor Kaburagi (天文月報、高瀬文志郎: 鏑木教授追悼). *Astronomical Herald, 81*(4), 106–107.

Taketani, M., & Hatanaka, T. (1956). Population and evolution of stars. *Progress in Theoretical Physics, 15*, 89–94.

Tokyo University. (1987). *One-hundred years of the Tokyo Astronomical Observatory*, Tokyo University (東京大学百年史—部局史—東京天文台).

Uchida, M. (1986). *Dictionary of calendar and time in Japan*, Uzankaku Pub. (内田正男: 暦と時の事典、日本の暦法と時法、雄山閣).

Unno, W. (2008). *One-hundred years of the Astronomical Society of Japan* (日本天文学会編、日本の天文学の百年)、Hyoronsha (評論社), Chapter 2, 13–25, A half century of astronomy in Tokyo (海野和三郎:東京における天文学の半世紀).

Yamashita, Y. (2006). Professor Osawa Kiyoteru and Okayama (天文月報、山下泰正: 大沢清輝先生と岡山). *Astronommical Herald, 99*(5), 278.

Yamashita, Y., Kozai, Y., et al. (2013). Reminiscences of Professor Fujita Yoshio, collection of Yamashita, Y., Kozai, Y., Hiei E. and Tsuji T. (天文月報、山下泰正, 古在由秀他: 藤田良雄先生追悼文集). *Astronomical Herald, 106*(4), 278–286.

Chapter 4
Astronomy in Early Showa. II. Kyoto 1926–1945

Abstract Since its establishment, Kyoto Imperial University has been imbued with a spirit of independence in all colleges of the humanities and natural sciences. This sprit was cultivated over the long history of Kyoto as the former capital of Japan for around 1000 years.

Astronomy was no exception. The first professor of astronomy, Shinjo Shinzo, developed his own view of astrophysics. His scholarly spirit was taken up by his disciples for about a half century (Kogure 2008).

In this chapter, the development of astrophysics is traced through the lives and works of Yamamoto Issei, Araki Toshima, Takeda Shin'ichiro, and Miyamoto Shotaro. In addition, the history of Chinese and Japanese astronomy is presented through the works of Noda Churyo, Yabuuchi Kiyoshi, and Watanabe Toshio, together with some postwar studies.

4.1 Yamamoto Issei and Variable Stars

4.1.1 Early Life and Observations of Novae

To the east of Lake Biwa extend fields and hills. On one of the hills, Kami-Tanakami Mura (上田上村、now the city of Otsu), Yamamoto Issei (山本一清, 1889–1959) was born into an old family from the area. In the Edo era, this family, though common people, was permitted to use the surname and wear a sword (Miyamoto et al. 1959, Kibe 1959).

Yamamoto entered Dai-San High School (old) in Kyoto in 1907 and then the Science and Technology College of Kyoto Imperial University in 1910. He studied astronomy under Shinjo in the Department of Physics. He was the first student of astronomy in Kyoto. In 1913 he graduated and entered the graduate school in astronomy. In the same year he married Kawasaki Eiko (川崎英子), who supported her husband's astronomy through the popularization of astronomy even after Issei's death (Fig. 4.1).

Yamamoto had been interested in observations of variable stars and novae. The first event was the appearance of a nova in the constellation Gemini in 1912. He

T. Kogure, *The History of Modern Astronomy in Japan*, Historical & Cultural Astronomy, https://doi.org/10.1007/978-3-030-57061-3_4

Fig. 4.1 Photograph of Yamamoto Issei standing by his Calver telescope. (Shinjo Bunko, Department of Astronomy, Kyoto University)

successively observed the novae in Gemini (1913) and in Lacerta (1914). After being appointed research assistant in the Department of Physics in 1914, he spent 2 years at the International Latitude Observatory, Mizusawa, studying positional astronomy. Thereafter he concentrated his work in Kyoto on the observations of variable stars. For photometric observations he used the unaided eye and binoculars, while for spectroscopic observations, he used an objective prism with a 13° apex attached to the Sartorius 18-cm refractor at the Kyoto University Observatory.

On June 9, 1918, a solar eclipse occurred at Torishima (鳥島, St. Peter's Island) in the Northwest Pacific Ocean. Although the eclipse was obstructed by clouds, Yamamoto discovered by chance a new star (Nova Aquilae No. 3) during the night of June 11. From that time he and his group conducted ongoing observations of this nova until December 7, 1918 (Yamamoto et al. 1919, Yamamoto 1919b). The light curve and some of the spectrograms are shown in Figs. 4.2 and 4.3. Yamamoto noticed that this nova declined rather rapidly and entered swiftly to the nebular stage, showing broad emission lines, in early July, as shown in Fig. 4.3.

Based on his observations of several novae, Yamamoto summarized the general properties of nova phenomena as follows (Yamamoto 1919a):

(a) After the maximum light, novae rapidly decline with periodic fluctuations to some degree in the light curves and gradually return to being original faint stars.
(b) In the spectroscopic variations of novae, some regularity can be seen in the order from a B-type star at the light maximum to an F-type in the declining phase, and then to the phase of emission-line stars.
(c) Emission lines generally demonstrate juxtapositions of bright and dark components in the red and violet sides, respectively. This indicates the existence of expanding atmospheres around the novae.

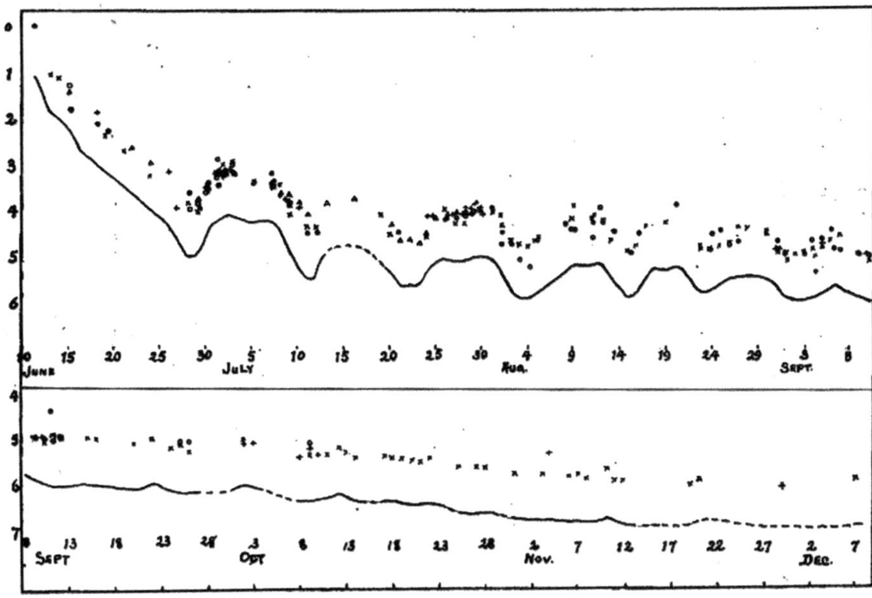

Fig. 4.2 Light curve of Nova Aquilae No.3 from June 11 until December 7, 1918. (Yamamoto et al. 1919)

Fig. 4.3 Samples of spectra of Nova Aquilae No.3, 1918. (Yamamoto et al. 1919)

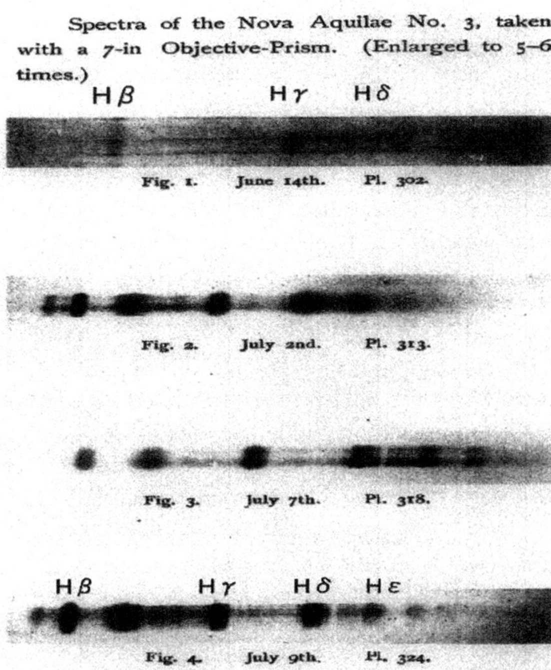

(d) Large widths of emission lines indicate large expanding motion of the atmospheres.
(e) Emission lines of hydrogen, helium, and calcium resemble those of solar prominence, suggesting the similarity of nova explosions with the prominence activities of the Sun.

From these behaviors Yamamoto conjectured that nova phenomena would be a tremendous gas outburst on a stellar surface, with velocity increasing to as high as 1000–2000 km s^{-1} or above.

4.1.2 Observations in USA

Yamamoto was sent to the USA and Europe by the government in 1922–1925. He enjoyed the journey with his wife and submitted his everyday diary to a magazine, *Tenkai* (天界, *The Heavens*), in some detail (Yamamoto 1922–1925).

In September 1922, the Yamamotos arrived at the Yerkes Observatory, welcomed by Director Edwin B. Frost, much like with Ichinohe Naozo in 1905.

When Yamamoto arrived, Yerkes Observatory was celebrating its 25th anniversary, and many special events were being held. This observatory was founded in 1897, as a first-generation modern observatory for astrophysical observations, like the Lick Observatory, established in 1888. The main instruments of the Yerkes Observatory were a 40-inch (102-cm) refractor and a 24-inch (61-cm) reflector. In the same year, a twin Bruce telescope (10- and 6.5-inch, or 25- and 12-cm) was added with the funds of Catherine Bruce of New York City. The main research staff at the time of Yamamoto's visit at Yerkes were E. B. Frost, Edwin E. Barnard, and G. Van Biesbroeck.

Yamamoto's work at the observatory was to get accustomed to the technique of photographic observations. He joined the group of Van Biesbroeck in the photographic search for faint asteroids. They measured asteroids on the accumulated photographic plates taken with the 24-inch reflector, and in 1922 they found five new asteroids of 15–16 magnitudes, which were the faintest ones at that time (Van Biesbroeck et al. 1923).

He also studied photographic techniques under Barnard throughout his stay. Barnard was not only a man of devoted to astronomical observations throughout his life; he was also a gentleman with grace and charm.

One day the Yamamotos were invited to the home of Barnard, who was in the latest stage of his life. On this visit Yamamoto wrote:

> He [Barnard] welcomed us with a face beaming with joy. He offered us chairs and a beverage. Feeding the fireplace, playing a phonograph, showing pictures and photos, he expressed his gentle welcome though in a piteous plight. We were entirely overwhelmed with gratitude.

Fig. 4.4 Snapshot of Barnard at Yerkes Observatory in 1922, photographed by Yamamoto. (Yamamoto Album, Department of Astronomy, Kyoto University)

Fig. 4.5 Snapshot of Shapley at tea time at HCO in November 1924, photographed by Yamamoto. (Yamamoto Album, Department of Astronomy, Kyoto University)

Soon after this visit, Barnard was hospitalized and received treatment; he passed away after 6 weeks at age 67. The Yamamotos attended the funeral held at the observatory (Figs. 4.4 and 4.5).

In November 1924, Yamamoto moved to the Harvard College Observatory (HCO), where he continued the photographic photometry of variable stars under collaboration with Harlow Shapley (1885–1972), the observatory's fifth director. The HCO is well known for two big projects, spectral classification and stellar photometry, promoted by E. C. Pickering and carried out by a group of women astronomers. When Yamamoto arrived, these works were almost finished. Annie J. Cannon had just published the last volume of the Henry Draper Catalogue, and Henrietta Leavitt was already deceased.

Cannon was a naturally sociable person and an optimist. He became fast friends with Mr. and Mrs. Yamamoto and invited them on driving tours and to musical concerts. They visited each other's homes and spent many relaxed hours together.

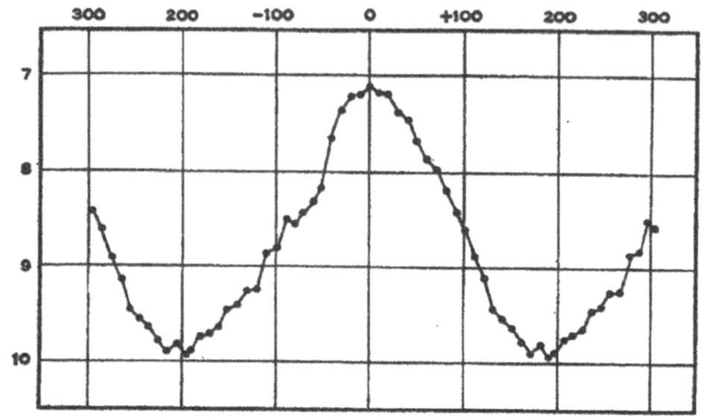

Fig. 4.6 Mean light curves of W Hydrae; ordinates are photographic magnitude, abscissas are days preceding and following maximum. (Yamamoto and Campbell 1924)

Yamamoto's first work at the HCO was the measurement of the light variation of long-period variable W Hydrae, which was known as an extremely red star with a period of around 380 days and (M8e in the HD Catalogue). The observatory housed a large collection of over 700 photographic plates of this star, taken between 1899 and 1923. The mean light curve measured by Yamamoto on the selected photographic plates is shown in Fig. 4.6 (Yamamoto and Campbell 1924). This was the first sample of light curves for very red long-period variables.

The second work of Yamamoto was the photographic photometry of Cepheid variables in the Small Magellanic Cloud (SMC) in collaboration with H. Shapley and H. H. Wilson. They had long been working on the large-scale structure of the Galaxy, based on the determination of distance to globular clusters, by making use of the period-luminosity relation of Cepheid variables. This relation was first derived by H. Leavitt for the 25 Cepheid variables in the SMC in 1912 (see appendix, Fig. A.2). Shapley attempted to revise this relation, and he invited Yamamoto to the further measurements of this relation in the same SMC. Yamamoto measured for additional 75 stars on the photographic plates accumulated in the HCO and derived the new period-luminosity relation. The results of his measurement are shown in Fig. 4.7, including the values of Leavitt (Shapley et al. 1925). In later discussions on the structure of globular-cluster systems, Shapley used this revised relation of Cepheid variables (Shapley and Sawyer 1929).

Yamamoto finished the up-to-date research on the photographic photometry at the HCO in August 1924. A month later he moved to Europe and came back to Kyoto after visiting Paris, Cambridge, the Vatican, and some other observatories.

4.1.3 Observations in Kyoto

In 1925, Yamamoto returned to Kyoto and was promoted to professor of astronomy in Kyoto Imperial University. He continued his observations started in 1918 with the Sartorius 18-cm refractor at the university observatory. Nakamura Kaname (中村要, 1904–1932) was hired as his assistant in 1921. He was a man of genius when it came

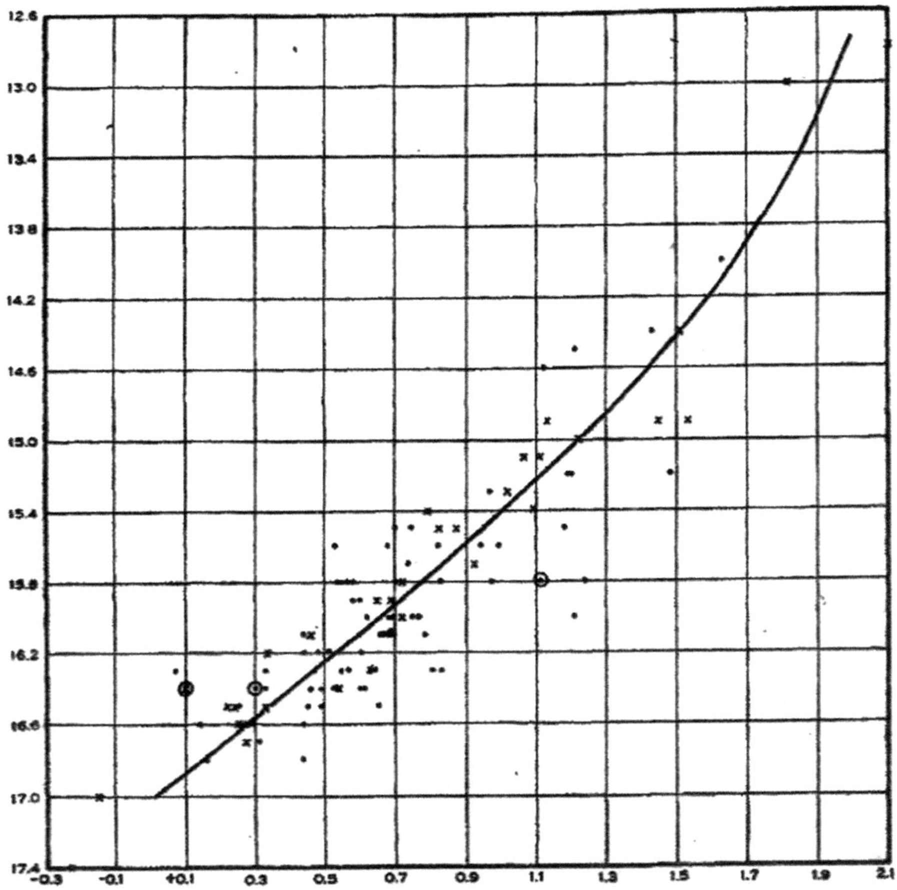

Fig. 4.7 Photographic period-luminosity curve. Ordinates are apparent magnitude; abscissas are logarithms of the periods. Crosses indicate measurements by Leavitt (1912) and dots by Yamamoto. (Shapley et al. 1925)

to observational technique including the making of various astronomical lenses and small telescopes. Yamamoto and Nakamura started photographic and spectroscopic observations. They collected around 2000 plates for stars and solar system objects up to 1930. These data included 90 spectrograms of 40 stars in various spectral types from B to M types and from luminosity classes I to V. From its rather homogeneous distribution of stars on the HR diagram, the purpose of observations might have been to produce a spectral atlas of bright stars, but the program was not implemented due to the premature death of Nakamura in 1932 at age 29 (Tomita and Kubota 2000).

In 1929, Kwasan Observatory was newly established at the summit of Mount Kwasan. Yamamoto served as the first director until his retirement in 1939 (Fig. 4.8).

Fig. 4.8 An overview of Kwasan Obsertatory. Left dome contains the Sartorius 18-cm refractor, and the right main dome is the Cooke 30-cm refractor. (Kwasan Observatory, its 80-year path, 花山天文台80年の歩み, p. 132)

Fig. 4.9 Exterior of Yamamoto Observatory. (Tomita 2012)

4.1.4 Popularization Activity

After retirement Yamamoto constructed a private observatory, Yamamoto Observatory, equipped with a 46-cm refractor made by Calver. The observatory was opened for amateur astronomers and for the education of young astronomers (Tomita 2012). Figure 4.9 illustrates the exterior of Yamamoto Observatory, the first and second floors to the left are rooms where astronomers can stay to conduct observations and astronomical experiments. To the right, the upper floor is the observing room with a sliding roof, where the Calver telescope is installed. The lower floor is an optical laboratory.

Yamamoto's other contribution in the popularization of astronomy was to found the Oriental Astronomical Society. While he was in the USA, he wrote a paper titled "Astronomy in Japan" (Yamamoto 1923) in which he compared the two astronomical societies:

There are two large astronomical societies in Japan:

1. The Astronomical Society of Japan, founded in 1908 by H. Terao, former director of the Tokyo Observatory, has about 600 members, and its publication was the *Astronomical Herald*, published monthly in Tokyo. Its office is in the Tokyo Observatory.
2. The Society of Astronomical Friends was founded in 1920 by I. Yamamoto. It has about 1200 members. It publishes the monthly *The Heavens* in Kyoto and the *Bulletin* (in English) at irregular intervals. Its office is the Kyoto University Observatory.

In this paper, the Society of Astronomical Friends was the original name of the Oriental Astronomical Society, and it was renamed to its present name in 1943. This society, along with the Astronomical Society of Japan, remains active in research and the popularization of astronomy.

Yamamoto died of cancer in 1959 at age 70. One of his disciples, Kibe Narimaro (木辺成麿), wrote in his reminiscences (Kibe 1959):

> He [Yamamoto] was essentially a virtuous man, so that he had a single-minded side. He sometimes thundered at us, but it cleared up after ten minutes. As it might be, he was a Christian, and he didn't like drinking or tobacco. We, his disciples, often sighed why he didn't like drinking even a little.

4.2 Araki Toshima and Astrophysics

4.2.1 Early Life and Foreign Study (Kiyonaga 1979)

Araki Toshima (荒木俊馬, 1897–1978) was born in Kutami Cho (来民町, present-day city of Yamaga, Kumamoto Prefecture) the son of Araki Takejiro (荒木竹次郎), principal of a middle school, who died while traveling when Toshima was 13 years old. Toshima entered Kyoto Imperial University and studied astronomy under Shinjo Shinzo in the Department of Physics. When the Department of Astronomy was established in 1921, he moved to this new department. After graduation in 1923, he was appointed lecturer. In the following year he married Shinjo Kyoko (新城京子), the eldest daughter of Shinjo Shinzo, and he was promoted to associate professor of astronomy.

Araki's first work in the Department of Astronomy was on the light variations of Cepheid and long-period variables in collaboration with his father-in-law (Shinjo and Araki 1924). These variables reveal some continuous period fluctuations. They tried to explain such fluctuations as caused by the existence of a third body in Shinjo's eccentric-nucleus model (Fig. 2.11). Their interpretation was not successful, and Araki finally abandoned the eccentric-nuclear model and accepted the pulsation theory of Harlow Shapley (Araki 1925).

In 1929, Araki was sent to the Astrophysical Observatory of Potsdam, where he studied observational astrophysics under Hans Ludendorff (1873–1941), who was

Fig. 4.10 Portrait of Araki Toshima at Kyoto University Observatory, ca. 1927. (Shinjo Bunko, Department of Astronomy Library, Kyoto University)

director of the Potsdam Observatory and was working on the physical nature of variable stars and globular clusters. When Araki was at Potsdam, Ludendorff was undergoing statistical studies of light curves of long-period variables.

Because Araki wished to study more theoretical astrophysics, he moved to the University of Berlin, where he learned quantum mechanics and relativity from Max. T. F. von Laue. Laue was working on the diffraction of X-rays by crystals. Although Araki studied theoretical physics under Laue, he showed little interest in Laue's X-ray physics. Instead, Araki was attracted by the works of E. A. Milne of the Victoria University of Manchester on the internal structure of stars. Stellar structure became his study subject after coming back to Kyoto (Fig. 4.10).

Araki loved traveling and sketching landscapes, and he visited many places in Europe. It is said that Araki sent more than 1800 picture postcards to his mother during his stay in Europe. These pictures were later collected and published under the title *Chuzan's Travel Pictures*, where Chuzan (疇山) was Araki's pen name. Figure 4.11 illustrates one of his sketches. He had a bit of a flamboyant character.

4.2.2 Internal Structure of Stars

After returning to Kyoto, Araki concentrated his work on the internal structure of white dwarfs as an extension of Milne's work.

In Milne's time, the generation of stellar energy was widely attributed to the annihilation of protons and electrons. Based on this theory, Milne (1930) considered the stellar structure and divided stars into two types. The first is the *centrally condensed* type, in which the energy source is confined to the central core; normal stars (giants and dwarfs) belong to this type. The second is the *collapsed* type, or white dwarf type, in which the mass is highly collapsed with a uniform distribution

Fig. 4.11 Landscape near the Observatory Capademonte in Napoli, sketched by Araki. (*Chuzan's Travel Pictures* 1929)

Table 4.1 Structure of white dwarf stars (Araki 1934)

	Milne's standard model	Araki's model	Observed value
Mean gas density (g cm^3)	4.79×10^3	7.66×10^4	6.1×10^4
Stellar radius (cm)	9.94×10^8	1.74×10^9	1.88×10^9
Effective temperature (K)	11,300	8330	8000–10,000

of the energy source. These two types are defined by the stellar luminosity, as shown in the following scheme (Milne 1930):

Luminosity L:	0	L_0	L_1 and large
Energy source:	Uniform source	Point source	No stable state
Stellar type:	Collapsed (white dwarf stars)	Normal (dwarf stars and Giant stars)	Nonexistent

This scheme was widely accepted in the years around 1930. Araki began to work on the structure of white dwarfs. As the standard model, Milne assumed that the absorption coefficient κ and the rate of energy generation ε were both constant inside stars, whereas Araki supposed that the energy production rate depended on gas density $\rho(r)$ in the form

$$\varepsilon = \kappa \left(1 - B \frac{\rho_0}{\rho}\right) \tag{4.1}$$

where K and B are constants. Araki constructed a stellar model in this case, and his numerical solution is given in Table 4.1, along with Milne's standard model and observational values (Araki 1934). Thus, Araki claimed that his model yields better agreement with observations.

After this work, Araki and Milne both switched their focus from the topic. Araki turned his interest to the physics of stellar atmospheres, and Milne moved to Oxford University and began work on relativistic cosmology.

4.2.3 Extended Atmospheres of Stars

In 1937–1942, Araki, collaborating with Kurihara Michinori (栗原道徳, 1903–1978), carried out theoretical studies on the formation of emission lines in extended atmospheres (or envelopes) in a state of expansion, as seen in Wolf-Rayet (WR), Novae, and P Cygni type stars.

The nature of WR stars was first investigated in 1929 by C. S. Beals (1899–1979) in the Dominion Astrophysical Observatory, Canada (Beals 1929). He paid special attention to the spectral line profiles of these stars. Their profiles are characterized by a strong and broad emission with an absorption edge on the violet side. Beals proposed for the first time a theory that the emission lines of these stars are formed in spherically expanding envelopes, whereas the dark violet edges represent the absorption of stellar radiation in the envelope lying between the star and observer. He also applied this theory to the P Cygni stars, which show similar line profiles (Beals 1934a, b).

Araki and Kurihara considered the emission-line profiles formed in an acceleratively expanding envelope (Araki and Kurihara 1937). Compared to planetary nebulae, stellar envelopes are characterized by higher electron densities that require the solution of radiative transfer in parallel with the equations of recombination processes.

To avoid solving such cumbersome processes, Araki and Kurihara introduced a new concept of *optically effective boundary* in the case of acceleratively expanding envelopes. Suppose a frequency ν in an emission line, and the radiation of this frequency is emitted from a limited region having an equal Doppler velocity inside the envelope. Araki and Kurihara called this region the *optically effective boundary*. By introducing this concept, they were able to calculate emission-line profiles. For a spherical envelope, they assumed that the expanding velocity $V(r)$ at radius r is given as a power function of index s as

$$V(r) = V^* \left(\frac{r}{R^*}\right)^s \tag{4.2}$$

where V^* denotes the outflow velocity at the stellar surface and R^* the stellar radius.

They calculated emission-line profiles for some cases of s, and the results are shown in Fig. 4.12, where the envelope size is assumed to be sufficiently large compared to the stellar radius.

Araki and Kurihara applied these theoretical profiles to the variation of the spectrum of Nova Aquilae 1918 observed by Beals, as shown in Fig. 4.13, where the profile of the Hβ line is observed in two epochs of August 23 (lower profile) and

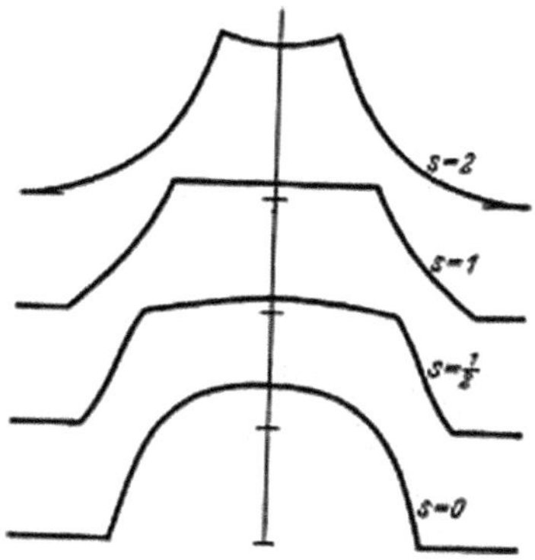

Fig. 4.12 Theoretical emission-line profiles formed in an acceleratively expanding envelope. The power index s is given in eq. (4.2). (Araki and Kurihara 1937)

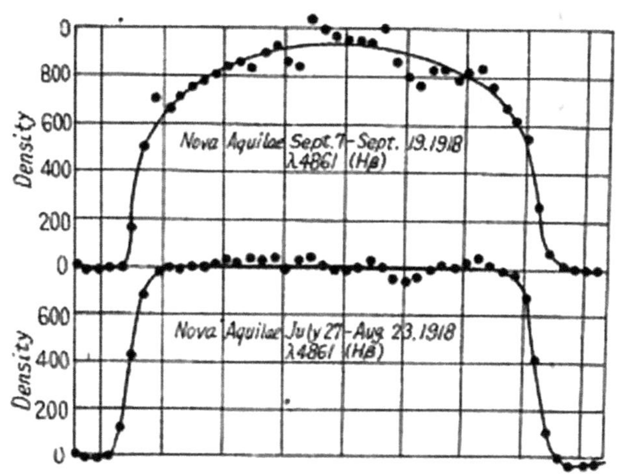

Fig. 4.13 Profiles of Hβ emission observed in Nova Aquilae in August (lower profile) and September (upper profile) in 1918, fitted with theoretical profiles. (Araki and Kurihara 1937)

of September 19 (upper profile) (Beals 1934a), along with the theoretical profiles for the fitting. Comparing these profiles, Araki and Kurihara argued that the state of expansion changed from $s = 1/2$ (root-mean-square law) in August to $s = 0$ (uniform expansion) in September.

The concept of optically effective boundary led to a new method for treating radiative transfer in acceleratively expanding envelopes. This method, however, requires lengthy numerical calculations, so it was not applied to other extended envelopes. Based on a similar idea, a much simpler method of *escape probability* was introduced by V. V. Sobolev (1915–1999) in the USSR, which has been widely applied to many types of expanding envelopes (Sobolev 1947).

After completing this work, Kurihara moved to Kyushu Imperial University as a professor of fluid mechanics at the College of Technology, and he also worked at the Natural Disaster Information Center of Western Japan attached to the university.

4.2.4 Postwar Activities

In 1941, Araki was appointed professor of astrophysics at Kyoto Imperial University. In 1945, World War II ended with the defeat of Japan, and Araki was purged from Kyoto University due to his nationalism in public opinions during the war. Following his resignation, he moved to the village of Kamiyakuno (上夜久野村) in northern Kyoto Prefecture, together with his disciples, Kiyonaga Kaichi (清永嘉一) and Takagi Kimisaburo (高木公三郎), and they devoted their days chiefly to writing books. In this period, they published *General Treatise of Astronomy* (天文学総論, seven volumes), *Encyclopedia of Modern Astronomy* (現代天文学事典), *A Lecture on Astronomical Chronology* (天文年代学講話), and others.

In 1954, after the purge at the university had ended, Araki returned to Kyoto as a professor of liberal arts at Ōtani University (Buddhist training university). After retirement from this university in 1964, Araki established Kyoto Sangyo University with two faculties of Economy and Science, and he was appointed president and general manager of this university. Three additional faculties, Industrial Administration, Law, and Foreign Languages, were added in 1969. The university had grown to the stature of a major private university in Kyoto.

Even in the busy work of the president, Araki took up lecturing in mathematics in the Faculty of Science. After several days of lecturing, he suffered a sudden heart attack and died in July 1978 at age 81. Sangyo University issued three publications by him: *Complete Set of Araki Toshima's Papers*, *Chuzan's Travel Pictures*, and *Chuzan's Anthology* in 1979 (Sangyo University 1979).

4.3 Takeda Shin'ichiro

Takeda Shin'ichiro (竹田新一郎, 1901–1939) was also a successor to Shinjo in theoretical astrophysics (Araki 1940, Kabumoto 1998).

Takeda was born in Tokyo, the eldest son of Takeda Sanshichi (竹田三七), a descendant of a samurai in the Wakayama clan. Shin'ichiro moved to Kyoto at a young age and was educated there. He entered the Department of Astronomy, Kyoto Imperial University, in 1923 and studied astrophysics under Shinjo. Upon graduation, he was appointed lecturer in 1927 and then promoted to associate professor of astrophysics in 1929.

Takeda's first work was the physical nature of comets in 1927, and he then moved to stellar structures and closed binaries in the 1930s. In 1936 he organized an expedition to observe a solar eclipse in Hokkaido. This expedition was hard on him. Also, his health was fragile from birth. It might have caused a chronic disease. Takeda died prematurely in 1939 at age 38. A manuscript of a popular book, *From Planet to Stars*, was found among his research papers. Araki and Noda published this book with reminiscences on him (Takeda 1940) (Fig. 4.14).

Fig. 4.14 Portrait of Takeda Shin'ichiro. (Takeda 1940)

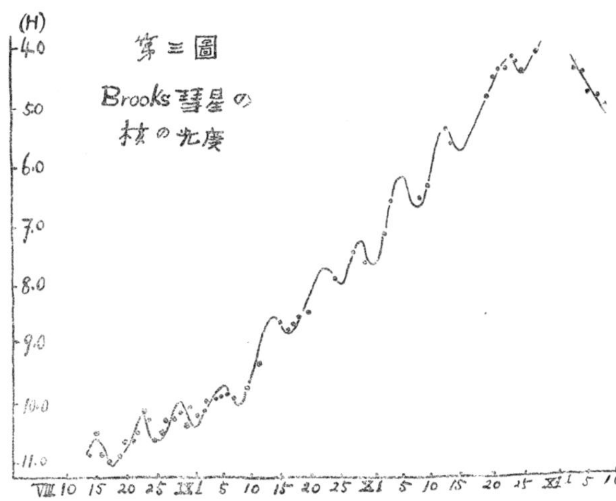

Fig. 4.15 Light curve of Brooks comet 1911c. (Takeda 1927)

4.3.1 *Physical Nature of Comets*

Takeda wrote a series of reviews on the physical nature of comets in the Japanese magazine *Tenkai* (*The Heavens*) in 1927. In it he introduced the contemporary state of cometary physics with his own criticisms. Among many research problems, he was particularly interested in the periodic variation of light curves of comets. He noticed the case of the Brooks comet (1911c), which showed a periodic variation of 7 days, as found by H. E. Lau (Lau 1912). Takeda collected data and drew a light curve, as shown in Fig. 4.15, where the abscissas are the month and day in 1911, and the ordinates are the magnitude (H) of the comet, as seen at one astronomical unit from the Sun (Takeda 1927).

Takeda regarded a comet as a gas sphere and applied Emden's gas-sphere theory. If a gas-sphere oscillates, the mean gas density can be derived from its oscillation period. Takeda obtained the mean density of the Brooks comet as

$$\rho = 3.6 \times 10^{-5}\,\rho_0, \tag{4.3}$$

where ρ_0 denotes the Earth's mean density.

Takeda thought, however, that this analysis was not definitive without additional observational data for other comets that exhibited periodic light variations. He expressed the need for further observations (Takeda 1927).

4.3.2 Homologous Contraction of Stars

Takeda accepted the mass annihilation hypothesis in connection with the energy source of stars, as did Araki, and he considered the evolution of stars (Takeda 1931). He supposed that stars gradually contracted, retaining a state of equilibrium due to the loss of mass and energy. He defined homologous contraction in the case of mass annihilation as a contraction that maintains mechanical equilibrium throughout the contraction process. When stellar mass is conserved, homologous contraction takes place under Lane's law, that is, the gas density and potential energy increase as R^{-3} and R^{-4}, respectively, and the temperature increases as R^{-1}, where R denotes the stellar radius. Takeda extended the homologous contraction to the case of mass-losing stars and derived the relation between stellar mass, M, and radius, R, in the form

$$\frac{M\beta}{R^{1-n}} = \text{constant}, \tag{4.4}$$

where β is the ratio of the gas pressure to the total (gas + radiation) pressure, which depends only on M, and n is a constant to be determined by observations. Takeda examined the loci of eq. (4.4) on the observed (log R vs. log $M\beta$) diagram, and he found good coincidence in two cases as shown in Fig. 4.16. One is the case of $n = 0$ for the main-sequence stars. In Takeda's theory, $n = 0$ corresponds to isothermal contraction. He supposed that the main-sequence stars were maintaining nearly the same central temperature, about 40 million K, as suggested by H. N. Russell (1925), and evolved along the main sequence losing mass and energy. Another case is $n = 0.55$ for the Cepheid variables. He argued that the Cepheids evolve from longer to shorter periods, under a homologous contraction of $n = 0.55$. Takeda's paper was contained in the *Astronomischer Jahresbericht*, but there was no international response. This may have been a result of the loss of popularity of the mass-annihilation hypothesis at that time.

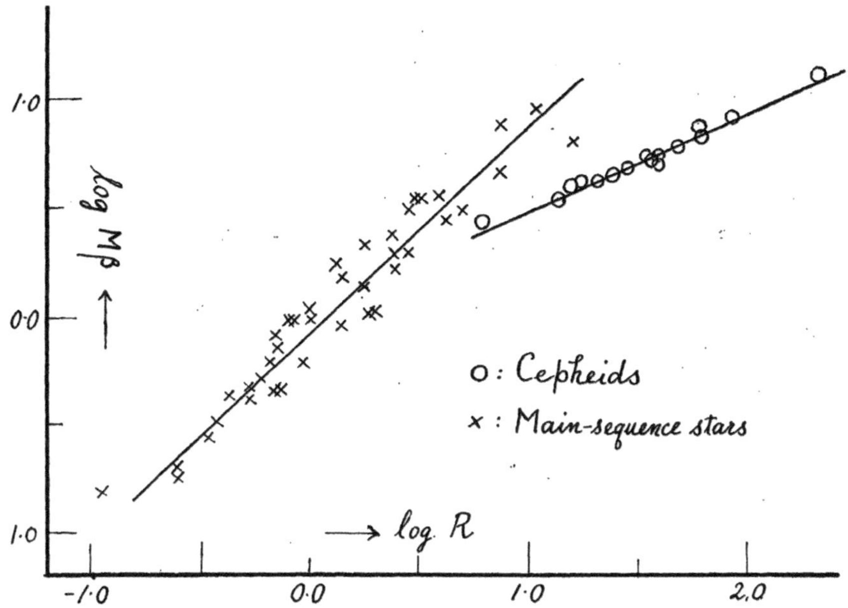

Fig. 4.16 Takeda's stellar evolution based on homologous contraction. (Takeda 1931)

4.3.3 *Distorted Outer Envelopes of Stars*

Takeda considered the rotational distortion of the outer layers of stars following Chandrasekhar's extended Roche model (Takeda 1934). Suppose a star of polytropic index $n = 3$, which corresponds to an ordinary main-sequence star. Chandrasekhar divided the interior of this star into two parts: the deep interior, which contains 90% of the mass within the sphere of about half the radius, and the outer envelope, which contains the remaining 10% of the mass (Chandrasekhar 1933).

When the star rotates, both the nucleus and outer envelope deviate from spherical symmetry. The deviation is small in the nucleus compared to the outer envelope. The oblateness, expressed by the ratio of the difference of equatorial and polar radii relative to the equatorial radius, grows with the increase in rotation velocity. While Chandrasekhar considered the case of slow rotation, Takeda carried out accurate numerical calculations and found that oblateness was enhanced at a much higher rate Chandrasekhar's case for an increase in rotational velocity.

Takeda also considered the steady configuration of a close binary whose orbit is circular, and the period of rotation is equal to that of revolution.

Binary interaction for the light curve in this system involves the effects of deformation of the outer layer and the effects of reflection of light. Takeda theoretically considered these effects and derived some general form of the deviation of the light curve. The effect of deformation reduces the brightness of stars, particularly at the epochs of light minima, whereas the effect of light reflection brightens up the surface of companion stars at the primary minimum, but no reflection effects appear at the secondary minimum. Both effects depend notably on the mass ratio of the binary components, and the maximum deviation appears at the primary minimum of the eclipse (Takeda 1934).

4.3.4 *Eclipsing Binary of β Lyra Type*

Eclipsing binaries are classified into two types: Algol type and β Lyra type. In the Algol type, eclipses are clearly defined by sharp depressions of light curves, whereas in the β Lyr a type, both primary and secondary eclipses overlap without showing any flat part of the light curve. As an example, the light curve of β Lyrae observed by Stebbins is shown in Fig. 4.17. The period of brightness variation is quite regular at 12.914 days.

In 1912, H. N. Russell derived the orbital elements of this binary by taking the ellipticity of the stars due to their mutual attraction and the effect of limb darkening of stellar surfaces into consideration (Russell 1912a, b). Excited by this work, Takeda analyzed the light curve of β Lyrae, based on his outer-layer theory for close binaries (Takeda 1937). He thus derived the theoretical light curve by which he obtained corrections due to distortion by tidal and rotational effects. The effects of

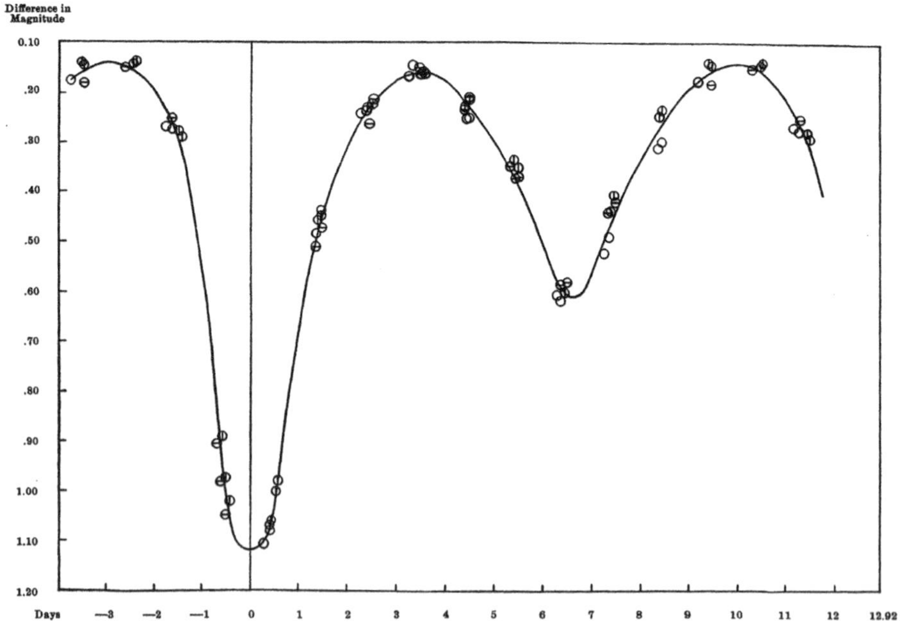

Fig. 4.17 Light curve of β Lyrae by photoelectric observations of J. Stebbins in 1915. Aabcsissas indicates days from primary minimum; ordinates indicates difference in magnitude. (Stebbins 1916)

light reflection and limb darkening were also taken into consideration. These results were applied to Stebbins's light curve (Fig. 4.17). Takeda stated that of his two models of uniform disk and limb-darkened disk, the latter yields a better fit for observed light curves.

Takeda's work was noticed by Russell. In his paper on the ellipticity of stars in eclipsing binaries, Russell wrote, "This problem has been discussed by Takeda in considerable detail. His work deserves careful consideration by anyone who pursues the question" (Russell 1939).

4.4 Miyamoto Shotaro, Astrophysics, and Planetary Science

Miyamoto Shotaro (宮本正太郎, 1912–1992) promoted astrophysics and planetary science in Kyoto (Hasegawa 1993; Miyamoto, S. 1954; Miyamoto, K. 1991).

4.4.1 Life and Works

Shotaro was born the eldest son of Miyamoto Ichisuke (宮本市助), a grain-import merchant, in Onomichi City (尾道市), Hiroshima Prefecture. He was an enthusiastic amateur astronomer in his youth, with special interest in the visual observations of Mars with a small telescope. During his amateur activity, he met with Yamamoto Issei of Kyoto Imperial University, who encouraged Shotaro to promote more interest in astronomy. Following Yamamoto's suggestion, Miyamoto entered Kyoto Imperial University in 1933 and studied astrophysics under Araki Toshima. After graduation in 1936, he served in the army but soon after was demobilized as a result of illness. Thereafter he spent his life as a lecturer (1940), associate professor (1943), and professor of astrophysics (1948) in Kyoto University until his retirement in 1976. Thereafter, he served as Honorary Director of the Kyoto School of Computer Science. When he died at age 79 in 1992, the computer school published and devoted to him a memorial book, *Complete Works of Miyamoto Shotaro* (Hasegawa and Sakka 1993).

The war years were difficult for him, and he lived isolated from the international community of astrophysicists. After the war, Araki and his main staff resigned from Kyoto University, Miyamoto found himself having to promote research and education in astrophysics alone.

The main topics in astrophysics and planetary science during his lifetime were as follows, with some overlap in time frame (Fig. 4.18):

1. Spectral theory of gaseous nebulae (1938–1950)
2. Stellar physics of neutron stars and white dwarfs (1942–1949)
3. Radiation field and stability of stellar envelopes (1942–1953)

Fig. 4.18 Photograph of Miyamoto Shotaro at Kwasan Observatory. (Hasegawa and Sakka 1993)

4. Solar physics: spectral theory of chromosphere and corona (1942–1951); profiles of Fraunhofer lines (1953–1957)
5. Planetary science (1958–1976)

4.4.2 Nebular Physics

In planetary and diffuse nebulae, photoionization and recombination are the basic processes in the formation of emission lines. Miyamoto paid attention to the effects of electron collisions with atoms for the excitation of energy levels. He considered the effects of collision in two cases of hydrogen and doubly ionized oxygen.

1. Hydrogen Balmer lines

In the visual spectral region, nebulae emit hydrogen Balmer lines in a manner whereby the intensity of Balmer emissions gradually decrease from Hα and Hβ to higher members of the Balmer series (Balmer decrement). The recombination process were investigated by D. H. Menzel and J. G. Baker at Harvard University in 1937–1938 and by Hagihara at the TAO, around the same time as Miyamoto's work in this field. Miyamoto considered the theoretical recombination process by considering the effects of collisional excitation, and he found that the Balmer decrement became steeper due to collisional excitation when electron temperature was higher than 20,000 K, as compared to the original recombination theory (Miyamoto 1938).

2. Doubly ionized oxygen [O III]

Emission lines at wavelengths λ4958Å, λ5006Å, had long been attributed to an unknown chemical element Nebulium. It was in 1926 that Ira Bowen identified these lines as the forbidden lines of doubly ionized oxygen [O III] (Bowen 1927). Miyamoto (1939) calculated the collisional transition probabilities, one year prior to the work of Hebb and Menzel (1940), and he applied to the estimation of electron temperature of planetary nebulae. Miyamoto's result showed that the electron temperature lies in a range of 10,000 to 25,000 degrees for most of the nebulae observed, in good agreement with the results of Menzel's group (Miyamoto 1939).

4.4.3 Neutron Stars and White Dwarfs

When a star is exhausted, its energy source collapses to some highly condensed state, as with white dwarfs or neutron stars. The existence of neutron stars was proposed by Lev D. Landau (Landau 1932), and the stellar structure was studied by Landau (Landau 1938) and J. R. Oppenheimer and G. M. Volkoff (Oppenheimer and Volkoff 1939).

In 1941, Miyamoto considered the structure of neutron stars, based on his own calculation of nuclear exchange force (Miyamoto 1941). He temporarily classified two types of neutron star: cold neutron stars (N stars), which are composed of pure neutrons with an absolute zero temperature, and proton-neutron stars (P stars), which are composed of the same abundance of protons and neutrons. According to his calculation, the radius and mass of N stars are 40–100 km and 0.001–0.010 solar mass, respectively, whereas P stars show 200–400 km and 1.1–1.2 solar mass, respectively.

For N stars, Miyamoto derived the internal density distribution for some case of parameter α, which represents the relative importance of the pressure due to the exchange force (P_e) and the pressure due to degeneration (P_o) in the form

$$\alpha \propto \frac{P_e}{P_o} \tag{4.5}$$

where $0 < \propto < 1$, and $\alpha = 0$ denotes the case of pure degenerate pressure, corresponding to white dwarfs with polytropic index 3/2, and $\alpha = 0.9$ for almost pure neutron stars. The results of his calculation are illustrated in Fig. 4.19, where the abscissa denotes the dimensionless radius and the ordinate the gas density relative to the central density.

According to his calculations, the density gradient becomes high with the increase of neutron abundance, suggesting the existence of highly compact cores. Supporting this result, current models also indicate the existence of highly condensed central cores (e.g., Katz 1992). Miyamoto's paper may be the first one treating nuclear physics in astronomy in 1940s Japan.

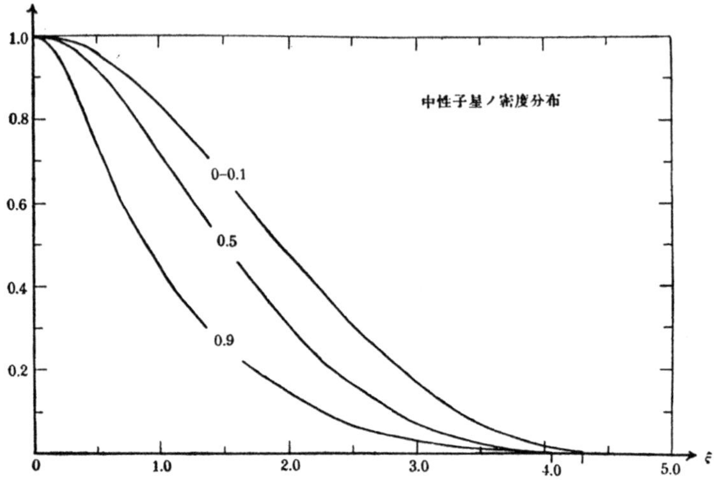

Fig. 4.19 Density distribution inside neutron stars. (Miyamoto 1941)

Table 4.2 Properties of planetary nebulae and stellar envelopes

	Stellar type	Outer radius R/R^*	Electron density number (cm^3)
Planetary nebulae	O, B	10^7–10^8	10^3–10^6
Stellar envelopes	O, B	10–50	10^{10}–10^{12}

Note: R^* denotes the stellar radius

In 1949, Miyamoto also considered the atmospheric structure of white dwarfs based on the theory of model atmosphere, particularly for the atmosphere of Sirius B (Miyamoto 1949). He supposed that the dominant source of opacity is the bound-free transitions of hydrogen, and he derived their atmospheric structure. He showed that due to pressure effects, the effective temperature of Sirius B is around 12,500 K, which is considerably higher than ordinary scales.

4.4.4 Early-Type Emission-Line Stars

1. Envelopes of Early-type stars

The physical parameters of the envelopes of early-type stars greatly differ from those of planetary nebulae, as shown in Table 4.2. The physical process of emission-line formation is also different for these objects. Under the assumption that the envelopes consist of pure hydrogen, Miyamoto distinguished two types of the radiation process: planetary nebula type (PN type) and Be star type (Be type) (Miyamoto 1942a, 1951).

In the PN type, the photoionization of hydrogen atoms occurs from the ground energy level and a recombination process takes place, forming emission lines by cascade transitions. The nebulae are generally transparent for the emission lines.

In contrast, in the Be type, photoionization takes place mainly from the first excited energy level and the emission lines are formed through a modified recombination process accompanying some self-absorption of emission lines.

Miyamoto showed that the envelopes of WR, Be, and P Cyg type stars are of the Be type. He applied this theory to the Balmer decrements of Be stars and found that the large variety of the observed decrement Hα/Hβ ranging from around 1 to 10 could be explained according to the variety of the envelope size and stellar temperatures.

2. Stability of stellar envelopes

In 1934, Boris P. Gerasimovič (1889–1937), a Soviet astrophysicist, first discussed the effect of radiation pressure for the stability of outer envelopes of stars. He showed that the radiation pressure from ultraviolet radiation could surpass the gravitational acceleration for supergiants such as P Cygni (Gerasimovič 1934). He thus developed the theory of a stellar envelope streaming out due to ultraviolet radiation pressure. His theory was based on the PN type of envelope.

In contrast, Miyamoto considered a similar problem for Be-type envelopes (Miyamoto 1952). He solved the equation of dynamical equilibrium in a simplified stellar envelope that is composed of pure ionized hydrogen and in an isothermal state. The main factors acting for their dynamical stability are gravitational acceleration, g, radiation pressure from ultraviolet radiation, $P_{rad,}$ and gas pressure due to electron scattering P_e. He derived the electron density, Ne, in the form

$$\frac{N_e}{N_{e0}} = \left[1 - \Gamma\left(1 - e^{-\xi}\right)\right]^{-1} e^{-\xi} \tag{4.6}$$

where ξ denotes a nondimensional radius of the envelope and Neo the electron density at the envelope base. Γ is given by

$$\Gamma = \frac{\kappa_1 F_1 N_{e0}}{cK\left(m_H\, g - \frac{\sigma F}{c}\right)} \tag{4.7}$$

where the right-hand side denotes the effect of radiation pressure due to the ultraviolet flux F_1, and its values are determined as the function of the MK spectral types, and Γ takes any value ($0 < \Gamma < \infty$) on the ($\log g - T_e$) diagram.

Miyamoto found that the stability of a stellar envelope depends on the value of Γ. He drew curves of equal values of Γ on the ($\log g - Te$) diagram for early-type stars, as shown in Fig. 4.20, where the three cases of $\Gamma = 0.1$, 0.5, and 1.0 are depicted. The first case indicates incipient swelling of the upper atmosphere. In the second case, the atmosphere may be almost unstable. And in the third case, it becomes definitively unstable. The case $\Gamma = \infty$ is also represented by the broken curve above which no star can exist. The positions of stars for each luminosity class and spectral type are schematically illustrated. It is shown that P Cygni, WR, and O*f* stars occupy the region above the curve of $\Gamma = 1.0$, and they clearly show outflowing phenomena.

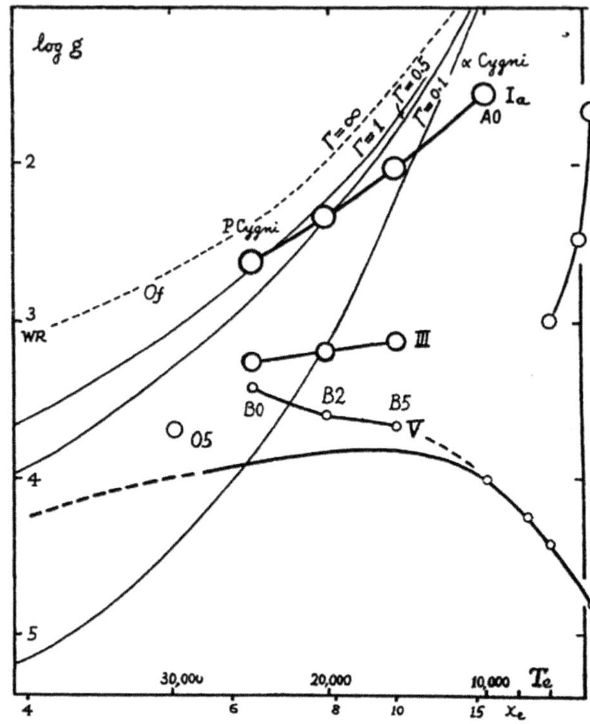

Fig. 4.20 Theoretical curves for different values of Γ on log g – effective temperature diagram. (Miyamoto 1952)

4.4.5 *Solar Physics*

Miyamoto contributed to solar physics, particularly on the physical state of the corona and chromosphere. The corona exhibits many emission lines and weak continuum. The origin of coronal emission lines had long been an enigma, and the lines were attributed to an unidentified chemical element Coronium. In 1941, Bengt Edlèn (1906–1993) attributed these lines to the multiply ionized iron, from the ninth order (Fe X at λ6374Å) to the thirteenth order (Fe XIV at λ5303Å) (Edlèn 1941), so the enigma of Coronium was clearly solved.

To explain the relative intensities of these ions, Miyamoto considered the effects of collisional ionization and formulated a new ionization formula, which was an extension of Saha's equation to the inclusion of collisional terms (Miyamoto 1942b, 1948). Miyamoto applied this formula and derived the electron temperature of the corona as high as one to two million degrees. The result of his calculation is shown in Fig. 4.21. The ordinate represents the logarithms of the relative abundance in units of the concentration of the ion of maximum abundance, and the abscissa denotes the series of Fe ions in its order. The electron temperature corresponding to each curve is shown in the upper part. Since the observed iron emission lines ranged from Fe X to Fe XIV, Miyamoto concluded that the coronal electron temperature should be one to two million degrees.

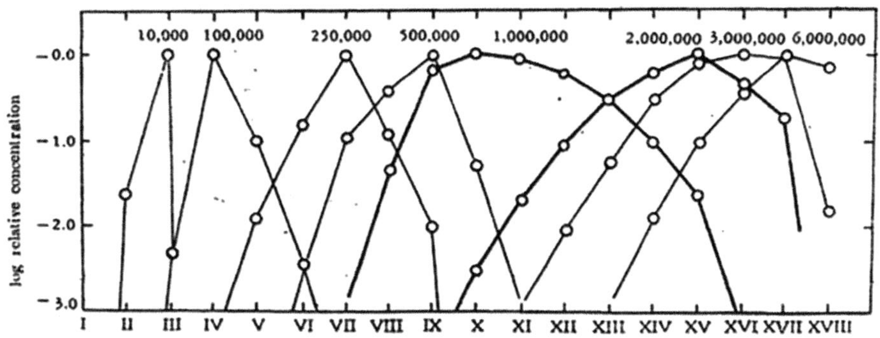

Fig. 4.21 Ionization of iron in solar corona. (Miyamoto 1942b, 1948)

A high electron temperature of the corona was also supported by large widths of emission lines observed by B. Lyot during the solar eclipse in 1935 (Lyot 1937). If the large widths were attributed to the thermal motion, the electron temperature should be as high as one million degrees.

Miyamoto's basic view was to take into consideration the effects of electron collisions in the formation of emission lines. In an article in the *Astronomical Herald* in 1948, he wrote:

> When I saw that the coronal emission lines had originated from the forbidden lines of highly ionized iron and nickel, I was quite at a loss what to do. Next day, however, I could build a new ionization theory based on electron collisions. (Miyamoto 1954)

Miyamoto also applied his collisional ionization formula to the estimation of the electron temperature of the chromosphere. On the electron temperature, higher value of 35,000 degrees had so far been proposed by R. O. Redman (1942) based on broad widths of emission lines and by R. Wildt (1947) based on density gradients of hydrogen, helium, and some metallic atoms along the height from the solar limb. In contrast, Miyamoto claimed that the electron temperature of the chromosphere had to be as low as around 6000 degrees, based on the metallic ionization and emission-line intensities of CaII H and K lines shown as follows (Miyamoto and Kawaguchi 1950):

(a) The concentrations of neutral, singly ionized, and doubly ionized metals were derived as a function of electron temperature from 5700 up to 35,000 degrees. The concentrations of neutral and singly ionized metals gradually decrease with temperature, whereas the concentration of doubly ionized metals drastically increases. The ionization state of these metals showed sharp differences between 6000 and 35,000 degrees. Comparison with observations has favored the former case.

(b) The emission-line intensities of the chromosphere decrease along the height from the solar limb at the solar eclipse. Miyamoto calculated the emission intensities of CaII K-lines as a function of height taking the thermal velocity

ν as a parameter. Best fit was obtained in the case of thermal velocity $\nu = 1.5$ km s^{-1}, which corresponds to an electron temperature of 6000 degrees.

With these considerations Miyamoto arrived at the low-temperature hypothesis, and his theory was later confirmed by Redman and Suemoto (Redman and Suemoto 1954).

4.4.6 Planetary Science

In 1958, Miyamoto was appointed third director of the Kwasan Observatory and began extensive work on planetary science through observational as well as theoretical studies. For observations, he used a Cooke 30-cm refractor (which was remodeled to a 45-cm aperture by Zeiss in 1968), and a Tsugami 60-cm reflector. In 1968, the Hida Observatory, equipped with a Zeiss 65-cm refractor, was created as a part of the Kwasan Observatory. Miyamoto served as the director of the Kwasan and Hida Observatories until his retirement.

Miyamoto's main interests in the planetary science were Martian meteorology and lunar geology.

1. Martian meteorology

Mars approaches the Earth at its opposition once every 2 years. Miyamoto observed Mars oppositions eight times from 1956 to 1973. In August 1956, he found a great yellow cloud that suddenly appeared in the Noachis highlands in the southern hemisphere of Mars (Miyamoto 1957). The cloud rapidly spread out and drifted in a southwestern direction along the isobar, and the polar ice cap was once covered by this cloud. A part of the cloud even spread into the northern hemisphere.

On the origin of the great yellow cloud, Miyamoto paid special attention to the rapid retreat of the southern polar cap.

The Martian atmosphere is characterized by two features: the low water vapor content and the low atmospheric density. These conditions usually make it difficult to form clouds over the Martian surface. To form clouds, water vapor and plentiful sunshine for heating are required. In Mars' summer, the hottest area appears not in the equatorial but in the polar region because sunshine reaches the pole without effective absorption and makes for long daylight hours. In midsummer, the polar ice cap rapidly melts and water streams flow out along some water courses, which are sometimes observed as temporal dark markings.

The retreat of the polar cap during the opposition in 1956 is shown in Fig. 4.22, where abscissas are the areocentric longitude of the Sun (lower axis) and the date (upper axis). The solid curve represents the diameter of the polar cap in units of the Martian diameter (left-hand ordinate). Crosses were observed by Saheki Tsuneo in Osaka and filled circles at Kwasan. The broken line indicates the area of the cap, A, and the dotted curve is the rate of water transfer from ice (both right-hand ordinates). The date of the first observed day of the great cloud is indicated by the upper mark.

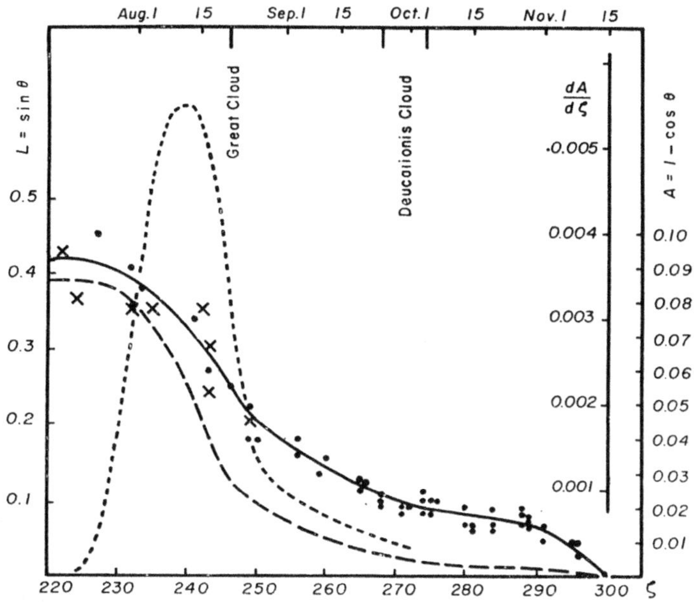

Fig. 4.22 Retreat of southern polar cap of Mars during 1956 opposition. (Miyamoto 1957)

Note that the cloud appeared at the time just after the mostly rapid melting of the polar cap.

Miyamoto suggested that the formation of the great cloud was the result of evaporation of water in a stream from the polar cap to the Naochis region. A large amount of water vapor was the motive power of the atmospheric temperature increase and a large-scale current to form a cloud. The yellow cloud was not yellow initially. After several days, it turned notably reddish. A strong ascending current on a large scale may have involved fine-grain desert sand to make the cloud yellow. A similar yellow cloud was also observed in the 1971 opposition.

In continuous observations of Mars at oppositions, Miyamoto traced the motion of the clouds and considered the general circulation of the Martian atmosphere (Miyamoto 1960a, 1963). According to his theory, cross-equatorial currents take place in summer and winter, as observed in the great yellow cloud in 1956. Such meridional currents had already been discovered by W. H. Pickering (1925). In equinoctial seasons (spring and autumn), no cross-equatorial currents appear. The circular pattern resembles the Earth's, viz., between the equator and pole. The prevailing wind over the middle latitudes is westerly. Miyamoto explained this pattern of general circulation in terms of the latitude variation of insolation. Since the Martian atmosphere is optically thin to sunlight, the hottest area appears near the pole in summer and winter, whereas the hottest area lies in the equatorial zone in equinoctial seasons.

2. Lunar geology

The lunar surface is characterized by bright highlands and dark maria. The craters are particularly concentrated in the mountainous regions. With the publication of the *Photographic Atlas of the Moon* (Miyamoto and Matsui 1960; Miyamoto and Hattori 1963), Miyamoto advanced the study of lunar geology.

On the origin of the characteristic structure of selenography, Miyamoto proposed a hypothesis according to which the differentiation of bright silicic mass from dark basaltic mass occurred during the formation of the lunar crust (Miyamoto 1960b, c). Due to the low surface gravity, the lunar crust was thin in its depth and rapidly solidified compared to the Earth, so that there had been no real orogeny on the Moon. The explosive nature of the silicic mass and the fluidity of the quiescent basaltic mass give rise to different aspects on the highlands and maria (Fig. 4.23). The volcanos in the silicic mass are explosive without accompanying lava flows. The lack of orogeny and explosive nature of silicic volcanos explain the distribution of craters without making any appreciable mountain ranges or any trace of lava flow in the highlands. In contrast, the mare basalts were formed by eruptions of rather quiet volcanos with wide lava flows, as seen on the Earth in the Hawaiian Islands in the Pacific Ocean (Miyamoto 1964, 1965).

On the origin of craters two hypotheses have so far been advanced: meteorite impact (Baldwin 1949) and volcanic (Spurr 1944). Baldwin argued that the kinetic

Fig. 4.23 Mountain site on Moon. (Miyamoto and Hattori 1963)

energy of impact is sufficient for the melting of the lunar crust, resulting in the formation of craters and maria. One of the unfavorable points for the impact hypothesis is, according to Miyamoto, the fact that the arrangement of numerous craters is far from a random distribution (Miyamoto 1960b). Thus, Miyamoto supported Spurr's volcanic origin.

4.5 History of Oriental Astronomy

Shinjo's work on the history of oriental astronomy was taken up by Noda Churyo, Yabuuchi Kiyoshi, and Watanabe Toshio.

4.5.1 Noda Churyo and Ancient Cosmology

Noda Churyo (能田忠亮, 1899–1987) was born in Nakajima (中島町), Ehime Prefecture, and entered Kyoto Imperial University in 1923. He studied astronomy under Shinjo in the Department of Astronomy (Saigusa et al. 1990).

In 1929, after graduation, he started his work on the history of oriental astronomy at the School of Oriental Culture, Kyoto Institute (東方文化学院 京都研究所) (presently Research Institute for Humanistic Studies 人文科学研究所 of Kyoto University). In 1950 he moved to Osaka Gakugei University (大阪学芸大) as a professor of astronomy, where he worked there until his retirement.

Noda worked mainly on ancient astronomy and cosmological thought in China. Noda's works are collected in his book *History of Oriental Astronomy* (Noda 1943) (Fig. 4.24).

Among Noda's works, two major topics are discussed in what follows.

1. Study of *Zhou Bi Suan Jing* (周髀算経, *Arithmetic Classics of the Zhou Gnomon*) (Noda 1933)

Zhou Bi Suan Jing is known as one of the oldest texts on mathematics and astronomy. It is a collection of ancient Chinese technical thought. Noda analyzed the astronomical arguments presented in the book in detail, and he concluded that this book was mostly written based on the knowledge of the Han Dynasty, but some parts date back to far more distant ages. Noda also noticed that the book was finally edited by Zhao Shuhang (趙君卿) in the late Han Dynasty. Zhou Bi (周髀) is a gnomon of the sundial used to measure shadow length, which varies with time as well as with distance along a longitude of the Earth. At night, the gnomon could be used to measure the angular height of the North Pole. Using these measurements, ancient people estimated the height of the sky, distance to the Sun and North Pole, and the extension of the Earth. Based on these measurements, they constructed models of the cosmos. According to Noda, cosmological thought was mostly developed in the Han Dynasty (202 B.C. – 220 AD), while its origins could be traced back even to the period of the Zhou Dynasty (1046–771 BC).

Fig. 4.24 Portrait of Noda Churyo in 1932. (Shinjo Bunko, Department of Astronomy, Kyoto University)

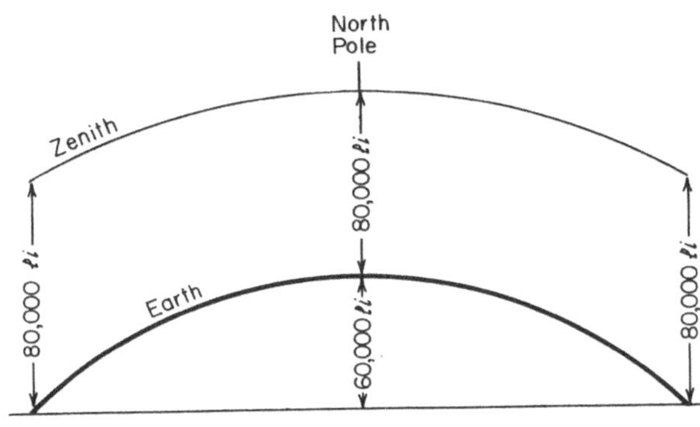

Fig. 4.25 Structure of sky and Earth in Kai t'ien theory. (Nakayama 1969a, b)

2. Cosmology in Han Dynasty (Noda 1943)

In the age of *Zhou Bi Suan Jing*, that is the age of observations using the gnomon, *Kai t'ien theory* (蓋天説, the sky as a cover) was widely accepted, according to which the sky and the Earth are both flat or both convex and parallel. This theory was developed and expanded into a cosmological theory in the Han Dynasty. Figure 4.25 illustrates a conception of Kai t'ien theory in the case of a convex sky and Earth. The distance scale one *li* (里) corresponds to around 500 m.

In the middle of the Han Dynasty, around the time of Christ, the spherical armillary was invented, and the idea of a spherical sky led to a new cosmology, *Hun t'ien theory* (渾天説, enveloping spherical sky). This theory might have been inspired by the image of a cosmic egg, the yolk being the Earth, but the spherical form of the Earth was not clearly recognized. Debate between both cosmological theories lasted a long time during the Han Dynasty. Noda critically traced the history of both theories from the Zhou to Han Dynasties. Noda argued that the original Kai t'ien theory in the Zhou Dynasty was the first scientific cosmology in China – scientific in the sense of being divorced from fantasy and folklore. On the other hand, Hun t'ien theory was based on observations with spherical armillary yielding a higher accuracy for the determination of heavenly bodies and for calendrical science.

These theories emerged as the traditional cosmology in China for a long time until the transmission of Western cosmology in the sixteenth century.

4.5.2 Yabuuchi Kiyoshi and Chinese Science

Yabuuchi Kiyoshi (薮内清 1906–2000) was born in Kobe and studied astronomy under Shinjo at Kyoto Imperial University. After graduation he worked at the Research Institute for Humanistic Studies (人文科学研究所) of Kyoto University. He made significant contributions to the history of Chinese science and technology (Hashimoto 1982). In 1972, he was awarded the George Sarton Medal from the History of Science Society of the USA. He was also a member of the Japan Academy.

Among his works on the history of China, two topics on astronomy and technology are presented in what follows (Yabuuchi 1949) (Fig. 4.26).

1. History of calendrical science

The *Mandate of the Heaven* (天命) was a principle used to justify the power of the emperor, as well as explain the auspicious circumstances for each dynasty in China. With this principle, astronomical observations and calendar making became the important national projects. Inspired by the importance of this principle,

Fig. 4.26 Portrait of Yabuuchi Kiyoshi. (Miyajima's private album 1982)

Yabuuchi insisted that the study of the history of calendrical science yields a key element in understanding the history of China from ancient to modern times.

In 1941, he analyzed calendrical history from the Yin Zhou (殷周) to Sui (隋) Dynasties and found that North and South Dynasties (南北朝) in the fourth and fifth centuries adopted different calendar systems that were unified into a single calendar in the Sui Dynasty, reflecting the political conditions in the late fifth century (Yabuuchi 1941). On this finding, Yabuuchi wrote in his recollections: "Through technical studies I could obtain a keystone to understand the history of China" (Yabuuchi 1982). Thereafter, the study of calendrical history in China became his life's work, along with the history of mathematics. In 1944, Yabuuchi published two books, the *Calendrical History of the Sui and Tang Dynasties* (隋唐暦法の歴史) and *Introduction to the History of Chinese Mathematics* (支那数学史概説). In these books, he not only summarized the great achievements of mathematics and astronomy in ancient China but also pinpointed developmental features of the sciences in China in these fields compared to the contemporary West.

2. Study of science and technology

From 1948, Yabuuchi led a research group on the study of *Tian Gong Kai Wu* (天工開物), an encyclopedia of Chinese science and technology authored by Song Yingxing (宋応星) in the later Ming (明) Dynasty (Yabuuchi 1954). This book consists of 18 topics, including grains, ceramics, iron wares and others. It served as the starting point for the systematic study of Chinese society. Yabuuchi and his colleagues studied this book closely. A Japanese translation and critical essays were published in 1954 (Yabuuchi et al. *Studies on Tian Gong Kai Wu*, 天工開物の研究). This was Yabuuchi's great contribution to the study of Chinese history from a sociological point of view.

4.5.3 Watanabe Toshio and History of Astronomy in Japan

Watanabe Toshio (渡辺敏夫, 1905–1998) contributed to the history of astronomy in the Tokugawa period.

He enrolled in Kyoto Imperial University and studied astronomy under Shinjo. After graduation in 1928, he continued his study of the history of astronomy and in 1955 was appointed professor of astronomy at the Tokyo University of Mercantile Marine (東京商船大学), where he remained until retirement in 1969. During his term as professor, he concentrated his attention on the collection of classical astronomical data of the Tokugawa period, including the discovery of many hidden data in the form of letters, diaries, memoranda, and others. These data reached around 5000 items, and they are now preserved at the National Diet Library as the Watanabe Collection.

Based on these extensive data, Watanabe studied the history of astronomy in the Tokugawa period, particularly the lives and works of Hazama Shigetomi and Asada Goryu (Watanabe 1943, 1983). The comprehensive history of astronomy in the

Tokugawa period was derived by Watanabe, which includes not only calendar making and the acceptance of Western cosmology (Chap. 1) but also a history of astronomical technology such as astronomical instruments, telescopes, clocks, celestial maps, and the records of observations of solar and lunar eclipses, comets, and planetary phenomena (Watanabe 1987).

4.5.4 History of Astronomy in Postwar Period

In the Kyoto region, the research style of Yabuuchi Kiyoshi, which treats the history of astronomy as part of science and technology, was taken up by the next generation, which included Yoshida Mitsukuni, Yamada Keiji, Hashimoto Keizo, Miyajima Kazuhiko, and others (Miyajima 2008).

Yoshida Mitsukuni (吉田光邦, 1921–1991) of the Research Institute for Humanistic Studies (人文科学研究所) of Kyoto University worked on science and technology in China and Japan in the pre-Meiji period (Yoshida 1955, 1972).

With respect to Chinese history in the Han Dynasty, Yoshida traced the social background of the development of science and technology with a main focus on the history of iron, ceramic, textiles, and military technologies. On astronomy, he argued that the spherical armillary and spherical globe were constructed against the technological background of the time and contributed to great improvements in calendar making (Yoshida 1972). According to him, the development of science is always due to social background, reflecting the research style of Yabuuchi.

Yamada Keiji (山田慶児) of Kyoto University contributed to Chinese history with respect to two topics. The first is the natural philosophy of the Zhu Xi school (朱子学), which covers the broad fields of astronomy, meteorology, and physics, in addition to classical Confucianism in the South Song period (南宋) (Yamada 1978). The other topic is the relation of science and social system in medieval China as a background on the formation of the Shou-Shih calendar (授時暦, 1281) (Yamada 1980). Hashimoto Keizo (橋本敬造) of Kansai University was interested in the paradigm theory of Joseph Needham and considered the production of the Chongzhen calendar (崇禎暦, 1635) by Xú Guangqi (徐光啓) as a scientific revolution in China partly introduced from European countries. Europe in those days was in a stage of transition from Ptolemy to Copernican cosmology, and Tycho Brahe's geocentric system was widely accepted as an interim step. Tycho Brahe's system was also incorporated into the Chongzhen calendar (Hashimoto 1972).

Miyajima Kazuhiko (宮島一彦) of Doshisha University worked on the history of East Asian countries (China, Korea, and Japan) with practical investigations of instrumentation. In 1998, a celestial map was discovered on the ceilling of the Kitora Tumulus (キトラ古墳) in the village of Asuka, Nara Prefecture. Miyajima identified Chinese constellations as shown in Fig. 4.27. The map covers a wide northern sky with three concentric circles around the North Pole and a circle presumed to be the ecliptic. The central circle denotes the celestial equator and the outer circle the southern boundary observable from China. This tumulus was

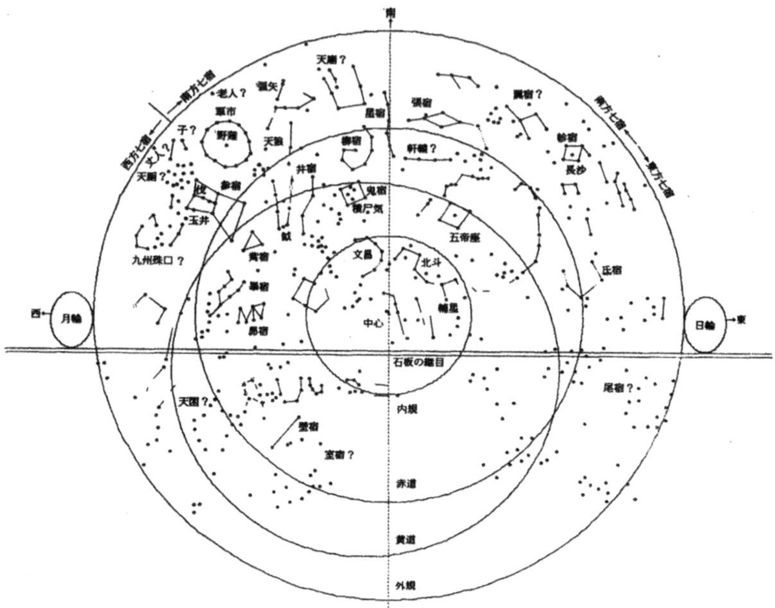

Fig. 4.27 Identification chart of Chinese constellations on astronomical map on ceilling of Kitora Tumulus. (Miyajima 1999)

supposedly constructed around 700 AD, and it is the oldest celestial map in East Asia (Miyajima 1999).

References

Araki, T. (1925). Uber die Erklarungshypothesen der Cepheidenerscheinung. *Memoirs of College of Science, Kyoto Imperial University, Series A, 8*, 75–129.

Araki, T. (1934). Zur Theorie des inneren Aufbaues der weissen Zwerge. *Zeitschrift für Astrophysik, 8*, 358–369.

Araki, T. (1940). Works of late Takeda Shin-ichiro on theoretical astrophysics (荒木俊馬: 理論天体物理学における竹田新一郎の業績). *Yusei kara Kosei he* (From Planets to Stars, 竹田新一郎: 遊星から恒星へ), 245–256).

Araki, T., & Kurihara, M. (1937). Zur Konturberechnung der Emissionslinien der expandierended Gashülle eines Sternes. *Zeitschrift für Astrophysik, 13*, 89–102.

Baldwin, R. B. (1949). *The face of the moon*. Chicago: University of Chicago Press.

Beals, C. S. (1929). On the nature of Wolf-Rayet emission. *Monthly Notice of the Royal Astronomical Society, 90*, 202–212.

Beals, C. S. (1934a). Spectrophotometric studies of WR stars and novae. *Publication of Dominion Astrophysical Observatory, 6*, 95–145.

Beals, C. S. (1934b). P Cygni and radial ejection. *Observatory, 57*, 378.

Bowen, I. S. (1927). The origin of nebulium spectrum. *Nature, 120*, 473.

Chandrasekhar, S. (1933). The equilibrium of distorted polytropes. *Monthly Notice of the Royal Astronomical Society, 93*, 539–574.

Edlèn, B. (1941). Identification of iron lines in coronal emission lines. *Swedish Journal of Physics, 28*, B. 1.
Gerasimovič, B. P. 1934, Monthly Notice of the Royal Astronomical Society, 94, 737–764, Non-static hydrogen chromospheres and the problem of Be stars.
Hasegawa, Y. (1993). Obituary of Professor Miyamoto ⟨長谷川靖子、宮本正太郎先生をお偲びして⟩. *Accume, 5*, 118–121.
Hasegawa, Y., & Sakka, K. (Eds.). (1993). *Complete works of Miyamoto Shotaro*, Kyoto School of Computer Science (長谷川靖子、作花一志編: 宮本正太郎論文集, 京都コンピュータ学院).
Hashimoto, K. (1982). A process of science revolution as seen in Chongzhen Calendar (橋本敬造: 崇禎暦書にみる科学革命の一過程、). *Science and Skill in Asia*, Dohosha, 370–380.
Kabumoto, K. (1998). The theoretical study on astrophysics by Shin-ichiro Takeda – On energy source of stars and stellar evolution (科学史研究, 株本訓久: 竹田新一郎の理論天体物理学研究-恒星の熱源と進化に関する研究を中心として一). *Journal of History of Science, Japan, 37*, 206–212.
Katz, J. I. (1992). Neutron stars: Physical properties and models. In S. P. Maran & C. Sagan (Eds.), *The astronomy and astrophysics encyclopedia* (pp. 761–763). Cambridge: Cambridge University Press.
Kibe, N. (1959). Reminiscence of Yamamoto Issei (天文月報、木辺成麿: 山本一清先生の思い出). *Astronomical Herald, 52*, 51–53.
Kiyonaga, K. (1979). Life and works of Professor Araki Toshima ("荒木俊馬論文集"、清永嘉一: 荒木俊馬教授の生涯と業績, 京都産業大学). In *Collected paper of Araki Toshima*, Kyoto Sangyo University (pp. 609–615).
Kogure, T. (2008). Dawn and tradition of astronomy in Kyoto (小暮智一:京都における天文学の草創と伝統). In *One-hundred years of the Astronomical Society of Japan* (日本天文学会編、日本の天文学の百年)、Hyoronsha (評論社) (Chapter 3, pp. 27–35).
Landau, L. (1932). On the theory of stars. *Physical Journal of the Soviet Union, 1*, 285.
Landau, L. (1938). Origin of stellar energy. *Nature, 141*, 333–334.
Lau, H. E. (1912). Über die Helligkeit der Kometen 1911 f (Quénisset) und 1911 c (Brooks). *Astronomische Nachrichten, 191*, 29–32.
Lyot, B. (1937). Quelques observations de la couronne solaire et des protuberances en 1935. *L'Astronomie, 51*, 203–218.
Milne, E. A. 1930, Monthly Notice of the Royal Astronomical Society, 91, 3–55, The analysis of stellar structure.
Miyajima, K. (1999). Ancient star map of Japan and astronomy in East Asia (京都大学人文学報、宮島一彦: 日本の古星図と東アジアの天文学). *Journal of Humanities, Kyoto University, 82*, 45–56.
Miyajima, K. (2008). History of Astronomy (宮島一彦: 天文学史). In *One hundred years of the Astronomical Society of Japan* (日本天文学会編、日本の天文学の百年), Hyoronsha (評論社) (Chapter 10, pp. 153–158).
Miyamoto, K. (Ed.) (1991). Obituaries of Miyamoto Shotato, Bright Starry Night, Private Issue (宮本周子、「星月夜」 — 宮本正太郎追禄集)、私版.
Miyamoto, S. (1938). On the Balmer emission of the planetary nebulae. *Memoirs of College of Science, Kyoto Imperial University, Ser. A, 22*, 173–202.
Miyamoto, S. (1939). Cyclic equations for O III and electron temperature of gaseous nebulae. *Memoirs of College pf Science, Kyoto Imperial University, Ser. A, 22*, 249–257.
Miyamoto, S. (1941). On the internal structure of neutron stars (宮本正太郎: 中性子星の内部構造). *Japanese Memoirs of Astronomy and Geophysics, 1*(3), 47–54.
Miyamoto, S. (1942a). On the radiation field of hydrogen envelopes of early emission-line stars (宮本正太郎: 特異星水素大気内の放射場について). *Acta Astronomica, 1*, 119–136.

Miyamoto, S. (1942b). On the emission lines of the Solar Corona ("天体の電離と輻射"、宮本正太郎: 太陽コロナの輝線について) Kouseisha-Koseikaku (恒星社厚生閣). In *Ionization and radiation of celestial objects* (pp. 179–194).

Miyamoto, S. (1948). Ionization theory of Solar corona (English version of Miyamoto 1942b). *Publications of the Astronomical Society of Japan, 1*, 10–13.

Miyamoto, S. (1949). Atmosphere of the white dwarfs. *Memoirs of Astrophysics, 1*, 17–26.

Miyamoto, S. (1951). On the radiation field of Be stars. I. Contribution from the Institute of Astrophysics and Kwasan Observatory, No. 19, pp. 17–25,

Miyamoto, S. (1952). Radiation pressure and stability of the atmospheres of early-type stars. *Publications of the Astronomical Society of Japan, 4*, 91–99.

Miyamoto, S. (1954). My life with astronomy (天文月報, 宮本正太郎: 天文学とともに). *Astronomical Herald, 47*(9), 139–142.

Miyamoto, S. (1957). *The Great Yellow Cloud and the atmosphere of Mars*. Contributions from the Institute of Astrophysics and Kwasan Observatory, No. 71, pp. 1–42.

Miyamoto, S. (1960a). *On the general circulation of Martian atmosphere*. Contributions from the Institute of Astrophysics and Kwasan Observatory, No. 88, pp. 27–33.

Miyamoto, S. (1960b). *A geological interpretation of the lunar surface*. Contributions from the Institute of Astrophysics and Kwasan Observatory, No. 90, pp. 1–8.

Miyamoto, S. (1960c). *Magnetic boiling and underground structure of the Moon*. Contributions from the Institute of Astrophysics and Kwasan Observatory, No. 96, pp. 1–6.

Miyamoto, S. (1963). Observational study on the general circulation of Mars. Contributions from the Institute of Astrophysics and Kwasan Observatory, No. 125, pp. 81–88.

Miyamoto, S. (1964). Morphological aspects of the Lunar crust. *Icarus, 3*, 486–490.

Miyamoto, S. (1965). Morphological aspects of the Lunar crust. II. *Icarus, 4*, 421–424.

Miyamoto, K. (Ed.). (1994). Private Issue (Obituaries of Miyamoto Shotaro). *Bright Starry Night* (宮本周子編: 星月夜).

Miyamoto, S., & Hattori, A. (Eds.). (1963). *Photographic Atlas of the Moon* (2nd ed.). Contributions from the Institute of Astrophysics and Kwasan Observatory, No. 137, pp. 1–136.

Miyamoto, S., & Kawaguchi, I. (1950). On the electron temperature of the chromosphere. *Publications of the Astronomical Society of Japan, 1*, 114–121.

Miyamoto, S., & Matsui, M. (1960). *Photographic Atlas of the Moon*. Contributions from the Institute of Astrophysics and Kwasan Observatory, No. 95, pp. 1–81.

Miyamoto, S., Ikaeda, T., & Doi, K. (1959). Memorial address of Professor Yamamoto Issei (天文月報, 宮本正太郎ほか: 山本一清教授追悼). *Astronomical Herald, 52*, 49–77.

Nakayama, S. (1969a). Chapter 4, Early Chinese cosmology. In *A history of Japanese astronomy*. Cambridge, MA: Harvard University Press.

Nakayama, S. (1969b). *A history of Japanese astronomy*. Cambridge, MA: Harvard University Press.

Noda, C. (1933). *A Study of Chou Pi Suan Ching* (A Study of arithmetric classic of the Chou gnomon), Academy of Oriental Culture, Kyoto Institute, Monograph Series No, 3 (東方文化研究所、能田忠亮: 周髀算経の研究).

Noda, C. (1943). *The oriental history of astronomy* (能田忠亮: 東洋天文学論叢) (Kosheisha, 恒星社).

Oppenheimer, J. R., & Volkoff, G. M. (1939). On massive neutron cores. *Physical Review, 55*, 374–381.

Pickering, W. H. (1925). Report on Mars, No. 31, Formation and melting of the polar caps. *Popular Astronomy, 23*, 576–587.

Redman, R. O., & Suemoto, Z. (1954). Temperature and turbulence in the chromosphere. *Monthly Notice of the Royal Astronomical Society, 114*, 524–539.

Russell, H. N. (1912a). On the determination of the orbital elements of eclipsing variable stars. I. *Astrophysical Journal, 35*, 315–340.

Russell, H. N. (1912b). ibid. II. *Astrophysical Journal, 36*, 54–74.

Russell, H. N. (1939). Notes on ellipticity in eclipsing binaries. *Astrophysical Journal, 90*, 641–674.

Saigusa, T., Hashimoto, et al. (1990). Obituaries of Noda Churyo (宇宙会報, 三枝利文、橋本敬三他: 能田忠亮先生追悼文集). *Uchukaiho, 5*, 11–15.

Sangyo University (Ed.). (1979). *Complete Set of Araki Toshima's Papers, Chuzan's Travel Pictures*, and *Chuzan's Anthology*, Private Issues (荒木俊馬論文集: 疇山旅日記, 疇山遺珠).

Shapley, H., & Sawyer, H. B. (1929). The distance of ninety-three globular clusters. *Harvard College Observatory, Bulletin* No. 869.

Shapley, H., Yamamoto, I., & Wilson, H. H. (1925). The Magellanic Clouds. VII. The photographic period-luminosity curve. *Harvard College Observatory, Circular, 289*, 1–8.

Shinjo, S., & Araki, T. (1924). On the periodic inequality in the light-elements of Cepheids and Mira-type variables. *Japanese Journal of Astronomy and Geophysics, 2*, 147–163.

Sobolev V. V. (1947). *Moving envelopes* (in Russian), (English translation published in 1961).

Spurr, J. E. (1944). *Geology applied to selenology, I and II*. Lancaster: Science Press Co.

Stebbins, J. (1916). A study of β Lyrae with a photo-electric photometer. *Lick Observatory Bulletin*, No. 277, 186–192.

Takeda, S. (1927). Physical nature of comets (天界、竹田新一郎: 彗星の物理的性質). *Tenkai (The Heavens), 7*(75), 233–239.

Takeda. (1931). Homologous Contraction in the case when mass is annihilated. *Japanese Journal of Astronomy and Geophysics, 9*, 90–98.

Takeda, S. (1934). Distorted outer layers of the stars. *Memoir of College of Science, Kyoto Imperial University, 17*, 197–217.

Takeda, S. (1937). Theoretical light curves of eclipsing variables of the β Lyrae type. *Memoir of College of Science, Kyoto Imperial University, 20*, 47–86.

Takeda S. (1940). *From Planets to Stars* (竹田新一郎: 遊星から恒星へ, Koseisha Kouseikaku (恒星社厚生閣).

Tomita, Y. (2012). *Data on Yamamoto Issei at Yamamoto Observatory* (冨田良雄: 山本一清資料の概要). Report of Archive project, Second, pp. 1–10.

Tomita, Y., & Kubota, J. (2000). *Nakamura Kaname and reflecting Telescopes*, Kamogawa Shobo (冨田良雄、久保田諄: 中村要と反射望遠鏡, かもがわ書房).

Van Biesbroeck, G., Struve, O., & Yamamoto, I. (1923). Observations of asteroids at the Yerkes Observatory. *Astronomical Journal, 35*, 45–46.

Watanabe, T. (1943). *Hazama Shigetomi and his family in the calendrical history of Japan* (渡辺敏夫: 天文学史上における間重富とその一家). Yamaguchi Book Shop (山口書店).

Watanabe, T. (1983). *Asada Goryu and the early modern history of science in Japan* (渡部敏夫: 麻田剛立と近世日本科学史), Yuzankaku (雄山閣).

Watanabe, T. (1987). *History of early modern astronomy in Japan* (渡部敏夫: 近世日本天文学史), 2 Vols. Koseisha Koseikaku (恒星社厚生閣).

Yabuuchi, K. (1941). China's calendrical history from Yin Zhou to Sui dynasty. *Journal of Oriental Studies* (東方学報), *12*(4).

Yabuuchi, K. (1949). *Astronomy in China* (薮内清、中国の天文学). Kouseisha-Kouseikaku (恒星社恒星閣).

Yabuuchi, K. (Ed.). (1954). *Studies on the Tian Gong Kai Wu* (薮内清編: 天工開物の研究). Kouseisha (恒星社).

Yabuuchi, K. (1982). Looking back on my study of Chinese history of science (薮内清科: 中国科学史研究をふりかえって). In *Science and skill in Asia*, Dohosha (東洋の科学と技術、同朋舎), (pp. 1–16).

Yamada, K. (1978). *Natural philosophy of Zhu Xi*, Iwanami Shoten (山田慶児: 朱子の自然学、岩波書店).

Yamada, K. (1980). *The road to the Shou-Shih calendar*, Misuzu Shobo (山田慶児: 授時暦への道、みすず書房).

Yamamoto, I. (1919a). Light curves of several recent novae and some notes on the general features thereof. *Memoirs of College of Science, Kyoto Imperial University* (京都大学理学部紀要), *4*, 13–23.

Yamamoto, I. (1919b). Nova Aquilae, No.3. *Popular Astronomy, 27*, 200–201.

Yamamoto, I. 1922–1925, Tenkai (天界), Vol. 2 – Vol. 5, Diary of travel abroad, No. 1 – No. 32. (天界、山本一清: 海外日誌).

Yamamoto, I. (1923). Astronomy in Japan. *Publications of the Astronomical Society of the Pacific, 35*, 199–203.

Yamamoto, I., & Campbell, L. (1924). The long-period variable W Hydrae. *Harvard College Observatory, Circular, 279*, 1–7.

Yamamoto, I., Ueta, Y., & Kudara, K. (1919). Observations of Nova Aquilae No.3. *Memoirs of College of Science, Kyoto Imperial University, 4*, 23–42.

Yoshida, M. (1955). *History of science in Japan*. Asakura Shotenn (吉田光邦: 日本科学史、朝倉書店).

Yoshida, M. (1972). *Collected Papers on Science-Technology in China*. Japan Broadcast Society (吉田光邦: 中国科学技術史論集、日本放送協会).

Chapter 5
Astronomy in Early Showa. III. Sendai 1926–1945

Abstract Sendai is a central city of the northeastern region of Japan's main island, and it was a capital of the Sendai clan, which continued from the early Tokugawa until the Meiji Revolution. In 1907 Tohoku Imperial University was established in Sendai as a new type of university opened for women and foreign students. Astronomy was promoted at first by physicists Kusakabe Shirota and Ishiwara Jun, then taken over by astrophysicists Matukuma Takehiko and Hitotuhanagi Zyuiti (Takeuchi and Seki 2008; Takeuchi 2014).

5.1 Early History of Astronomy

5.1.1 Sendai in the Tokugawa-Meiji Period

Around 1670, the Sendai clan sent Eshi Tomotatsu (江志知辰, 1649–1714) to Kyoto to study astrology and calendar making under Shibukawa Harumi (Chap. 1). In Sendai, Eshi's disciples, Aoki Nagayoshi (青木長由, 1669–1740) and Toita Yasusuke (戸板保祐, 1708–1784), worked at the Tenmonkata (天文方, Bureau of Astronomy) of the clan and built a base of astronomy in Sendai. Toita constructed an armillary sphere in 1733 and began to make accurate time observations of celestial phenomena, including solar eclipses, comets, and stellar motions. Whereas the purpose of his observations was high-precision calendar making, he also paid attention to peculiar phenomena, such as comets, meteors, parhelion, planetary motions, and others. In his book *Iseishō* (Digest of Peculiar Stars, 異星鈔, 1759), Toita described numerous particular phenomena with an astrological interpretation, such as what type of disaster might happen in the near future given a particular sky event.

In the middle of the nineteenth century, Furuyama Makoto (古山誠, or Furuyama Toshisada, 古山利貞, 1824–1887) began to work on astronomy and geodesy in Sendai (Takeuchi 2014). Furuyama was born in Sendai as a samurai of the Sendai clan. From his youth he studied mathematics, astronomy, and geodesy in Western style, as well as classical Shintoism. He was hired as an assistant astronomer in 1840 and then promoted as a Tenmonkata of the clan in 1855. His work was the annual

T. Kogure, *The History of Modern Astronomy in Japan*, Historical & Cultural Astronomy, https://doi.org/10.1007/978-3-030-57061-3_5

creation of a local calendar in the Sendai region up to the Meiji Revolution. In 1857, the Navy of the Sendai clan constructed a warship, *Kaisei-Maru* (開成丸), for the defense of their domain, and Furuyama held an additional post as the navigating officer of the ship.

In 1868, the political state of Japan completely changed in the Meiji Revolution. All the feudal clans vanished, and the capital of Japan moved from Kyoto to Tokyo. The Tenmonkata of the Sendai clan also vanished, and Furuyama was appointed astronomer of the Seigaku Kyoku (星学局, Department of Astronomy) in 1870 by the Meiji government in Tokyo. He contributed to the calendar reform, where the country swtiched from the Tempo calendar to the Gregorian calendar, which was promulgated in 1873. Since Seigaku Kyoku was closed in 1874, Furuyama moved to the Ministry of War and worked there for 14 years. He died in 1887 at age 63.

In Sendai, there was a long tradition of astronomical observations and calendar making in collaboration with astronomers in Kyoto and Edo. This tradition, however, was not adopted by the new astronomy at Tohoku University, which was founded in the Meiji era (Takeuchi 2014).

5.1.2 Tohoku University and Astronomy

In 1907, Tohoku Imperial University was established by the Meiji government as the third Imperial University following Tokyo and Kyoto. The Ministry of Education, under the advice of Nagaoka Hantaro, selected the first professional staff, which included Honda Kotaro, Kusakabe Shirota, and Ishiwara Jun, and sent them to Europe to study physics. From the beginning the university advocated open-door policies. It was the first national university in Japan to accept female students and foreign students from 1913.

The Faculty of Science was founded in 1911 with four departments: mathematics, physics, chemistry, and biology. The first staff of the Department of Physics were Honda Kotaro (本多光太郎) (experimental physics), Aichi Keiichi (愛知敬一) and Ishiwara Jun (石原純) (both theoretical physics), and Kusakabe Shirota (日下部四郎太) (geophysics).

Education and research in astronomy were among the main subjects taught from the beginning. Astronomy education was successively promoted by professors at three departments as given in Table 5.1.

Table 5.1 Professors of astronomy at Tohoku University in the period 1911–1950 (Takeuchi 2014)

Department	Years	Professor of astronomy
Physics	1911–1919	Kusakabe Shirota (日下部四郎太)
Geophysics	1920–1933	Okada Takematsu (岡田武松)
		Kusakabe Shirota
		Nakamura Saemontaro (中村左衛門太郎)
Astronomy	1934–1950	Matukuma Takehiko (松隈健彦)

In 1911, Mukaiyama Observatory for geophysical and astronomical observations was constructed near the university campus. It was equipped with a Bamberg 8-cm meridian circle and a Zeiss 13-cm equatorial refractor for astronomical observations. It was also equipped with a seismometer and meteorological instruments for geophysics. Mukaiyama Observatory served as the meteorological station of Sendai as well.

5.1.3 Kusakabe Shirota and Geophysics

Kusakabe Shirota (日下部四郎太, 1875–1924) was born into the family of a farmer in a small village, now part of the city of Yamagata, Yamagata Prefecture. He spent several years as a novice monk in a Buddhist temple near his house before entering elementary school. He was a troublesome boy in his youth (Tsugane 1973).

After 3 years at Dai-Ni High School (old) in Sendai, he enrolled in Tokyo Imperial University, where he studied physics under Tanakadate Aikitsu and Nagaoka Hantaro. After graduation in 1900, he continued his study of physical lithology at Nagaoka's laboratory. This included measurements of elasticity of numerous kinds of rocks and the propagation velocity of seismic waves in various strata. To support himself, for several years he held a second job as timekeeper at Tokyo Astronomical Observatory in Azabu. There he studied positional astronomy under special circumstances. His lithological works were recognized by the Japan Academy in 1914 (Fig. 5.1).

In 1908, he was appointed professor candidate of physics at Tohoku University and sent to Europe by the government for 3 years. Though he stayed mainly at Berlin

Fig. 5.1 Portrait of Kusakabe Shirota. ⟨Source: 東北大学史料館⟩⟨要使用許可⟩

and Göttingen Universities, he visited many other universities and observatories in Western Europe.

After returning home, he moved to Sendai as a professor of physics in charge of geophysics and astronomy. In Sendai, he continued his study of physical lithology along with making astronomical observations at Mukaiyama Observatory. He took on a heavy workload as the Dean of the College of Science and died from erysipelas suddenly in 1924 at age 50.

Kusakabe was widely known as a pious physicist. His faith and focus were the extermination of civilian superstitions. He traveled to many places around the country from south to north in order to collect lithological data and earthquake records. In parallel with these works, he also collected superstitious beliefs among local people and published an instructive book entitled *Pious Physics – Zodiac and Auspicious Days* on the rational interpretations of astrological phenomena (Kusakabe 1924).

In this book, Kusakabe argued that all superstitions and legends had reasonable origins in ancient times, but later on they were interpreted literally or religiously. For example, the auspicious days (吉日) in the calendar were selected based on astrological phenomena that were related to systematic motions of planets in the sky. Another example was the Tanabata Festival (七夕祭), which is celebrated on July 7 every year around the country. Tanabata originated in a legend of China about two stars at both side of the Milky Way. One is the Cowherd star (Altair) and the other is the Weaver star (Vega). While the cowherd boy loved the Weaver, they were permitted to meet each other across the Milky Way only once a year on July 7. This story of lovers meeting had become celebrated as the Tanabata Festival in Japan. On this day people show off bamboo branches decorated with many ornaments and strips of colored paper on which they write their wishes. While this festival is based on a Chinese legend, people do not think about its origins. Kusakabe suggested some astronomical reasoning behind this legend. One of its plausible origins is a legend in the region of the Han River (漢江) in China. According to Uozumi, the Milky Way in the sky flows in parallel with the Han River once a year on a night in early July. This astronomical event was dramatized as a love story and handed down from generation to generation (Uozumi 1982). Kusakabe cited many examples and gave them reasonable explanations to superstitions and legends in Japan.

5.1.4 Ishiwara Jun and Theoretical Physics (Nishio 2011)

Ishiwara Jun (石原純, 1881–1947) was born in Tokyo, the eldest son of Ishihara Ryo (石原量, Christian priest, descendant of a samurai). His mother died when he was 6 years old. His father was in poor health and retired early from the priesthood. As a result, Jun's life was difficult. After a 3-year course of study at Dai-Ichi High School (old) in Tokyo, he entered Tokyo Imperial University thanks to financial support of relatives. He studied theoretical physics under Nagaoka Hantaro in the

Fig. 5.2 Portrait of Ishiwara Jun. (After Tohoku University 東北大学史料館 利用許可必要)

Department of Physics. After graduation in 1906, he continued his studies at Nagaoka's laboratory. In the following year he married Hashimoto Itsuko (橋元逸子). They had three children together. In 1908, he was appointed lecturer of physics at the Military Academy of the Army. In those days, his interest was directed at Einstein's relativity, and his first work, in 1909, was on optics in a moving medium (Ishiwara 1909).

Einstein's relativity was based on two presuppositions: (a) the relativity of motion and (b) the constant velocity of light. Ishiwara began to work on relativity based on the first presupposition. He studied optical phenomena, such as the Doppler effect, the dispersion and absorption of light, and radiation density in optical spectra from the viewpoint of relativity. However, he could not accept presupposition (b), since he thought that the velocity of light might change in a moving medium (Fig. 5.2).

In 1911, he moved to Sendai as an associate professor of physics at Tohoku Imperial University. Half a year later, he was sent to Europe by the government for 3 years. He studied theoretical physics under, successively, Arnold Sommerfeld at Munich, Max Planck at Berlin, and Albert Einstein at Zürich.

In Sendai, he was promoted to professor after returning from Europe. He began to work on quantum theory. He proposed the generalized quantum condition at the same time as A. Sommerfeld and W. Wilson, along with some theoretical models of hydrogen and helium nuclei (Ishiwara 1915, 1916).

With his works on relativity and quantum physics, he was recognized by the Japan Academy in 1919.

Ishiwara was known as a *waka* poet for his *waka* anthologies from his student days. In 1917–1919, he fell in love with Hara Asao (原 阿佐緒), a painter as well as a fellow *waka* poet. Due to this love affair he was suspended from the university.

Two years later he finally resigned from his post and moved with Asao to Yasuda (安田), Chiba Prefecture. Thereafter, he devoted his life to science journalism.

From November to December 1922, Einstein made a lecture tour to Sendai, Tokyo, and Kyoto, under the sponsorship of the publisher Kaizo Sha (改造社). Ishiwara acted as an interpreter during Einstein's stay in Japan and published a book entitled *A Collection of Einstein's Lectures*, which was coauthored with a painter named Okamoto Ippei (岡本一平). Einstein and Ishiwara became very close friends during this tour (Ishiwara, H. 1971) (Figs. 5.3 and 5.4).

Fig. 5.3 Photograph of Einstein and the staff of the Department of Physics. Left to right: Honda Kotaro (本多光太郎), Einstein, Aichi Keiichi (愛知敬一), and Kusakabe Shirota (日下部四郎太), Tohoku University, December 1922. (After Tohoku University東北大学史料館 要使用許可)

Fig. 5.4 Image of Einstein and Ishiwara at Chion-In Temple (知恩院) in Kyoto, sketched by Okamoto Ippei. (Ishiwara, H. 1971)

In 1945 Ishiwara suffered a serious injury in a traffic accident and died in1947 at age 66.

5.2 Matukuma Takehiko and Stellar Astronomy

5.2.1 Life and Works

Matukuma Takehiko (松隈健彦, 1890–1950) was born in Karatsu (唐津) in Saga Prefecture, Kyushu, and showed an interest in astronomy in his youth. After 3 years at Dai-Roku High School (old) in Okayama, he entered Tokyo Imperial University and studied astronomy under Terao Hisashi. Upon graduation in 1913, he continued his studies on a scholarship at Terao's laboratory. After several years, he became a professor at the Navy Academy at Edajima in Hiroshima and taught at Dai-Roku High School (old) in Okayama and then at Dai-Ichi High School (old) in Tokyo, successively.

In 1924, he moved to Sendai as an associate professor at Tohoku Imperial University. The following year he was sent to Europe by the government for 3 years. He stayed mainly in Cambridge and studied general relativity and astrophysics under A. S. Eddington. Eddington was an early advocate of Einstein's general theory of relativity and his book *The Mathematical Theory of Relativity* was just published in 1923.

Matukuma's study in the early 1930s was directed to two subjects: celestial mechanics and the structure of globular clusters.

When the Department of Astronomy in Tohoku Imperial University was established in 1934, Matukuma was appointed as the first professor. While his main interest was in theoretical astrophysics, he also promoted observational astronomy as well. When a total eclipse was observed in Hokkaido, Japan, he organized an expedition of solar eclipse for the purpose of confirming Einstein's effect.

During the World War II (1941–1945), Matukuma worked on the optical system of the Schmidt camera and celestial mechanics on the motion of planets. The results of his works were barely published in the late 1949 due to social confusion and economic difficulties immediately after the end of the war.

In this way, Matukuma worked hard during and after the war. His work, however, was broken by his sudden death by ileus in 1950 at age 59 (Hagihara et al. 1950).

He was also an esperantist in Sendai, and he served as the President of the Esperantist Society of Sendai for a long time. The title *Contributions from the Department of Astronomy* was named *Sendai Astronomaj Reportaj* (Fig. 5.5).

Fig. 5.5 Portrait of Matukuma Takehiko. ⟨After Tohoku University, 東北大学史料館 要許可⟩

5.2.2 Celestial Mechanics

In celestial mechanics, Matukuma was interested in the numerical solutions of the restricted three-body problem for the orbit of a third body (Matukuma 1930, 1932). G. W. Hill considered the numerical solution of periodic orbits in some cases (Hill 1902a, b). Matukuma extended Hill's work and derived the periodic orbits as shown in Fig. 5.6, where X and Y are nondimensional axes, and the Y axis is on the line connecting the two main bodies. He calculated two-dimensional orbits of a small body moving on the (X-Y) plane. The orbits are characterized by the so-called Jacobi constant 2C, which appears as an integration constant of the equations of motion for the square of velocity V, as given by

$$V^2 = 3x^2 + \frac{2}{r} - 2C \tag{5.1}$$

where x denotes the distance on the X axis and r the distance from the center. The integration constant 2C can take any numerical value. The orbits in Fig. 5.6 present some different groups of 2C taken as a parameter.

Matukuma classified orbits into several families: Family B (looped orbit), given as B1 (2C = 2.0), B2 (1.3), B3 (0.8), and B4 (0.4), and Family F (liberation orbit), given as F1 (2C = 4.0), F2 (3.7), and F3 (3.4). Family F represents the oscillating motion around the equilibrium points (Lagrange's equilateral triangles). Family B is contained between a cusped orbit (2C = 2.55778) and the ejection orbit (2C <0.4). No ejection orbit is given in Fig. 5.6.

In addition, Matukuma proposed the existence of periodic orbits in the case of 2C < 0. He examined several cases from 2C = −3, −10 and −100,000 and proved the possible existence of periodic and nonperiodic orbits, but no detailed numerical

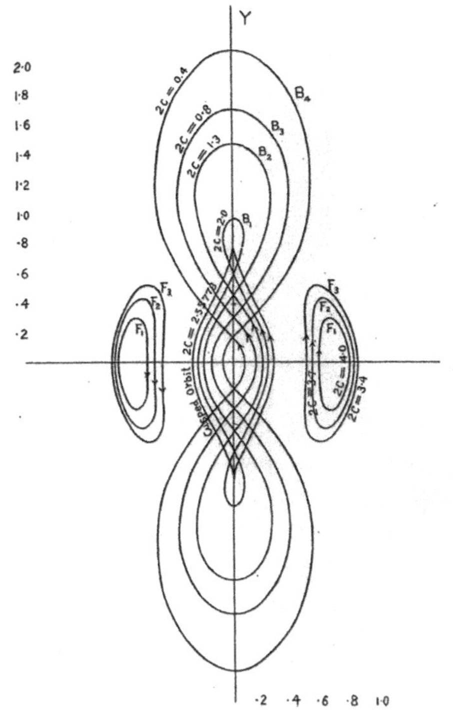

Fig. 5.6 Periodic orbits in restricted three-body problems. (Matukuma 1930)

calculations were performed. The orbits of these cases remained as future problems (Matukuma 1933).

In the early 1940s, Matukuma was also working on the theoretical study of the motion of perihelion and node of the planet Venus, Earth, and Mars, with lengthy numerical calculations (Matukuma 1949).

5.2.3 *Structure of Globular Clusters*

When Matukuma began working on the structure of globular clusters (Matukuma 1930), he adopted the Kapteyn model of the Milky Way galaxy. In this model, the Galaxy was a rotating ellipsoidal stellar system; the radius of the galactic plane was 9000 pc, the vertical height at the center was 1800 pc, and the Sun was located near the center. In contrast, the distance to the nearby globular cluster, the Hercules cluster (M13), was 11,000 pc and that of the most distant cluster, NGC 7006, was 67,500 pc. Matukuma accordingly argued that globular clusters were almost extra-galactic objects, like spiral nebulae. On the concentration of globular clusters in the direction of the constellation Sagittarius, he supposed that it might be related to the structure of the whole Universe.

Matukuma's aim on the study of globular clusters was to derive the three-dimensional distribution of stars in the clusters. Jeans (1915) considered this problem under the assumptions of spherical symmetry and steady state, whereas Eddington (1930) added the assumptions of polytropic distribution of stars and velocity ellipsoid of Karl Schwarzschild. Referring to these previous works, Matukuma adopted the three assumptions of steady state, spherical distribution, and velocity ellipsoid of stellar motions.

Based on these assumptions, Matukuma derived a new differential equation on space stellar density:

$$\frac{d^2\varphi}{dr^2} + \frac{2}{r}\frac{d\varphi}{dr} + \frac{\varphi^n}{a^2 + r^2} = 0. \tag{5.2}$$

Here, φ denotes the potential at each point in the cluster, a is a constant to be determined from observations, and n is a constant related to the structure of the cluster as follows:

	Radius of cluster	Total mass
$n < 3$	Finite	Finite
$n = 3$	Infinite	Finite
$n > 3$	Infinite	Finite

When $n = 3$, the star density ρ is given in the form

$$\rho = \rho_0 \frac{\varphi_1^3}{1 + r^2} \tag{5.3}$$

where ρ_0 is the central star density, φ_1 is a solution of eq. (5.2), which is generally solved numerically. Matukuma did not give numerical solutions, so the problem remained for future work.

5.2.4 *Observations of Einstein Effect*

In 1919, A. S. Eddington (1882–1944) observed a solar eclipse on May 29 at Principe Island off the west coast of Africa. According to general relativity, stellar light passing near the Sun should be curved by its gravitational field. Eddington measured the displacements of some stars during and outside of the eclipse, and he showed that the observed shift could be interpreted as the effect of general relativity. Eddington's announcement was reported in newspapers all over the world as a confirmation of Einstein's theory. However, it was soon claimed that Eddington's observations were of poor quality, and the simultaneous observations of the same eclipse at Sobral, Brazil, failed to detect the Einstein effect. Thus, further observations were required for confirmation.

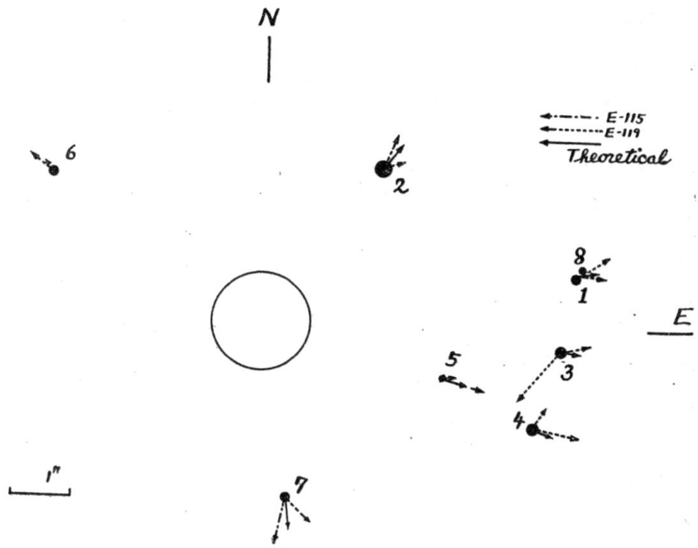

Fig. 5.7 Displacement of star image during total solar eclipse compared with theoretical prediction observed by Matukuma's group in 1936. (Matukuma et al. 1940)

Matukuma organized a solar eclipse expedition, observed on the northeastern coast of Hokkaido Island on June 19, 1936, for the confirmation of the Einstein effect (Matukuma et al. 1940).

During the solar eclipse Matukuma's group obtained some photographic plates with different exposures. After half a year, they observed the same sky area and measured the positions of stars on the two plates, E-115 and E-119, The displacements are shown in Fig. 5.7, along with a theoretical prediction.

According to the theory of relativity, the deflection of light, D, can be expressed as a function of distance from the solar center by the formula

$$D = E\,(R_*/r), \tag{5.4}$$

where R_* denotes the solar radius, r the distance from the solar center, and E a constant. Einstein's theory gives a value of $E = 1''.745$. Matukuma's group measured the positions of eight stars around the Sun and derived a value of $E = 2''.134 \pm 1''.14\,6$. Thus, they derived reasonable values for the confirmation of the Einstein effect.

5.2.5 *Optical System of Schmidt Camera*

During World War II (1941–1945), Matukuma was working on the optical system of the Schmidt camera. The fields of view of reflecting telescopes are generally small compared to refracting telescopes. Bernhard V. Schmidt (1879–1935) at the Hamburg Observatory invented a new optical system, called the Schmidt camera, in 1932, to widen the field of view of reflecting telescopes. The principle is to put a nonspherical correcting plate in front of the main spherical mirror. With this combination, the fields of view can be widened up to several degrees in angular diameter without spherical aberration. Matukuma took up this problem to derive the exact form of a correcting lens. According to his memoirs, he began his work at the very beginning of the war, and the most of the results were obtained near the end of the war. He carried out lengthy numerical calculations and finally published the results in 1949 (Matukuma 1949). In 1948, however, a large Schmidt telescope was already completed at the Mount Palomar Observatory. Matukuma's study indicates Japan's isolation during the war.

5.3 Hitotuyanagi Zyuiti and Astrophysics

5.3.1 Life and Works

Hitotuyanagi Zyuiti (一柳寿一, 1910–1998) contributed to the progress in theoretical astrophysics at Tohoku University (Suda 1998; Seki, 2014). He was born in Tokyo, though his family was from Adono (安曇野), Nagano Prefecture. After graduation from Tokyo High School (old) in 1928, he enrolled at Tokyo Imperial University and studied astronomy under Hagihara and Hirayama Kiyotsugu. The only other student in his class was Fujita Yoshio (Chap. 3). Upon graduation in 1931, he was invited by Matukuma to be his assistant at Tohoku Imperial University, and he was promoted to associate professor and then professor. He spent more than 40 years at the university until his retirement in 1973. Thereafter he continued his research work as professor emeritus at Tohoku University.

In Sendai, he started his work on the stellar model in the early 1930s. After completing this work, he came down with tuberculosis and was affected by it for several years. After recovery, his interest was directed to the theoretical study of the solar atmosphere in the early 1940s. During the war, Sendai suffered severe damage from US air attacks, and, as a member of the board of trustees, he spent significant amounts of time on the recovery of the education and research system at Tohoku University.

After the 1960s, Hitotuyanagi became interested in galactic astronomy. He joined the research group of Taketani Mitsuo, Hatanaka Takeo, Hayakawa Sachio, and others. His main subjects were the structure of spiral and barred galaxies.

Hitotuyanagi was a lifelong bachelor. He loved music, the arts, and skiing. Every winter he went mountain skiing at a hot-spring resort. He died in July 1998 in Sendai at age 88 (Fig. 5.8).

Fig. 5.8 Portrait of Hitotuyanagi Zyuiti. (Takeuchi 1972)

5.3.2 Stellar Model

In 1926, A. S. Eddington considered the theory of stellar pulsation for Cepheid and Mira types, based on the subatomic energy source operating in the central part of the stars. He showed that if the rate of energy generation, ε, is proportional to the nth power of gas density, ρ, the central region of the stellar core enters a state of overstability when n is larger than 5/3. This suggests that the amplitude of a small disturbance in the central region gradually rises up and propagates to the stellar surface. This large value of n is the condition of the maintenance of pulsation phenomena.

Hitotuyanagi's work in 1934 on the interior structure of stars began with an examination of star stability against the pulsation (Hitotuyanagi 1934). He accepted the annihilation hypothesis for the generation of stellar energy and assumed that the rate of energy generation ε was proportional to the square of gas density, ρ. Based on these assumptions, he calculated anew the stellar model making use of H. A. Kramers' absorption coefficient and S. Rosseland's theory for the stellar structure. Hitotuyanagi made numerical calculations for a star of 6.4 solar mass and 290 solar luminosities (1.1×10^{36} erg s^{-1}), which corresponds to around a B5 main-sequence star, and he found that this star had a stable state for pulsation, contrary to Eddington's theory, according to which this star should be a pulsation star by its large value of $n = 2$. On this disagreement, Hitotuyanagi supposed that one of the reasons might be the assumption of the mass annihilation at the stellar center, but this was inconclusive in the days of the early 1930s.

5.3.3 Limb Darkening of Solar Photosphere

The brightness of the solar surface is not uniform. It gradually decreases from the disk center toward the limb. This is called solar limb darkening. In the early 1940s, the solar atmosphere was widely assumed to be in radiative equilibrium. Hitotuyanagi pointed out that the limb darkening derived on this assumption clearly deviates from the observed darkening. He considered the existence of a convection layer under the solar atmosphere in radiative equilibrium. In the convection layer, since a part of radiative energy is consumed by the kinetic energy of convection flow, the temperature declines faster than the case of pure radiative equilibrium. The existence of a convection layer was first recognized in 1930 by A. Unsöld, and its effects on the Fraunhofer lines and continuous spectrum were investigated by E. A. Milne in 1930 and by P. C. Keenan in 1938.

Hitotuyanagi examined the effects of the convection layer in detail, and he derived theoretical limb darkening. The results of his numerical calculations in the

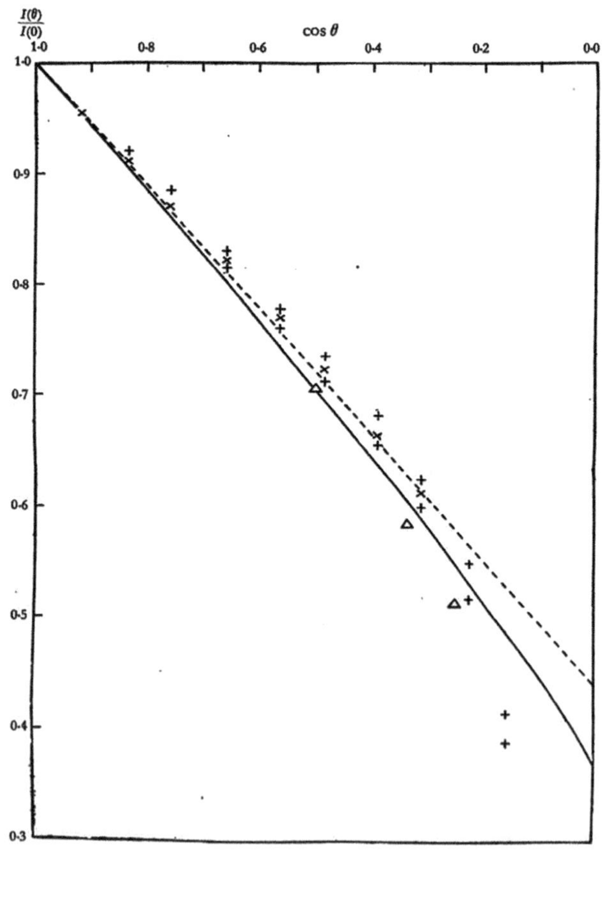

Fig. 5.9 Theoretical limb darkening and comparison with observations. The dotted and real curves represent the first and second approximation, respectively. (Hitotuyanagi 1941a)

first and second approximations are illustrated in Fig. 5.9, along with observed data for comparison (Hitotuyanagi 1941a, b).

In addition, based on his theory of model atmosphere (Hitotuyanagi 1944, 1947), Hitotuyanagi and Inaba derived the center-limb variation of the sodium D line. The line profile is sensitive to the distribution of stellar temperature and sodium abundance. By comparing these findings with previous works of C. W. Allen (1940) and W. Priester (1953), Hitotuyanagi derived a new plausible abundance of sodium in the solar atmosphere (Hitotuyanagi and Inaba 1954).

5.3.4 Galactic Astronomy

After the war, Hitotuyanagi, together with his young colleagues, promoted theoretical studies of galaxies.

In the 1960s, the main subjects were the formation of spiral arms in galaxies and the dynamical structure of barred galaxies (Oki et al. 1965).

1. Formation of spiral arms in Galaxies

In 1958–1962, various models were proposed on the formation of spiral arms based on the magneto gas dynamical point of view. A. G. Pacholczyk and J. S. Stodolkiewicz (1958), F. Hoyle and J. G. Ireland (1961), J. Tassoul (1962), and A. Elvius, N. Herlofson, and P. O. Lindblad (1961) proposed different models. Following the publication of these works, Hitotuyanagi's group considered the formation of spiral arms. They assumed that the initial gravitational field was determined by stars alone and unchanged thereafter, since the total mass of gas is too small compared to the mass of stars. They considered a two-dimensional model of a galaxy and approximately solved the magneto-hydrodynamic equations of gas and magnetic field (Oki et al. 1965).

As the initial conditions, they supposed that the gas was distributed in a circular, homogeneous thin layer and that the magnetic field was also homogeneously extended through galactic and intergalactic space. Under regular galactic rotation, the formation of spiral arms begins and develops as delineated in Fig. 5.10, where the structure of arms is represented at four time intervals t, measured in units of 2×10^8 years, which corresponds to around one rotation period of the galaxy. Arms are traced as the curves of maximum density ρ_{max} relative to the mean density ρ_0. In Fig. 5.10, full lines represent the region for $\rho_{max}/\rho_0 > 1.5$ and dotted lines for < 1.5.

A serious difficulty arose in this solution concerning the time scale of spirals. When winding curves in Fig. 5.10 are compared with observations, the age of arms can be estimated such as $t \approx 6$ for our Galaxy, and $t \approx 4$ for M81.Thereafter arms wind up toward the galactic center. These values were evidently much smaller than the actual age of the galaxies, $t \approx 100$ or even more. To avoid this difficulty, they assumed that once a spiral arm formed, the arm and other physical conditions in a galaxy remained nearly stationary state for a time interval comparable with the age of the galaxy. The formation of spiral arms will be considered again in Chap. 7.

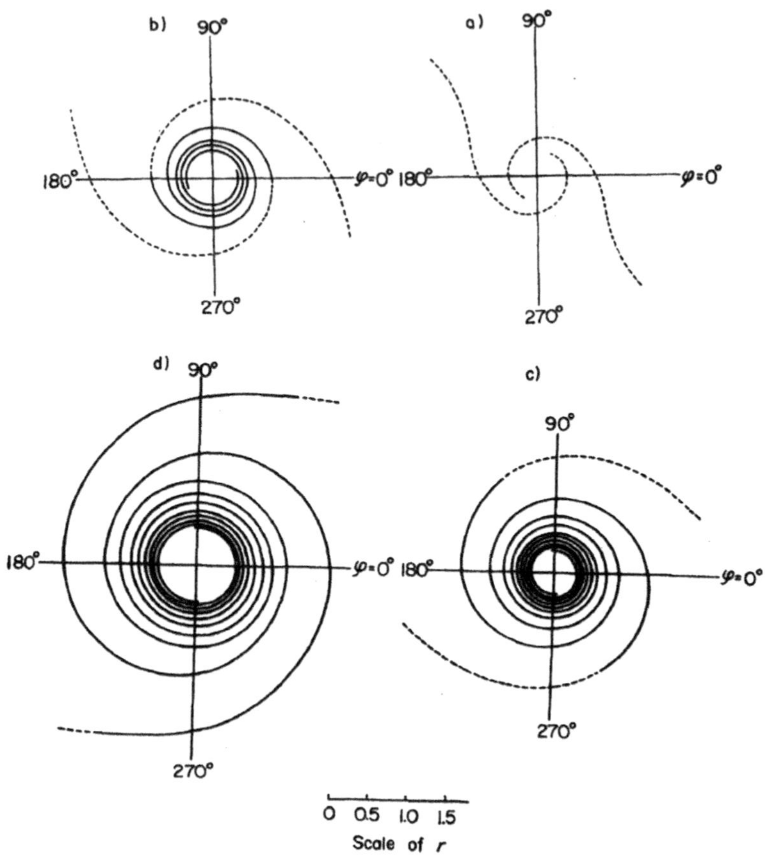

Fig. 5.10 Formation of spiral arms. Forms of arms at time t in units of 2×10^8 years. (**a**) $t = 1.0$, (**b**) $t = 3.2$, (**c**) $t = 5.6$, and (**d**) $t = 10.0$. (Oki et al. 1965)

2. Dynamical structure of the barred galaxies

In the early 1960s, E. M. and G. R. Burbidge and K. H. Prendergast proposed a prolate spheroid model of barred galaxies with uniform stellar densities. Another model was proposed by S. J. Aarseth, who constructed a model consisting of a spherical central region and cylindrical bars. G. and M. de Vaucoulours showed the existence of expanding motion of gas in barred galaxies.

Hitotuyanagi's group considered the two-dimensional motion of gas confined to the galactic plane in Aarseth's sphere-cylinder model. The model is shown in Fig. 5.11, where R_1 and a denote the radii of the central sphere and cylindrical bar, respectively. $R_2 - R_1$ is the length of the bar (Aarseth 1960). Under the assumption that the gas is ejected from the surface of the central sphere at velocity V_0,

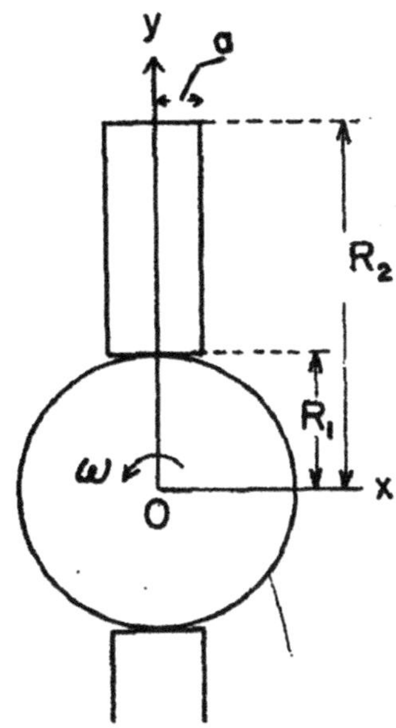

Fig. 5.11 Sphere-cylinder model of barred galaxy seen from pole. (Oki et al. 1965)

Hitotuyanagi's group solved the magnetohydrodynamic equations of gas flow, and as a result, they showed that gas flows out in parallel with a bar or forming curved patterns. After arriving at the end of the bar, the gas causes the subsequent formation of trailing spiral arms from both ends of the bar. Based on this model, they considered in depth the dynamical structure of barred galaxies.

References

Aarseth, S. J. (1960). The rotation of a barred galaxy under gravitational forces. *Monthly Notice of the Royal Astronomical Society, 121*, 525–529.

Hagihara, Y., Hitotsuyanagi, Z., & Kaburaki, M. (1950). Memorial addresses for Matukuma Takehiko (天文月報、萩原雄佑、一柳寿一ほか: 松隈健彦追悼文集). *Astronomical Herald, 43*, 21–24.

Hill, G. W. (1902a). Illustrations of periodic solutions in the Problem of three bodies. I. *Astronomical Journal, 22*, 33–37.

Hill, G. W. (1902b). Illustrations of periodic solutions in the Problem of three bodies. II. *Astronomical Journal, 22*, 117–121.

Hitotuyanagi, Z. (1934). Bemerkungen über ein Sternmodell mit der Energierzeugung $\varepsilon \sim \rho^2$. *Japanese Journal of Astronomy and Geophysics, 12*, 113.

Hitotuyanagi, Z. (1941a). Über die Randverdunkelung der Sonne, I. Zweite Nöherungslösung in der Milne – Lindbladoken Theorie von Blanketing Effekt der umkehrenden Schicht. *Sendai Astronomiaj Reportaj, II*(1), 99–111.

Hitotuyanagi, Z. (1941b). Über die Randverdunkelung der Sonne, II. Der Einfluss des Fraunhoferschen Linien auf des Kontinierliche Spektrum. *Sendai Astronomiaj Reportaj, II*(2), 114–134.

Hitotuyanagi, Z. (1944). Zur Theorie der Fraunhoferschen Linien in der Sonnenatmoosphären. *Sendai Astronomiaj Reportaj, II*(3), 87–23.
Hitotuyanagi, Z. (1947). Über die Adiabatengleichung in Sternatmosphären. *Japanese Journal of Astronomy and Geophysics, 21*, 119–124.
Hitotuyanagi, Z., & Inaba, H. (1954). Model solar atmosphere and center-limb variations of the sodium D lines. *Sendai Astronomiaj Reportaj, 38*(3).
Ishiwara, J. (1909). Zur Optik der bewegten ponderablen Medien. *Proceedings of Tokyo Mathematico-Physical Society, 5*, 150–180.
Ishiwara, J. (1915). Die universelle Bedeutung des Wirkungsquantums. *Proceedings of Tokyo Mathematico-Physical Society, 8*, 106–116.
Ishiwara, J. (1916). Über die Struktur des Atomkerns. *Proceedings of Tokyo Mathematico-Physical Society, 8*, 540–550.
Ishiwara H. (Ed.). (1971). *A Collection of Einstein's Lectures*, edited by Ishiwara Jun and Okamoto Ippei (石原紘編、石原純、岡本一平: "アインシュタイン講演録" Tokyo Tosho (東京図書).
Kusakabe S. (1924). *Pious Physics — Zodiac and Auspicious Days*, (日下部四郎太: "信仰物理 - 黄道吉日"), Iwanami Shoten (岩波書店).
Matukuma, T. (1930). On the periodic orbits in Hill's case., I. *Sendai Astronomiaj Reportaj, 1*, 6–8.
Matukuma, T. (1932). On the periodic orbits in Hill's case., II. *Sendai Astronomiaj Reportaj, 1*(6), 147–450.
Matukuma, T. (1933). Periodic orbits in Hill's case, Third paper, Periodic ejective orbit. *Sendai Astronomiaj Reportaj, 1*(7), 364–366.
Matukuma, T. (1949). Theory of Schmidt Camera – Preliminary Report. *The Science Report of the Tohoku University, Ser. 1, 33*(1), 43–50.
Matukuma, T., Onuki, A., Yoshida, S., & Iwane, Y. (1940). On Einstein effect derived from the observations of the total Solar eclipse of June 19th in 1936. *Sendai Astronomiaj Reportaj*, 1(12) (*Japanese Journal of Astronomy and Geophysics,* 18(1), 51–72).
Nishio, S. (2011). *Biography of Ishiwara Jun – A Pioneer of Science Journalism*, (西尾成子:"評伝石原純 -科学ジャーナリズムの先駆者—") Iwanami Shoten (岩波書店).
Oki, T., Fujimoto, M., & Hitotuyanagi, Z. (1965). Formation of spiral arms and dynamical model of barred galaxies. *Progress in Theoretical Physics,* Supplement, No. 31, 77–115.
Seki, M. (2014). *Reminiscence of Professor Hitotuyanagi*. Private communication.
Suda, K. (1998). Obituary of Professor Hitituyanagi Juiti (天文月報、須田和男: 一柳寿一先生を悼む). *Astronomical Herald, 91*(10), 514.
Takeuchi, M. (Ed.). (1972). *Proceedings of the symposium: The theory of the structure and evolution of galaxies* (Retirement Memories of Prof. Hitotuyanagi) (竹内峯編: "銀河の構造と理論シンポジュウム集録"、一柳寿一教授退官記念会).
Takeuchi, M. (2014). *Tohoku University and Astronomy – First half century –* (竹内峯: "東北大学と天文学—最初の半世紀—), Private Issue.
Takeuchi, M., & Seki, M. (2008). *In one hundred years of astronomy in Japan* (日本の天文学の百年), Chapter 4, 37–42, New breath in the woody city (竹内峯, 関宗蔵; 杜の都に新たな息吹).
Tsugane, S. (1973). *A pious physicist – Kusakabe Shirota* (津金仙太郎: "信仰物理学者—日下部四郎太") Chuou Shoin (中央書院).
Uozumi, T. (1982). *Legend of the Milky Way* (魚住孝義:"古事記天の川伝説"), Kadensha Publisher (花伝社).

Chapter 6
Postwar Development of Astrophysics, 1946–2000 (Part I: Instrumentation)

Abstract In the history of postwar astronomy in Japan, constructions of large instruments is remarkable. This chapter is devoted to the development of observational facilities in optical, infrared, radio, and space observations. Their capabilities in observations are also noted, with examples.

6.1 Social and Economic Background

The early Showa period ended in 1945 with the defeat of Japan in World War II. Japan was occupied by the Allied Forces for several years. Most Japanese were awakened from an attitude of militarism to democracy. This change could be regarded as the second turning point in modern Japanese history after the Meiji Restoration (Bartholomew 1989).

Japan was destroyed, and confusion persisted for several years in social and economic affairs. Thereafter, the Japanese economy expanded at a stunning pace from the late 1950s through the early 1970s. At unprecedented speed, Japan emerged from a state of destruction and poverty and moved to a place of prosperity. Japan's gross national product (GNP, the total value of goods and services produced in a year) expanded by an average annual rate of more than 10%. Such a high growth rate is often called the Japanese miracle. Nominal GNP of four major nations are compared in Table 6.1.

In the period 1947–1949, the educational system was greatly transformed: compulsory education was expanded from 6 to 9 years, and the university system was also changed. Seven Imperial Universities were upgraded to National Universities, along with the establishment of 60 more National Universities across the country. These new universities were established in the wake of reform of preparatory schools (formerly high schools) or normal schools for training teachers. Large numbers of private universities and technical schools were also established. This new educational system contributed to the popularization of higher education and to the acceleration of industrial development and business activities.

Remarkable developments occurred in information technology (IT) in business, industry, and science after around the 1960s. The electric computer was superseded

T. Kogure, *The History of Modern Astronomy in Japan*, Historical & Cultural Astronomy, https://doi.org/10.1007/978-3-030-57061-3_6

Table 6.1 Nominal GNP of four major postwar nations, 1951–1980 (GNP in $US billion) (Gordon 2003)

Year	Japan	USA	West Germany	UK
1951	14.2	328.3	28.5	41.4
1955	22.7	398.0	43.0	53.9
1960	39.1	503.8	70.7	71.9
1965	88.1	688.1	115.1	100.2
1970	203.1	992.7	184.6	124.0
1975	498.2	1549.2	415.2	234.5
1980	1040.1	2633.3	816.5	585.5

by the electronic computer, and computer centers were established in 1965 at Tokyo and Kyoto Universities. The nation as a whole entered the computer age in the 1970s.

These social, economic, and technological developments enabled the accomplishment of national projects for constructing large scientific facilities, including observatories and orbiting satellites. Astronomy and space science grew up together into the Big Sciences.

6.2 Tokyo Astronomical Observatory

In 1949, the Tokyo Astronomical Observatory (TAO) changed from being a classical engineer-technician system to a department-section system. In the same year, the National University system was reformed; the Tokyo Imperial University changed to the University of Tokyo, and the TAO became an integral part of the university.

The TAO was composed of 6 departments and 19 sections, with each department and section composed of research staff of professors, associate professors, lecturers, and assistants. The number of staff increased to 123, and thus the TAO became a research institute. The names of the directors in the postwar era are given in Table 6.2. Since 1950, the status of the TAO increased gradually into a core institute of the interuniversity astronomical community (TAO 1977).

As shown in Table 6.2, the TAO was reorganized in 1988 as the National Astronomical Observatory of Japan (NAOJ), together with the Latitude Observatory in Mizusawa and Research Institute of Atmospherics of Nagoya University. For a time, the NAOJ came under the jurisdiction of the Ministry of Education and Science, and it was operated by a wide astronomical community across universities.

In 2004, the NAOJ was incorporated as the Inter-University Research Institute Corporation, the National Institutes of Natural Sciences (NINS), and the National Astronomical Observatory of Japan. The NINS (自然科学研究機構) includes five institutes: NAOJ (国立天文台), National Institute for Fusion Science (核融合科学研究所), the National Institute of Physiological Science (生理学研究所), National Institute of Basic Biology (基礎生物学研究所), and the National Institute of Molecular Science (分子科学研究所). NINS are a member of the Inter-University Research Institute Corporation (大学共同利用機関法人), which covers a broad

Table 6.2 Past directors in postwar period (1946–2000) (TAO 1977)

Order	Name	Tenure	Specified field
Tokyo Astronomical Observatory			
5.	Hagihara Yusuke (萩原雄佑)	1946–1957	Ap
6.	Miyaji Masashi (宮地政司)	1957–1963	As
7.	Hirose Hideo (広瀬秀雄)	1963–1968	As, Ap
8.	Furuhata Masaaki (古畑正秋)	1968–1973	Ap
9.	Osawa Kiyoteru (大沢清輝)	1973–1977	Ap
10.	Suemoto Zenzaburo (末元善三郎)	1977–1981	Ap
11.	Kozai Yoshihide (古在由秀)	1981–1988	Ap
National Astronomical Observatory of Japan (NAOJ)			
1.	Kozai Yoshihide (古在由秀)	1988–1994	Ap
2.	Kodaira Keiichi (小平桂一)	1994–2000	Ap
3.	Kaihu Norio (海部宣男)	2000–2006	Ap

Note specified field: *Ap* Astrophysics, *As* Astronomy

range of fields, including humanistic and natural sciences, high-energy physics, and information sciences. The Inter-University Research Institute is a *research institute operated by the research community*, that is, a type of world-class organization unique in Japan.

The astronomical facilities of the NAOJ at the end of the 1990s were as follows:

NAOJ Headquarters, Mitaka City, Tokyo
Dodaira Station, Tokigawa, Saitama Prefecture
Mizusawa VERA Observatory, Ōshu, Iwate Prefecture
Nobeyama Radio Observatory (NRO) and Nobeyama Solar Radio Observatory, Minamimaki, Nagano Prefecture
Okayama Astrophysical Observatory, Asaguchi, Okayama Prefecture
Hawaii Observatory, Hilo and Mauna Kea, Hawaii

6.3 Optical and Infrared Observatories

6.3.1 Okayama Astrophysical Observatory

Astrophysical observations in Japan passed a milestone in 1961 with the construction of the Okayama Astrophysical Observatory (OAO) for the physical observations of celestial objects. The OAO is the first facility built for the astronomical community in Japan.

The OAO has three major telescopes: 188-cm reflector, 91-cm reflector, and 65-cm solar Coudé telescope, the last having been constructed in 1968. Of these, the 188-cm reflector was made by Grubb-Parson of England, whereas the 91-cm and 65-cm telescopes were made by Nihon-Kogaku of Japan. The 188-cm reflector

Fig. 6.1 The 188-cm reflector at OAO in 1962. (Courtesy NAOJ)

(Fig. 6.1) has three foci: Newton (Astro-camera), Cassegrain (photoelectronic photometer, prism spectrograph), and Coudé (High dispersion grating spectrograph).

In the early phase, the main observations were made by Fujita's and Osawa's groups for the spectroscopic and photometric observations of stars, both using the 188-cm and 91-cm reflectors (Chap. 3). In parallel, proposals for the observation of stars, galaxies, nebulae, and Solar System objects were also submitted from many users all over the country. This suggests that the expectation for large telescopes had already been fully met among Japanese astronomers in the 1960s.

How the OAO met the expectations of the astronomical community is shown by the number of proposals, which increased every year for each focus of the 188-cm reflector, as shown in Fig. 6.2 for the first 23 years of operation.

Note that the total number of proposals rapidly increased from 40 in the first half of the 1960s to 50–60 in the next decade, and as high as to 70 in the last decade. The total allocation days of the OAO is about 42 weeks per year, so that the mean allocation days for each proposal were around 1 week in the early 1960s, whereas it reduced to 3 or 4 days in the last decade. This suggests that proposals for observations on one telescope were almost completely filled in the last decade and even thereafter. The need for the next large telescope had already greatly increased in the 1970s. The strong demand for the next large telescope was the driving force behind the promotion of the Japanese National Large Telescope project (JNLT), which was realized finally as the Subaru telescope in the late 1990s (Sect. 6.3.5).

In addition, Fig. 6.2 shows in the trend of observations that the proposals for the Cassegrain focus notably increased starting in the 1970s. This trend was spurred by the attachment of new powerful detectors, such as an image intensifier (1969), an infrared (IR) detector (1971), and a multichannel photometer (1975).

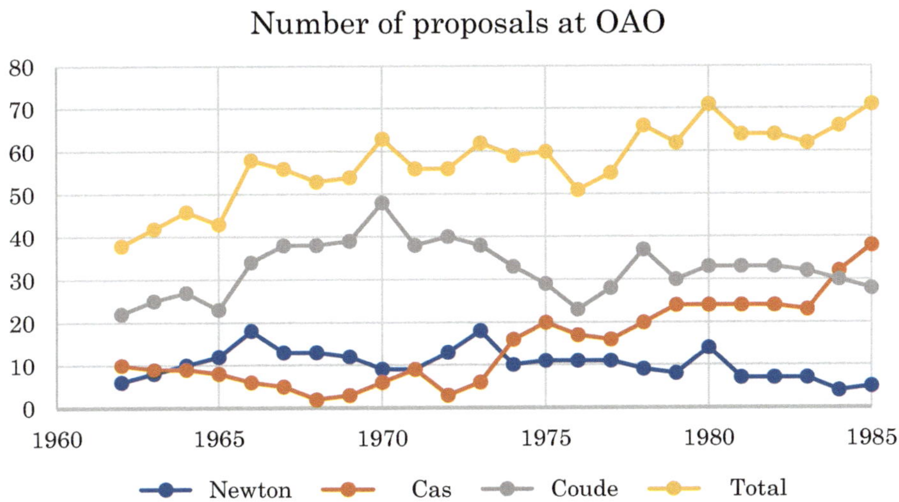

Fig. 6.2 Number of proposals per year at OAO for first 23 years (OAO)

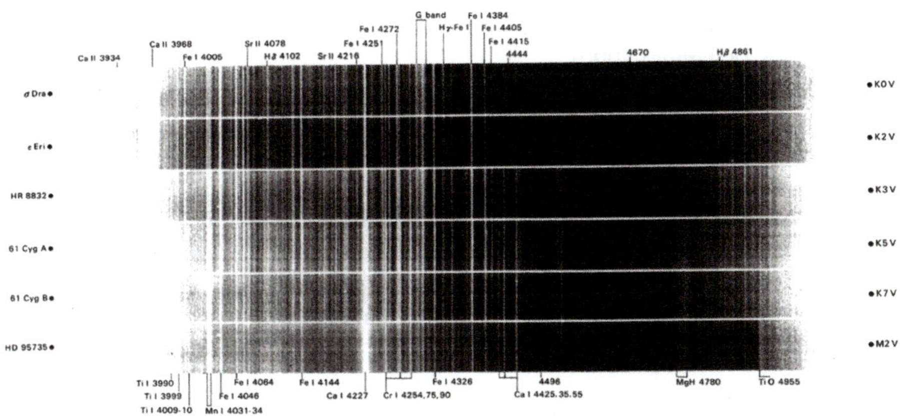

Fig. 6.3 Spectral atlas for K0V to M2V stars. (Yamashita et al. 1976)

The 91-cm Cassegrain reflector, equipped initially with a photoelectric photometer and a prism spectrograph, and then an IR photometer (1968), photoelectric spectral scanner (1967), and an image tube spectral scanner (1980), had also been in use. With intensive use of the 188-cm reflector, many observational results had been produced even from the 1960s.

With this telescope, *An Atlas of Stellar Spectra, Taken with the Prism-Spectrogram of the 91-cm Reflector* was published by Yamashita Yasumasa, Nariai Kyoji, and Norimoto Yuji in 1976. A sample page is illustrated in Fig. 6.3.

Advances in the detector came about with the invention of the charge-coupled device (CCD) by W. Boyle and G. E. Smith of Bell Telephone Laboratories in 1969.

CCD experiments started at the OAO by Kawakami Hajime and his group in the early 1980s. The first observations with a CCD camera were made in 1985 for both the 188- and 91-cm telescopes (Kawakami et al. 1985). The age of photographic observations ended in around 1990. Thereafter, CCD became the main detectors of observations and rapidly spread to every observatory in Japan.

6.3.2 Dodaira Station of TAO

As an attached station of the Mitaka Campus of the TAO, Dodaira Station was established in 1962 atop Mt. Dodaira in the Chichibu Mountains of Saitama Prefecture (TAO 1977, pp. 89–92).

The main telescope was the 91-cm equatorial reflector made by Nihon Kogaku in Japan. A photographic camera was attached in the primary focus, and the photometric photometer was first prepared in the Cassegrain focus, and later a prism spectrograph was installed. Kitamura Masatoshi (Chap. 7), who participated in the construction of the station, used a photometric photometer for the 91-cm telescope and observed variable stars and binary stars. According to his memoirs, sky transparency during the winter in the period 1960–1970s was excellent for photometric observations (Kitamura 2000).

A Schmidt telescope with a 50-cm aperture (50, 65 cm, $f = 100$ cm) was moved from the Mitaka campus and used to observe comets, asteroids, and artificial satellites. An automatic photographic camera attached to a prism spectrograph was constructed in 1964 by Hirose's group for observations of meteor showers (see section on Hirose in Chap. 3). Other facilities included a polar tube for the observations of the North Pole, an observation room with a night sky light.

In the 1980s and 1990s, the night sky of Dodaira was becoming bright because of the city lights of the Tokyo area. Finally, Dodaira Station was closed at the end of 2000.

6.3.3 Kiso Observatory and Schmidt Telescope

A large Schmidt telescope was built following long discussions in an interuniversity group on galactic astronomy (SAM, see Sect. 3.6). The SAM meeting in 1965 was devoted to the design of a Schmidt telescope as well as to the future research programs to be pursued with the new telescope.

In 1974, the Schmidt telescope was installed at the Kiso Observatory on a peak on the Kiso Mountains, Nagano Prefecture. The site was selected for its ample dark sky as a result of its location far from the two big cities of Tokyo and Nagoya.

The telescope was made by Nihon-Kogaku, and it has a correcting plate (105 cm) and main spherical mirror (150 cm), with a focal length of 330 cm (the focal ratio is 3.1). The 30 × 30-cm photographic plate covers a sky field of 5° × 5°. Compared to

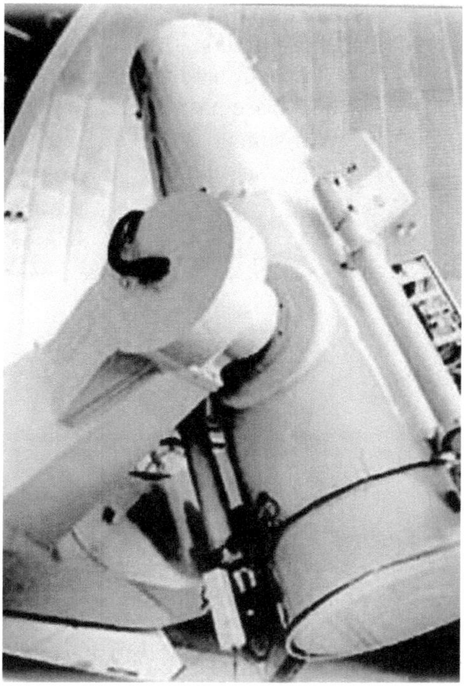

Fig. 6.4 Schmidt telescope at Kiso Observatory. (Courtesy Kiso Observatory, University of Tokyo)

the 122-, 150-, and 305-cm Palomar Schmidt telescope (focal ratio 2.5), the Kiso Schmidt, though it has a smaller aperture, has a limiting magnitude comparable to the Palomar Schmidt because of its larger focal ratio. The Kiso Schmidt also has two objective prisms of apex angles 2° and 4° as well as some basic color filters (Fig. 6.4.).

The observation system is the same as that of the OAO, based on proposals and programs. Notable observations in the first 20 years are summarized in what follows.

1. Survey of Ultraviolet-excess galaxies

By making use of three-band color filters, U (3600 Å), G(4800 Å), and R (6500 Å), the group of Takase Bunshiro conducted a survey of ultraviolet-excess galaxies (Kiso ultraviolet galaxy = KUG). KUGs are generally Markarian-type emission-line galaxies. The emission lines include the Balmer lines and some forbidden lines such as [OIII] and [NII], and their relative intensities differ from HII regions and planetary nebulae, rather resembling some spectra of Seyfert galaxies. By 1993, a total of 8968 KUGs had been identified in some 5100 square degrees in the sky field (Takase 1980; Takase and Miyauchi-Isobe 1993).

2. Surface photometry and quantitative classification of galaxies

Surface photometry and quantitative classification of galaxies were carried out in the period 1974–1983 by Takase Bunshiro, Kodaira Keiichi, and Okamura Sadanori. Some of their observations were published as *An Atlas of Galaxies* in 1990, which

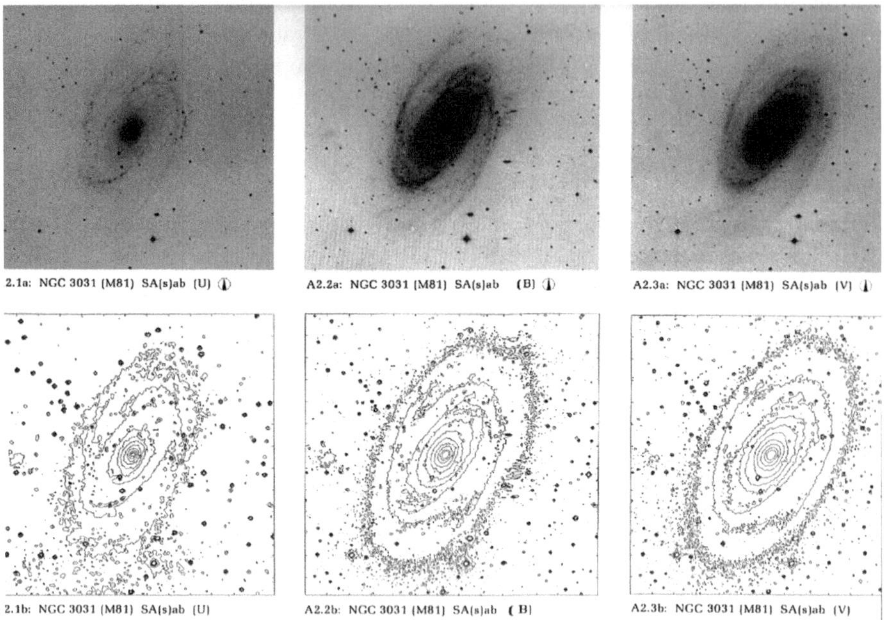

Fig. 6.5 NGC 3031 (M81) in U (left), B (center), and V (right) colors. (Takase et al. 1983)

contains 145 galaxies and clusters of galaxies with surface photometry for some galaxies. One page of this Atlas is shown in Fig. 6.5.

3. Survey of carbon stars

Carbon-rich late-type stars were classified into nine types (C0–C8) in order of effective temperature by W. W. Morgan and P. C. Keenan in 1941. Maehara Hideo of the TAO carried out a deep carbon star survey with the 105-cm Schmidt telescope using a 4° objective prism in the V and IN spectral bands. He selected cool carbon stars (C4–C8), which present three strong CN molecular bands in the λ8000 Å region almost independently of the subclass of carbon stars. From 1987 through 1999, Maehara and Soyano Takao surveyed 1565 square degrees along the galactic equator and identified 1069 carbon stars, among which 198 stars were new detections. He cataloged all carbon stars by stellar position (l, b) and V-magnitude, along with a finder chart for each star (Maehara and Soyano 1987; Soyano and Maehara 1999). This catalog updated the previous carbon-star catalog with precise stellar positions and V magnitudes in the epoch of 1999.

4. Survey of Hα emission-line stars in the Orion Region

The Orion molecular cloud region is an active star formation site composed of OB associations, HII regions, emission-line stars, and IR regions. In the 1980s, Suhardja D. Wiramihardja, Kogure Tomokazu, Nakano Makoto, and their group conducted a

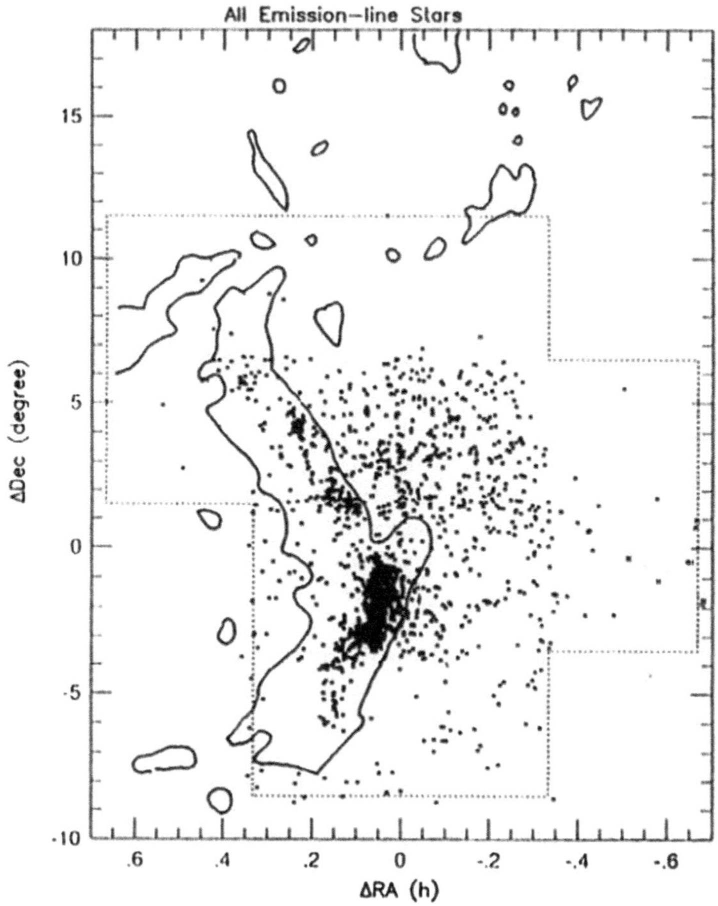

Fig. 6.6 Distribution of all Hα emission-line stars in 1995, overlapped on the contour of molecular clouds. (Nakano et al. 1995)

broad survey of Hα emission-line stars in the Orion region by making use of the 4° objective prism of the Schmidt telescope (Wiramihardja et al. 1989). They surveyed a sky area of 325 square degrees and identified 1220 $H\alpha$ emission-line objects, including 804 new detections and 29 nonstellar objects. These stars are widely distributed on the western side of the Orion molecular clouds. The distribution of all Hα emission stars, including both previously and newly identified stars, is illustrated in Fig. 6.6, along with the contours of the molecular clouds (Nakano et al. 1995). In this figure, the aggregation of emission-line stars on the western side of the main molecular cloud is remarkable. The emission-line stars in this part are mostly identified by the Kiso survey, and this group is free from molecular clouds and OB stars in the Orion region. For around 30 stars in this group, spectroscopic observations were carried out by Kogure, Ogura Katsuo, and others at the

Mt. Stromlo Observatory, Australia, and it was shown that all of the sample stars were of the T Tauri type (Kogure et al. 1992). They thus found that the west side aggregation is a simple T-association, which is older than the present Orion star-forming region.

Photographic observations at the Kiso Observatory were finished in 1993 by the end of production of large Schmidt plates at the Kodak Company. After that, the observations have been taken over by single or multiple CCD detectors for deeper sky surveys.

6.3.4 Agematsu IR Observatory and Infrared Astronomy

Infrared radiation lies between optical and radio radiations in a wavelength range of 0.8–1000 μm, which is subdivided into near (0.8–5 μm), mid (5–30 μm), and far (30–1000 μm) IR radiation. The terrestrial atmosphere is generally opaque to IR radiation with several windows, observable from ground-based telescopes.

The first IR observations in Japan were made at the OAO in 1968 with the PbS detector for the lunar surface by the group of Hayakawa Sachio and Okuda Haruyuki of Nagoya and Kyoto Universities (Hayakawa et al. 1968). Thereafter, they started polarization observations in the near-IR spectral region. Polarization is robust against the unstable sky in Japan. With a polarimeter mounted on the 91-cm reflector, Hashimoto Jun'ichiro, Maihara Toshinori, and others observed a bright late-type supergiant VY Canis Majoris at H and K bands (1.6 and 2.2 μm) and found a disk structure around the star (Hashimoto et al. 1970).

In 1973, the Agematsu Infrared Observatory (AIRO) was constructed at an altitude of 1200 m in the Kiso Mountains equipped with a 100-cm IR telescope (Fig. 6.7). At the AIRO, observational conditions have been substantially improved with the reduction of background radiation owing to its site at high altitude. By adopting a second mirror chopping system, high efficiency of observations was obtained. Moreover, plenty of observational time became available compared to at the OAO (Okuda 2018).

With this telescope, novae, comets, and the Galactic Center were observed in the near-IR spectral regions. Noguchi Kunio and his group made narrow-band spectro-photometric observations of 22 carbon stars and 23 M-type stars between 2.85 and 4.1 μm, and they found that all carbon stars exhibited a broad absorption feature centered at about 3.05 μm, whereas no trace of absorption was found in the M stars. They suggested that some polyatomic molecules in the carbon-star atmospheres were possibly responsible for the absorptions (Noguchi et al. 1977).

Sato Shuji and his group carried out broadband IR photometry for Nova Vulpeculae 1976 (NQ Vul) for a period covering two hundred days after its discovery. After a gradual decline in optical light, the IR radiation suddenly increased by about 3.5 mag at L band (3.2–4.1 μm). By comparing optical and IR spectra, they traced the process of the dust formation and its growth rate in the nova envelope (Sato et al. 1978).

Fig. 6.7 Agematsu 100-cm IR telescope. (Okuda 2018)

For a long time, the Agematsu IR Observatory had been the home base of IR astronomy in Japan. Activity in observations and development of instruments lasted until 2003.

In 2000, an IR telescope with 1.4-m aperture was constructed at Sutherland, South Africa, by the group of Sato Shuji and Nagata Tetuya (Sato et al. 2001). Taking advantage of good weather and the good location, the telescope started a variety of observations, particularly the observations of the Magellanic Clouds.

In parallel with ground-based observations, observations with a balloon-borne telescope were carried out in the 1970s by Okuda Haruyuki and Matsumoto Toshio. They have mapped the galactic plane in the near-IR radiation in the 1970s. The Nagoya group led by Hayakawa Sachio also started a series of rocket experiments, through which they observed the cosmic background radiation in the near- and mid-IR range. In the 1990s, Japan entered the age of space observations including a wide range of IR observations (Okuda 2018).

6.3.5 Subaru Telescope and NAOJ

1. Large telescopes in the 1970s

There were two international trends in the construction of large telescopes in the 1970s: one was the construction of 4-m-class telescopes, for example:

1973: 3.8-m reflector at Kitt Peak National Observatory (USA)
1974: 4.0-m reflector at Cerro Tololo Inter-American Observatory (CTIO, Chile)

1974: 3.8-m reflector at Siding Spring Observatory (Australia)
1977: 3.57-m reflector at the La Cilla Observatory (ESO)

The other trend was the development of the new-technology telescope (NTT), which consisted of a lightweight main mirror and computer-controlled mirror to maintain a correct surface.

The first NTT was the 3.58-m telescope constructed at ESO in 1989. The Keck 10-m telescope I at Mt. Mauna Kea followed in 1993, and four 8.2-m very-large telescopes (VLTs) at ESO in Chile were built in 1998–2000.

2. Large telescope plans in Japan

Discussions on the next telescope after the OAO started in the 1970s and various plans were proposed. A study group with a membership of around 150, Group of Optical and Infrared Astronomers (GOPIRA), was established at the end of 1980. After long discussions, the NTT plan with an aperture larger than 5 m was accepted by the National Committee for Astronomy (NCA) of the Science Council of Japan (学術会議). Afterward GOPIRA undertook a feasibility study of the so-called Japanese National Large Telescope (JNLT), and, finally, a plan for a NTT with an aperture of 8.2 m with a single mirror was accepted (Kogure and Tokunaga 1989).

In 1988, the TAO was reorganized as the NAOJ, and the project of the JNLT was endorsed in 1991. Construction began in 1992 on the summit of Mauna Kea, Hawaii, under cooperation with Hawaii University, and later that year a public contest was held to officially name the telescope, and it became known as *Subaru Telescope* (Subaru = *Pleiades* in Japanese). Construction was completed in 1998 as one of the telescopes grouped on the summit of Mauna Kea (Figs. 6.8).

3. Subaru Telescope

The Subaru telescope is one of the NTTs and uses various updating technologies to achieve outstanding performance.

The primary mirror has an effective diameter of 8.2 m, against its small thickness 20 cm. This mirror is made of ultralow thermal expansion (ULE) glass with a weight of 22.8 metric tons and a focal length of 15 m. To support this thin mirror during observations, 261 robotic fingers were inserted into holes drilled into the back of the mirror to keep it in perfect condition regardless of where the telescope is pointing in the sky.

The telescope structure has an altazimuth mounting, with four foci (Fig. 6.9). The focal ratio and the field of view (FOV) of these foci are as follows:

Focus	Focal ratio	FOV
Primary focus	2.0	30 arcmin
Cassegrain focus	12.2	6 arcmin
Nasmyth focus (Optical)	12.6	6 arcmin
Nasmyth focus (IR)	13.5	6 arcmin

Fig. 6.8 Telescopes on summit of Mauna Kea (4200 m above sea level) at time of completion of Subaru telescope. Along ridge from Subaru (8.2 m, 1998) (center left), Keck I and II (10 m, 1992 and 1996), IRAF (NASA Infrared Telescope Facility, 3.0 m, 1979), CFHT (Canada-France = Hawaii telescope, 3.6 m, 1979), Gemini (8.1 m, 1999), UKIRT (United Kingdom Infrared Telescope, 3.8 m, 1979), UH (University of Hawaii telescope, 2.2-n, 1976), and UH 0.6 m (University of Hawaii telescope, 1968). The two telescopes in the lower right are CSO (Caltech Submillimeter Observatory, 10.4 m, 1987) (left) and JCMT (James Clerk Maxwell telescope, 15 m, 1987). (Courtesy NAOJ, after pamphlet "8.2-m Subaru telescope")

These foci are equipped with various types of observational instruments that are constantly updated to increase their observational capacity. What follows are examples of instruments:

Primary focus	Fiber multiobject spectrograph (FMOS)
	Subaru prime focus camera (Supreme Cam)
Nasmyth (Op)	High-dispersion spectrograph (HDS)
Nasmyth (IR)	IR camera and spectrograph (IRCS)
Cassegrain	Multiobject IR camera and spectrograph (MOIRCS)
	Faint object camera and spectrograph (FOCAS)
	Cooled Mid-IR camera and spectrograph (COMICS)

The telescope enclosure is also shaped to improve the quality of astronomical images by minimizing the effects of atmospheric turbulence.

The Subaru telescope was completed in the end of 1998 and first-light images were released in January 1999. Open-use observations began in December 1999. Since then, Subaru has begun to operate as the basic telescope in Japan and in the astronomical community of the world.

A number of projects and programs had been proposed and successfully carried out in many fields and subjects including the following:

1. Solar System: small objects such as cometary nuclei, small asteroids, small object orbiting the Edgeworth-Kuiper Belt.

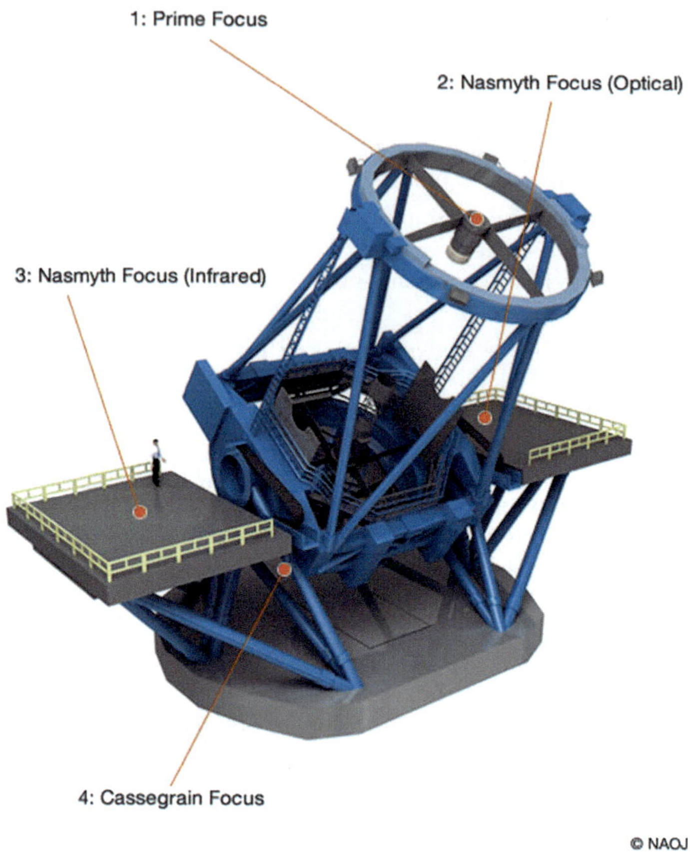

Fig. 6.9 Subaru telescope and its four foci; primary focus, two Nasmyth foci, and Cassegrain focus behind main mirror. (Courtesy NAOJ)

2. Star-forming regions: Orion region and others. Figure 6.10 illustrates an IR image of the Orion Nebula, taken as the first-light image in December 1998 by IR camera at the Cassegrain focus.
3. The Milky Way and the Local Group: The structure of our Galaxy, Andromeda Galaxy, and the associated dozens of dwarf galaxies are observed, along with the activities of their nuclei.
4. Realm of galaxies: the diversity of galaxies in shapes and sizes, active galaxies and active nuclei of galaxies, dynamical structure of colliding galaxies and mergers.
5. Distant galaxies with high redshift z making use of Subaru prime focus camera (Supreme-Cam) and spectrograph (FMOS).

Figure 6.11 demonstrates the most distant galaxy (red object in enlarged picture) in October 2007. The redshift of this galaxy is $z = 7$, which means it lies about 13 billion light-years away.

Fig. 6.10 One of the first images of Subaru telescope. IR image of Orion Nebula, with Trapezium stars lying near center. (Kaifu et al. 2000)

Fig. 6.11 The most distant galaxy (central red object of enlarged picture) in October 2007, discovered by Subaru. (Courtesy NAOJ)

6.4 Solar Observations

6.4.1 Tower Telescope of TAO

In 1929, a 65-cm coelostat (called Tower Telescope) was built at the top of a tower 15 m high on the Mitaka campus of the TAO. Solar light is sent to an underground observation room for high-dispersion analysis of the solar spectrum. After construction, Tower Telescope was used for a short time, but it was not active for a long time.

Active observations began in 1950 by Suemoto Zenzaburo on the solar chromosphere. His group, Hirayama Tadashi, Hiei Eizaburo, and Makita Mitugu, continued solar observations with this telescope in the 1960s.

6.4.2 Coronagraph at Norikura Solar Station, TAO

Norikura Solar Station was constructed in 1948 near the top of Mt. Norikura, 2876 m above sea level (Fig. 6.12). It has three coronagraphs (apertures: 10, 10, and 25 cm). With these coronagraphs, K-corona (continuous emission due to electron scattering) and 5303-corona (the highly ionized iron green line at λ5303 Å) have been monitored.

In 1964, short-term variation of K-corona within a single day was observed as shown in Fig. 6.13 by Nishi Keizo and Nagasawa Shingo. Remarkable variation was observed in K-corona, while no change was observed in the 5303-corona (Nishi and Nagasawa 1964).

Fig. 6.12 Norikura Solar Station, TAO. (Tokyo Astronomical Observatory)

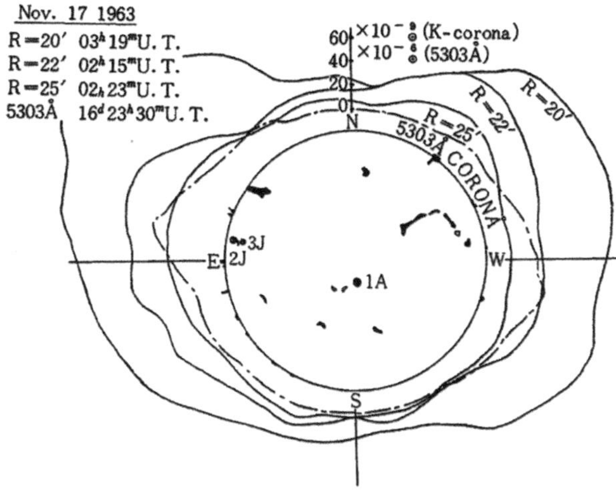

Fig. 6.13 Short-term variation of coronal structure observed in November 1963. (Nishi and Nagasawa 1964)

6.4.3 Solar Coudé-Telescope at OAO

In 1968, a 65-cm Coudé telescope was constructed at the OAO for solar observations. With this telescope, Tower Telescope at Mitaka was practically retired from service. Suemoto's group used this Coudé telescope for observations of active regions, flares, and magnetic activities. Kawaguchi Ichiro of Kyoto University was also a regular user of this telescope, mostly for observations of prominences.

In 1979, a newly designed polarimeter was installed. With this polarimeter, Makita Mitsugu and his group determined the magnetic field of the solar penumbra by analyzing Zeeman-line profiles. Sunspot spectrograms obtained with this polarimeter in May 1971 are shown in Fig. 6.14. Analyzing these spectrograms, they derived the magnetic field of penumbra, which was inwardly directed, making an angle of 34° from the solar surface, with a strength of 1500 Gauss. An outward Evershed flow of 5.6 km s^{-1}, parallel to the lines of magnetic field, was also observed (Makita 1979).

6.4.4 Domeless Solar Telescope and Hida Observatory

The Hida Observatory was established on a peak of Mt. Ōamami (1290 m above sea level) in Gifu Prefecture, as a part of the Kwasan and Hida Observatories of Kyoto University. The main telescopes are the 65-cm refractor (1972) and the domeless solar telescope (DST, 1979) both made by Zeiss (Kawaguchi 1982).

1. Domeless Solar Telescope

The DST is a 64-cm reflector that is installed at the top of a 20-m-high tower in order to avoid atmospheric scintillation. (Fig. 6.15). Inside the tower are two

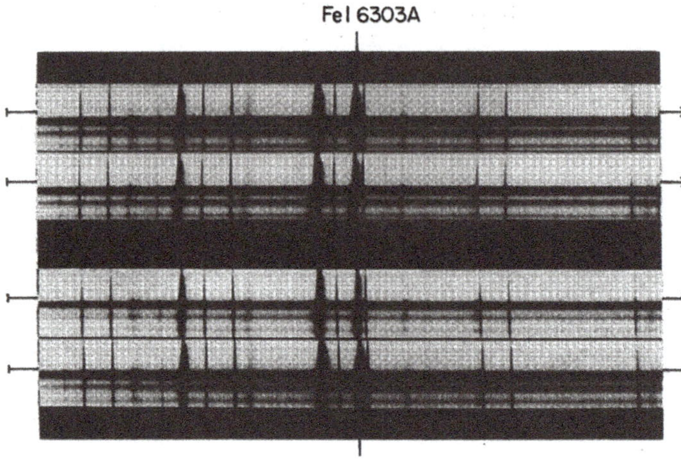

Fig. 6.14 Sunspot spectrograms obtained with polarimeter. The two upper ones are for circular polarization and the lower two are for linear polarization. Arrows at both sides denote the position of penumbras. (Makita 1979)

Fig. 6.15 Domeless Solar Telescope (DST) at Hida Observatory of Kyoto University. (Courtesy Kwasan and Hida Observatory)

observation rooms. On the first floor a vacuum vertical spectrograph is mounted giving high-dispersion spectra and monochromatic solar images at the Hα line and G-band. On the second floor a horizontal spectrograph gives the entire visible solar spectrum.

The DST was designed to acquire solar surface images at the highest spatial (0.5 arcsec) and temporal (2 sec) resolutions, along with high-dispersion spectrograms. Figure 6.16 demonstrates a monitor image of the Sun at the Hα line, where solar activities of bright flares and dark filaments are seen on the solar disk. The solar surface is monitored with this image, and any phenomena can be observed in detail. In Fig. 6.17, Hα images of flares and eruptive prominences are shown as examples of solar active phenomena.

2. 65-cm Refractor and Mars observation

One of the main subjects at the Hida Observatory was the observation of the planetary system. The 65-cm refractor was constructed for this purpose. This telescope has a 65-cm aperture, 1050-cm focal length (focal ratio 16), and 0.18-arcsec resolving power. With this telescope, Akabane Tokuhide, Iwasaki Kyosuke,

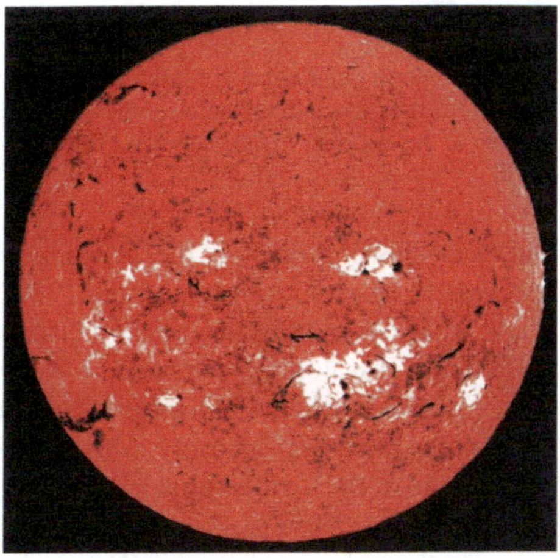

Fig. 6.16 Total image of Sun at Hα line obtained byDST. (Courtesy: Kwasan and Hida Observatories, Annual Report 2003, p. 16)

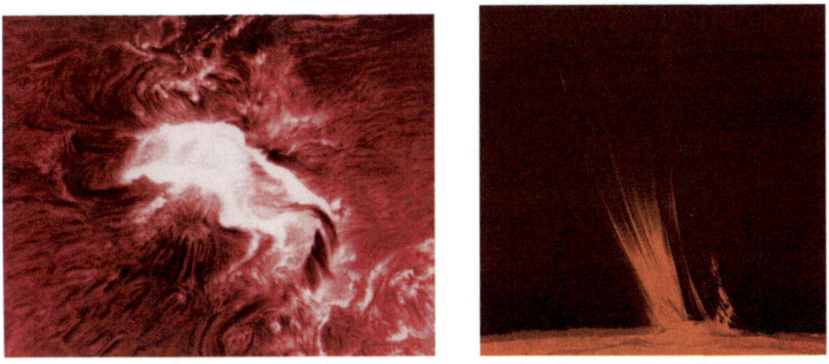

(a) Solar flare (b) Eruptive prominence

Fig. 6.17 Examples of solar phenomena. (**a**) Solar flare. (**b**) Eruptive prominence. (Courtesy Kwasan and Hida Observatories)

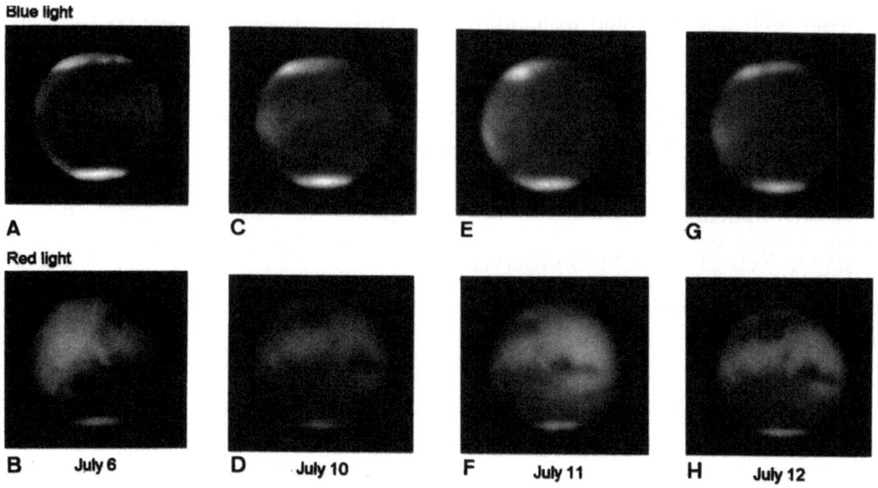

Fig. 6.18 North polar hood of Mars in 1986. Upper row: blue-filter photos; lower-rows: red-filter photos. North is top, and the dates of the observations are given at the base. Mars was in the early autumn in the northern hemisphere, and the polar cap is seen at the southern pole. (Akabane et al. 1995)

and their group promoted the imaging and photometric observations of planets, particularly Mars and Jupiter, in the 1980s and 1990s.

An example of Mars observation is shown in Fig. 6.18, which illustrates the time variation of the north polar hood (a kind of cloud) observed in 1986 at an early autumn in the northern hemisphere. It was observed in two colors, blue and red. The north polar hood appears only in the blue-light photos (upper panel) reflecting the high albedo of the hood in blue light. The southern polar cap can be seen in both blue and red photos. Note that the north polar hood is not uniform in brightness and in size. The boundaries of the hood extend to a latitude of 30°–40°N of the Martian surface in early morning and recede poleward to around 60°N in the afternoon (Akabane et al. 1995).

6.4.5 *Solar Activity Telescopes*

1. The Solar Flare Telescope was constructed at Mitaka, NAOJ, in 1992. This instrument is composed of four telescopes for the respective observations of (a) Hα image, (b) continuum image, (c) vector magnetic fields, and (d) velocity fields in photosphere. The instrument aims at the study of energy built-up and released in solar flares, and it has been operated as a ground-based support for the X-ray satellite Yohkoh launched in 1991.
2. The Solar Magnetic Activity Research Telescope (SMART) was constructed at Hida Observatory of Kyoto University in 2003. SMART comprises four

Fig. 6.19 Solar Magnetic Activity Research Telescope (SMART) at Hida Observatory. (Courtesy Kwasan and Hida Observatories)

telescopes for total Hα solar images and for a high-precision magnetic-field map. The main purpose of this system is to predict the occurrence of flare activities on the solar surface through continuous observations of the solar surface and to predict the influence of flares on the Earth. This research field is called cosmic meteorology (Fig. 6.19).

6.5 Radio Observations

Cosmic radio waves were discovered in 1932 by K. G. Jansky of the Bell Telephone Laboratories as cosmic noise at 20.5 MHz coming from all parts of the Milky Way. In 1939, G. Reber, an amateur radio enthusiast and amateur astronomer, detected radio waves similarly coming from the Milky Way at 160 MHz. He surveyed the wide sky and delineated a rough radio map of the Milky Way in 1944. This map has attracted major attention from astronomers around the world, and it opened up a new research field, radio astronomy.

6.5.1 Early Radio Observations

In Japan, radio astronomy began with the observations of solar noise in the late 1940s at the following four institutes (Orchiston and Ishiguro 2017):

1. Osaka University. Oda Minoru (see Sect. 6.6) and Takakura Tatsuo observed solar noise at 3300 MHz using a handmade horn in 1949–1951. Later they moved to Osaka City University.
2. The Research Institute of Atmospherics (空電研究所), Nagoya University, Toyokawa. Tanaka Haruo and Kakinuma Kiyotaka began solar radio observations in 1950 with a 2.5-m dish operating at 3700 MHz. In 1953, they constructed a five-element interferometer consisting of a 1.5-m dish at 4000 MHz. The interferometer was later expanded to eight elements and was used for observations of the brightness distribution of radio emissions on the solar surface, along with observations of dynamic solar bursts.
3. Tokyo Astronomical Observatory. Hatanaka Takeo (Chap. 3) and Moriyama Fumio constructed a 5 × 5-m radio telescope at 200 MHz and began observations of solar radio bursts in 1949. They successively constructed a 10-m parabolic dish (200–3000 MHz) in 1953 and a 12-m parabolic dish (9500 MHz) in 1957. These telescopes were used for solar radio observations, particularly for solar radio bursts.
4. Hiraiso Station, Radio Research Institute of the Ministry of Posts and Telecommunications (郵政省電波研究所、平磯電波観測所). In 1950, a broadside array was erected by Obayashi Tatsuzo operating at 61.2 MHz for solar noise. It was replaced in 1952 by a new 200-MHz array for the regular monitoring of solar noise, and this was operated during the 1950s.

The Sun often demonstrates violent burst phenomena that are usually classified into Type I to Type V based on the structure of the frequency space that changes with time. Each type is also associated with some active phenomena such as flares, eruptive prominences, and proton emissions. In the 1960s, Morimoto Masaki, Kai Keizo, Takakura Tatsuo, and their group at the TAO observed extensively every type of radio burst and carried out theoretical analysis for the mechanisms of bursts. Figure 6.20 illustrates an example of a dynamic spectral structure of the Type IV solar burst observed on September 11, 1957 (Takakura and Kai 1961).

Radio observations for galactic objects started with the construction of a 24-m spherical-dish telescope at the Mitaka campus of the TAO and a 30-m parabolic telescope at the Kashima Branch of the Radio Wave Institute of the Telecommunication Ministry (郵政省電波研究所鹿島支所) in the early 1960s. These telescopes have been used mainly for HII regions and supernova remnants at 1420 and 4170 MHz.

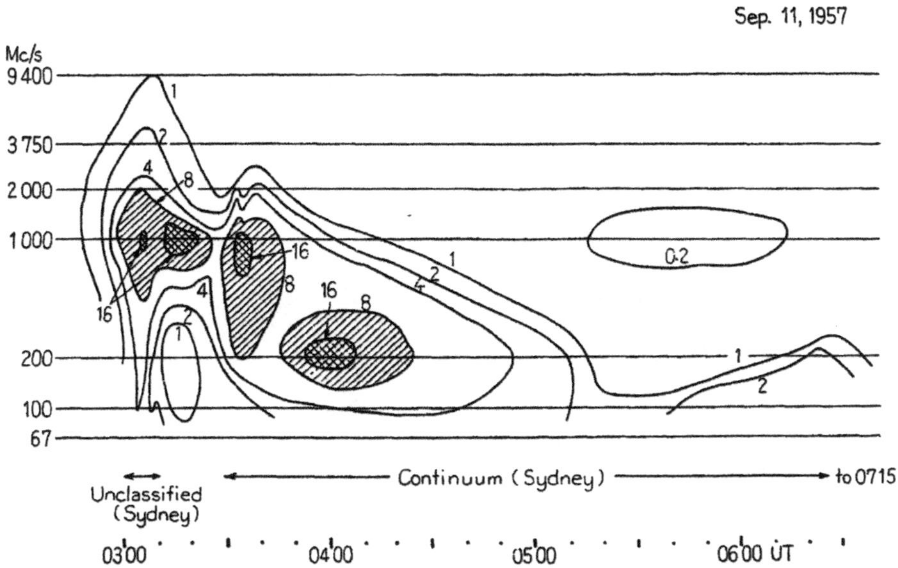

Fig. 6.20 Example of a dynamic spectrum of a Type IV solar burst (Duration: hours to days; frequency range: 20 MHz to 2 GHz). (Takakura and Kai 1961)

6.5.2 Nobeyama Radio Observatory and Molecular Clouds

1. Dawn of millimeter-wave astronomy

Molecular-line astronomy had begun with the discoveries of an OH line (λ18 cm) in 1963, NH_3 (anmonia, λ1.23 cm) in 1968, and H_2CO (formaldehyde, λ6.21 cm) in 1969. In the 1970s, discoveries of molecular lines moved to millimeter and submillimeter lines, such as CO (monoxide carbon, λ0.260 cm) and CH_3OH (methyl alcohol, λ0.820 cm). These lines were observed with an11-m millimeter-wave telescope at the Kitt Peak National Radio Observatory in the USA.

In Japan, a millimeter-wave radio parabolic telescope with a 6-m aperture was constructed by Akabane Kenji, Morimoto Masaki, and Kaifu Norio at the Mitaka campus of the TAO in 1972. With this telescope they discovered some interstellar molecules such as $(H_2CO)_n$ (para-folmaldehyde, λ0.412 cm) and CH_3NH_2 (methylamine, λ0.411, 0.348 cm) in the Orion and Sagittarius nebulae (Morimoto and Kaifu 1972; Akabane et al. 1974).

With this success, the construction program of the 45-m millimeter-wave radio telescope became a reality.

2. Nobeyama Radio Observatory

The NRO, in Nobeyama, Nagano Prefecture, was thus established in 1982 with a 45-m millimeter-wave telescope and an array telescope system. This array is composed of five 10-m telescopes for interferometric observations. The 45-m telescope is the largest one by aperture in the world; its surface accuracy reaches 0.01 mm, and its tracking accuracy is 0.1 arcsec at a 180° angle. The angular resolution reaches

Fig. 6.21 Nobeyama Radio Observatory. The 45-m parabola and the Nobeyama Millimeter Array. A 10-m parabola was later added to the array system. (Courtesy NAOJ, after "Thirty-year Anniversary of the Nobeyama Radio Observatory")

0.004 arcsec (maximum) (corresponding to a visual acuity of around 4). This telescope operates in the millimeter and submillimeter range (10–230 GHz) (λ3 cm–0.13 cm) (Fig. 6.21).

Observation programs are divided into two projects. One is for observatory projects and the other is assigned to a proposal project opened for interuniversity users. Observatory projects consist of two programs. One is a large-scale survey of the entire galactic surface, which is part of an international project overseen by NRO to cover the highest radio frequencies of 10 GHz (λ 3 cm) and 43 GHz (λ 7 mm). Another project is the systematic survey of interstellar molecular lines. In the first 3 years, more than 200 new lines were discovered. An example of a line survey is illustrated in Fig. 6.22, which is a spectrum of the Orion KL object taken in the 45-GHz band of the 45-m telescope.

The proposed projects have been opened to astronomers for free observation programs. The main objects of observation include, but are not limited to, the following:

Formation and evolution of molecular clouds
Structure of molecular clouds around young stars
The nucleus of the Milky Way
Supergiant black holes in galaxies
Structure of active galaxies

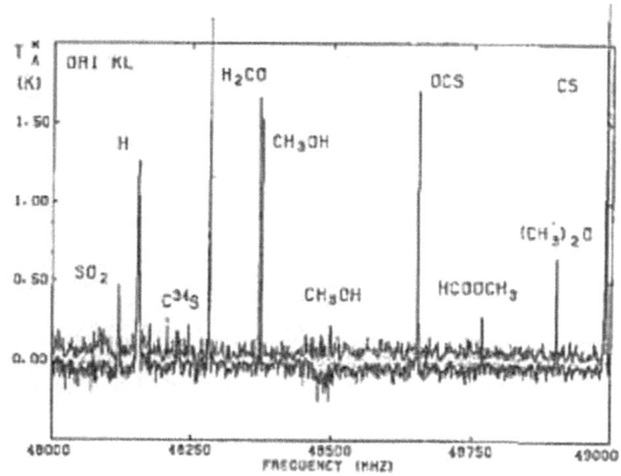

Fig. 6.22 Spectrum of IR object Orion KL in 45-GHz band observed with 45-m radio telescope (Suzuki and Ohishi 1985)

Some results will be given in Chap. 7.

The Nobeyama Millimeter Array (NMA) is a telescope array initially consisting of five 10-m parabolas and another similar parabola added later. The observable frequency ranges from 80 to 230 GHz. Its ability to produce high spatial resolution is equivalent to a 600-m diameter parabola, and the maximum angular resolution reaches 0.0003 arcsec, which corresponds to a visual acuity of around 60.

6.5.3 Nobeyama Solar Radio Observatory

In 1975,the Nobeyama Solar Radio Observatory was established, equipped with a radio heliograph. It is composed of eight-four 80-cm-aperture parabolic antennas arranged in a clove. This heliograph started regular observations in June 1992. It observes the Sun at 17 GHz in a field of view that covers the entire solar disk, with spatial and temporal resolutions of 12 arcsec and 1 sec, respectively. In the 1990s, simultaneous observations in radio waves (Nobeyama Radio Heliograph), soft X-rays (Yohkoh satellite), and Hα lines (Hida DST) were carried out for the analysis of flares, eruptive prominences, coronal arcades, and other solar activities.

Figure 6.23 illustrates an example of a solar image at 17 GHz observed with the Radio Heliograph on July 6, 1992.

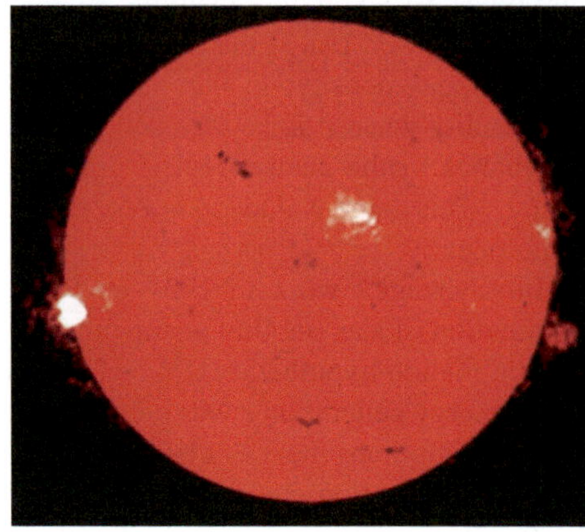

Fig. 6.23 Image of Sun at 17 GHz (pseudo-colors) observed with radio heliograph on July 6, 1992. (Courtesy NAOJ)

6.5.4 Millimeter-Wave Telescopes of Nagoya University

While construction of the NRO was under way, Fukui Yasuo, Kawabata Kin'aki, and Ogawa Hideo of Nagoya University built a 4-m millimeter-wave radio telescope for molecular-line observations in 1982. This telescope allows for fast tracking of the sky due to its lightweight antenna (~700 kg), with a beam width of 3 arcmin. It is suitable for survey observations of a wide sky field, like a Schmidt telescope in optical astronomy. Fukui's group started survey observations of molecular clouds in radio bands, 110 GHz (CO) and 90 GHz (HCO^+, HCN), as a telescope to complement the Nobeyama 45-m telescope (Kawabata et al. 1984).

In 1996, Fukui Yasuo, Abe Rihei, and their group constructed a second 4-m radio telescope at Las Campanas Observatory in Chile, and it was named Nanten Telescope (meaning Southern Sky). With the Nanten they carried out a CO-line survey of the southern sky, particularly the Magellanic Clouds. At that time there were two millimeter-wave telescopes in the southern hemisphere: a 1.2-m Smithsonian telescope and a 1.5-m EWO telescope. Fukui's 4-m telescope was the most powerful one in the southern sky at the time (Fukui et al. 1999).

Fig. 6.24 shows a CO map of the Large Magellanic Cloud (LMC) in the area around 30 Doradus. Fukui's group found 55 CO clouds coexisting with HII regions and stellar clusters. They showed that about 47% of CO clouds are associated with stellar clusters and large HII regions, suggesting active, ongoing formation of stellar clusters. The remaining CO clouds show almost no sign of star formation, that is, they have no HII regions and stellar clusters. Some parts of CO clouds are associated with small HII regions but have no stellar clusters. From these distributions Fukui's group found that young clusters coexist with CO clouds, whereas in old clusters their surrounding CO clouds have already dissipated (Fukui et al. 1999).

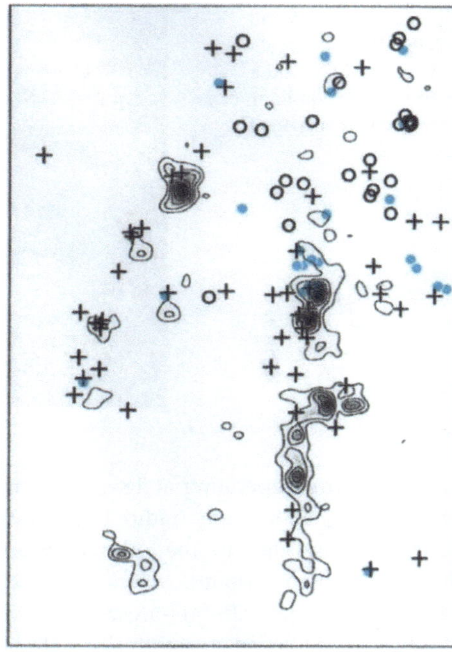

Fig. 6.24 Molecular clouds near 30 Dor, in LMC. Real line profiles denote CO cloud, and other symbols are as follows: + (HII region), ○ (stellar clusters older than 10 Myrs), and • (stellar clusters younger than 10 Myrs). (Fukui et al. 1999)

6.5.5 VSOP and Vera Projects

The spatial resolution of telescope increases with the increase of telescope aperture in any wave-length telescope. Large telescope gives high resolution, so that when two parts of both ends of large telescope is cut-down to make some smaller ones, and the cosmic image is focused just like at one large telescope, then the images of two small telescopes yield the same spatial resolution, though light gathering power significantly reduced. In order to get high spatial resolution, two or several telescopes are combined into one focus at some telescope center. In order to get the fine image, the wave fronts sent from distant telescopes must be reserved. The method to make this reservation possible belong to the technique of interferometer. So that such telescope systems are called radio or optical interferometers, and when separation of telescopes is very long (over 100 km or even several 1000 km), the system is called very-long baseline interferometer (VLBI). Remarkable VLBI in Japan is presented here in two cases.

1. VSOP

In the late 1990s, the Institute of Space and Aeronautical Sciences (ISAS) started a VLBI project connected with a space radio telescope, that is, the VLBI Space Observatory Program (VSOP). For this purpose, a satellite named HALCA was launched in February 1997. HALCA has an apogee height of 21,000 km and a perigee height of 580 km, with an orbital period of 6 h 20 min. It has an antenna with

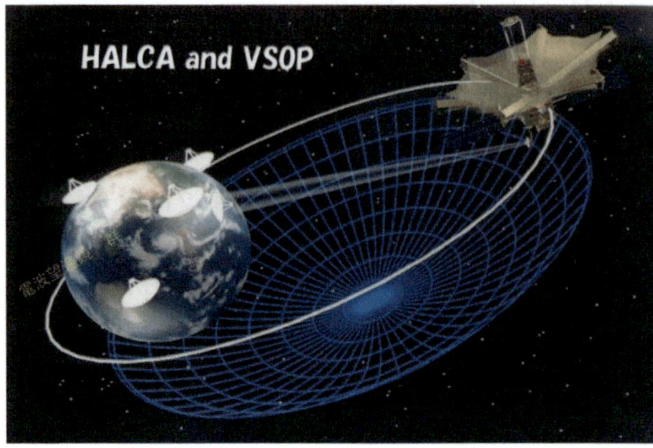

Fig. 6.25 HALCA and VSOP. VLBI observations between HALCA and a radio telescope on Earth can be made with very long baselines

an 8-m aperture operating at 18- and 6-cm wavelengths. HALCA can make VLBI observations with many radio telescopes around the Earth under international collaborations. Due to the orbital motion of HALCA, the baseline can reach 30,000 km at maximum, when the space resolution becomes 0.0001-arcsec on 18-cm waves and 0.00003-arcsec on 3-cm waves. This high space resolution of VSOP may be compared with the 0.05 arcsec of the Hubble Space Telescope in optical regions (Hirabayashi 1998; Hirabayashi and Edwards 1999) (Fig. 6.25).

More than 30 radio observatories in the world joined these VSOP observations. The main objects of VSOP observations are quasars, nuclei of active galaxies, and other active objects. Since HALCA provides the measurement of circular polarization, the direction of magnetic fields can be derived, in addition to the submilliarcsecond structure of the objects.

One of the most active observers in VSOP was Denise C. Gabuzda of the University College of Cork, Ireland (Gabuzda 2009). She stayed in Japan at the early phase of VSOP project. Under collaboration with Gabuzda, Asada Keiichi and their group carried out VSOP observations of the bright radio source 3C 84 in the Seyfert galaxy, NGC 1275, in 2006. They found the expansion of a young radio lobe associated with 3C 84. The apparent expansion velocity was measured as 0.50 ± 0.09 *light years* and the age of this lobe was estimated as 45 ± 8.9 years in 2001. In this way they revealed the very active nature of 3C 84 (Asada et al. 2006).

2. VERA

Observations of latitude and longitude had been the responsibility of the Mizusawa International Latitude Observatory since the Meiji era. In the 1980s, it became clear that the accuracy of optical observations was much lower than radio observations for geodetic purposes. Also, the Global Positioning System (GPS) proved the superiority for time keeping compared to optical observations. As a consequence, the International Astronomical Union (IAU) and International Union of Geodesy and Geophysics (IUGG) established the International Earth Rotation

Service for global geodetic observations. The Mizusawa was reorganized as the Mizusawa VLBI Observatory and became a branch of the NAOJ in 1988 (Yokoyama 2008).

As the new establishment, the Mizusawa VLBI Observatory launched the VLBI Exploration of Radio Astronomy (VERA) project. This project is composed of four 20-m radio telescopes at Mizusawa Station in Iwate Prefecture, at Kagoshima Station in Kagoshima Prefecture, at Ogasahara Station in Ogasahara Islands, and at Ishigaki Station in Okinawa Prefecture (Fig. 6.26). These were successively constructed by 1998. Then the VLBI observations connecting these four stations started as a VERA project in 2001. VERA has a pointing accuracy of several 10 micro-arcsecond angles for millimeter-wave sources and water masers, so that the main roles of VERA involve the radio astrometry of the Galaxy, for the solar neighbor sources, large-scale structure of the Galaxy, period-luminosity relation of Mira-type variables, and even astrometry of outer galaxies. Figure 6.27 illustrates the 10-year achievements of parallax and proper-motion observations for galactic objects (Honma et al. 2010).

The next goal of VERA is the observations of hundred or even thousand objects. VERA is expected to contribute to the deep understanding of the structure and dynamics of the Galaxy.

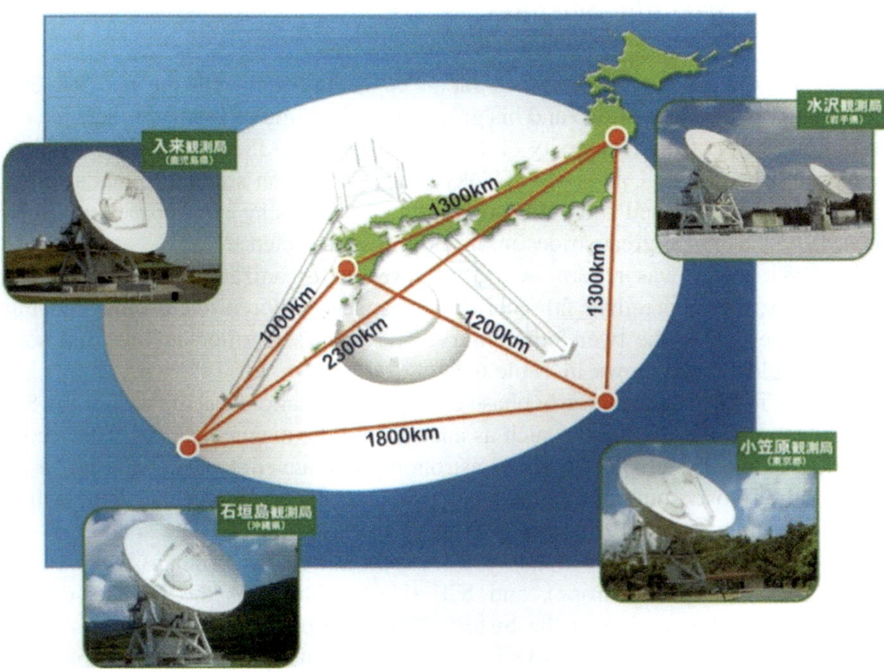

Fig. 6.26 Four VERA stations. (Courtesy NAOJ, NAOJ News No. 205, 2010)

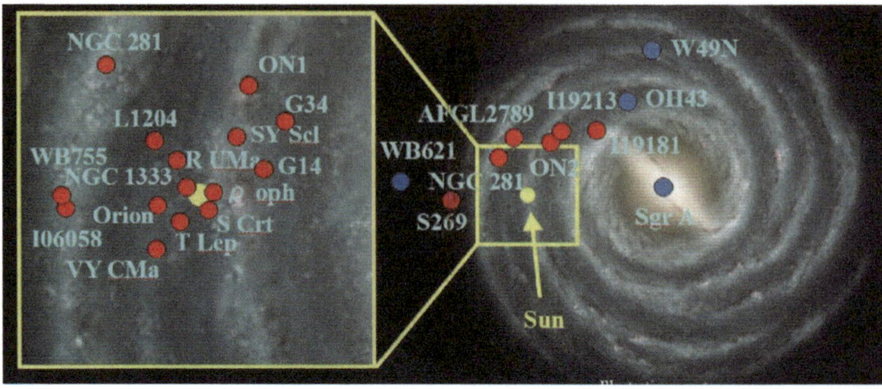

Fig. 6.27 Image of Galaxy seen from VERA. The left panel shows a solar neighbor. Red marks indicate the objects used to measure both parallaxes and proper motions, while blue marks are the objects used to measure the proper-motions. (Honma et al. 2010)

6.6 Space Observations and High-Energy Astrophysics

6.6.1 Institute of Space and Astronautical Science and Scientific Satellites

The history of space observation in Japan began in 1955 with the pencil-rocket experiment by Itokawa Hideo and his group at the Institute of Industrial Science (生産技術研究所) of the University of Tokyo. In 1964, the ISAS (宇宙科学研究所) was established at the University of Tokyo. In 1970, Japan's first artificial satellite, OHSUMI, was launched and put into orbit by a rocket using solid propellant. Since then, ISAS has made great strides in promoting space science.

In 1981, ISAS was reborn as a joint research organization among Japanese universities, and more than 20 satellite and space explorers were launched into orbit by the end of the twentieth century. The main scientific satellites and solar system explorers are shown in Table 6.3.

As seen in Table 6.3, the early phase of space observation was directed at the Sun and the terrestrial environment such as auroras and the magnetosphere. The second decade was mostly devoted to X-ray astronomy conducted by Oda Minoru. In the 1980s, the most stunning event was observations of Comet Halley, which approached the Sun in 1987 after 76 years of orbital motion. The spacecraft that participated in Halley observations were SAKIGAKE, SUISEI (Japan), VEGA I and II (USSR), GIOTTO (Europe), and ICE (USA). This was the largest scale of international collaboration in the history of space observation, and it was called the Halley fleet of spacecraft. SAKIGAKE was originally the first test explorer to

Table 6.3 Scientific satellites launched from Japan by 2000

Name	Launch–end year	Observations
SHINSEI (New star)	1971–1973	First scientific satellite
TAIYO (The Sun)	1975—1980	Solar X-ray, solar UV-radiation, etc.
KYOKKO (Polar light)	1978–1982	Aurora, magnetosphere
JIKIKEN (magnetosphere)	1978–1985	Aurora, magnetosphere
HAKUCHO (Swan)	1979–1985	X-ray astronomy
HINOTORI (Sunbird)	1981–1991	X-ray solar flares, X-ray bursts
TENMA (Pegasus)	1983–1988	X-ray astronomy
OHZORA (The sky)	1984–1988	Middle atmosphere, magnetosphere
SAKIGAKE (Forerunner)	1985–1999	Test of interplanetary orbit, Halley's Comet, solar wind
SUISEI (Comet)	1985–1992	Halley's Comet
GINGA (Galaxy)	1987–1991	X-ray and γ-ray astronomy
AKEBONO (Dawn)	1989–2015	Magnetosphere, aurora
HITEN (Angel)	1990–1993	Lunar swing-by experiments
YOHKOH (Sunlight)	1991–2004	Solar activity and corona
ASCA	1993–2001	X-ray astronomy of Universe
HALCA	1997–2005	Radio astronomy satellite
NOZOMI (Hope)	1998–2003	Japan's first Mars explorer

ISAS News Series: Past spacecraft/ISAS

follow the solar orbit in interplanetary space. It observed the solar-wind magnetic field and plasma activities near Comet Halley at its near-perihelion stage. SAKIGAKE continued the observations of solar-wind plasma waves for 14 years until it ceased operation in 1999. SUISEI was launched for the observations of Comet Halley. It was able to determine the rotation period through ultraviolet imaging at close approach to the comet of 151,000 km. SUISEI had a heliocentric orbit. SUISEI completed its final mission, an Earth swing-by, in1992.

In the 1990s, ISAS promoted the technological development of lunar and planetary missions as in the cases of SAKIGAKE, HITEN, and NOZOMI.

In 2003, an independent administration agency, the Japan Aerospace Exploration Agency (JAXA, 宇宙航空研究開発機構), was established, merging three aerospace organizations: the ISAS, the National Aerospace Laboratory (NAL), and the National Space Development Agency (NASDA) of Japan. It was designated as a core performance agency to support the Japanese government's overall aerospace development and utilization programs. The promotion of space science has remained at ISAS as the central task of the institute.

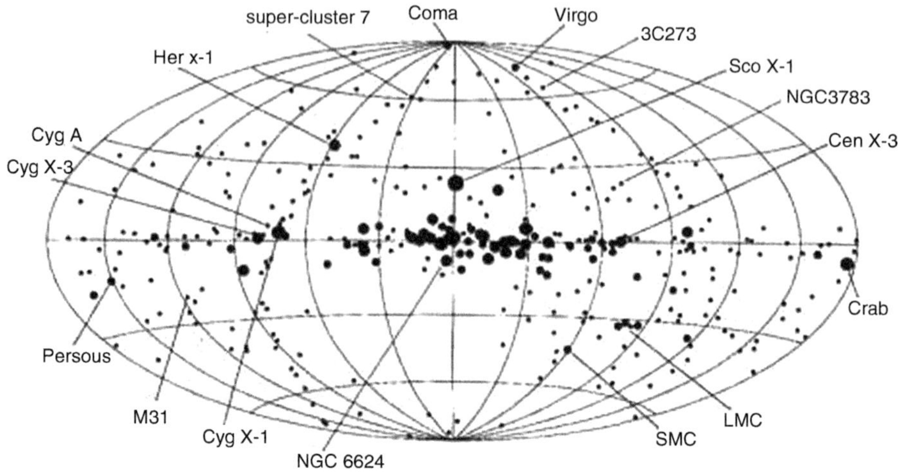

Fig. 6.28 Distributions of X-ray sources in galactic coordinates observed by UHURU. The sizes of the sources are proportional to the logarithm of the peak intensity. (Forman et al. 1978)

6.6.2 Oda Minoru and X-Ray Astronomy

1. Uhuru and X-ray observations

In 1970, the first X-ray satellite, UHURU, was launched from NASA's San Marco launch platform in Kenya. UHURU means *freedom* in Swahili. UHURU loaded the *sudare collimator* (the modulation collimator, a device of two or more layers of parallel wires), invented by Oda Minoru (Oda 1965). Since the sudare collimator had a high angular resolution, UHURU identified 339 X-ray sources in the whole sky, as shown in Fig. 6.28, and this satellite laid the foundation of present-day X-ray astronomy.

2. Life of Oda Minoru (Maddox 2001; Ogawara 2001)

Oda Minoru (小田稔1923–2001) was born in February 1923 in Sapporo, Japan, the elder son of a physician. He studied physics at Osaka Imperial University (now Osaka University). Fortunately, he escaped war service. After graduation in 1944, he embarked on the study of cosmic physics, which he studied for the next 2 years. In 1950 he moved to Osaka City University as an associate professor of physics.

In 1953–1956, Oda stayed at the Massachusetts Institute of Technology (MIT) as a visiting staff of the cosmic-ray group of Bruno Rossi (1905–1993). They followed up extensive cosmic-ray air-shower experiments with new methods of density sampling and fast timing. This marks the beginning of a lifelong friendship between Oda and Rossi, both among the most influential scientists in the development of cosmic-ray and space physics.

Oda's 3-year stint at MIT changed his life. He learned that research, meritocratic though it must be, should be a democratic process, too. The research system in place

Fig. 6.29 Oda Minoru in March 2000. (Proceedings of the Symposium "Strategy for Human Survival" held at Tokyo University of Information Science)

in Japanese universities was still traditional, such that apprenticeships at universities lasted into the Taisho and early Showa eras. Therefore, Oda's attitude to science had a significant influence on the modernization of Japanese science in the 1960s (Fig. 6.29).

Oda returned not to Osaka but to Tokyo, where he became one of the leaders of projects to build and launch Earth orbiting satellites. This was realized as the establishment of the ISAS in 1964, and Oda was appointed professor and project director at the institute. Oda's democratic management was on vivid display at the ISAS, and this was one of the reasons why X-ray astronomy at ISAS developed so successfully.

Oda Minoru was the founder of space astronomy in Japan, as well as a key participant in the exploratory work in the USA that initiated the field of X-ray astronomy. With his achievements in space science, he was awarded the Order of Culture (文化勲章) by the Japanese government in 1993. He retired from ISAS in 1988 and died in 2001 of heart failure at age 78 (Clark et al. 2001).

6.6.3 *X-Ray and* γ*-Ray Observations*

Under Oda's initiative, the ISAS updated the *sudare collimator* in spatial and spectral resolutions for wide-ranging purposes from sky survey to high-dispersion spectral and imaging observations of various X-ray and γ-ray objects. The sensibility of X-ray detectors had been remarkably increased also.

Among the scientific satellites listed in Table 6.3, the notable X-ray satellites are as follows (ISAS News No. 169, 1995, Special Issue, "Forty Years from Pencil"):

1. HAKUCHO (1979–1991): This was the first Japanese X-ray satellite for the observations of X-ray stars, especially X-ray burst sources. Moreover, HAKUCHO observed various time variations of X-ray sources, such as X-ray pulsars and black-hole candidates.

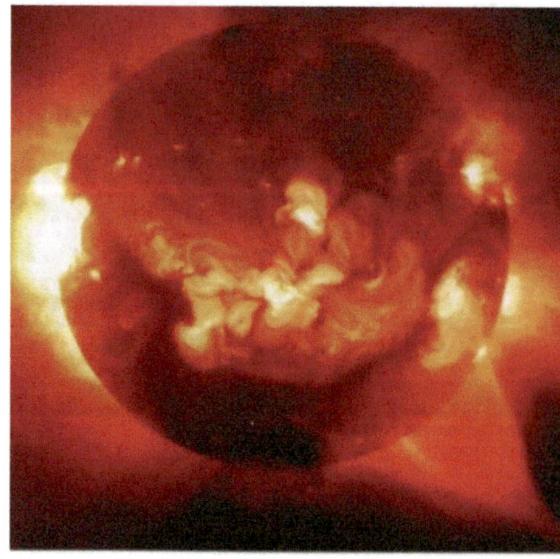

Fig. 6.30 Soft X-ray image of Sun on January 25, 1992, taken with soft X-ray telescope on board Yohkoh. (Ogawara et al. 1992)

2. TENMA (1983–1988): With the X-ray spectral analyzer boarded, TENMA obtained X-ray spectra of various X-ray stars with high energy resolution. The first discovery was the distribution of hot plasma at temperatures of several tens of millions of degrees along the galactic plane. The satellite also discovered and analyzed iron absorption lines in the X-ray spectra in many low-mass X-ray binaries (LMXRB), high-mass X-ray binaries (HMXRB), and active galactic nuclei (AGN).
3. GINGA (1987–1991): Just after the launch, a supernova appeared in the LMC. GINGA successfully observed X-ray emissions from the supernova. In addition, GINGA conducted a wide survey of X-ray objects in the Galaxy, including black-hole candidates, high-temperature plasma inside dark nebulae, flare bursts in binary systems, extragalactic X-rays in the nuclei of Seyfert galaxies, and so on. GINGA also was installed with a gamma-ray burst detector (GBD) and discovered more than 30 events of γ-ray bursts.
4. ASCA (1993–2001): With an X-ray CCD camera and an imaging spectrometer, ASCA carried out precise X-ray observations of various celestial objects in the Universe, including a supernova that appeared in 1993 in the M81 galaxy, X-ray observations of γ-ray bursts, and a distant cluster of galaxies located at a distance of 7000 million light-years.
5. Yohkoh (1991–2004): Yohkoh contributed to the observations of solar phenomena. It was said that the world's solar community was rocked by the launch of Yohkoh in its high spatial (2.5 arcsec) and temporal (0.5–2 sec) resolutions both in a soft X-ray telescope (SXT) and hard X-ray telescope (HST). Figure 6.30 exhibits an SXT image that shows active regions with a thread-like structure, as well as many X-ray bright areas in dark coronal holes.

These activities made Japan's space astronomy program one of the most advanced in the international space community. Progress in X-ray astrophysics will be presented in connection with certain topics in Chap. 7.

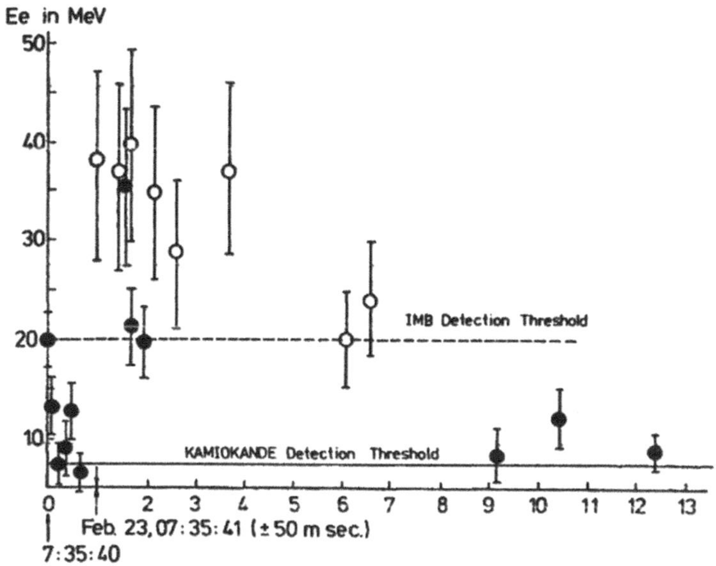

Fig. 6.31 Neutrino observations at Kamiokande and IMB. (Koshiba 1987)

6.6.4 Kamiokande and Cosmic Neutrinos

Kamiokande was constructed in 1983 in the deep interior of the Kamioka mine, 1000-m below sea level. Kamiokande means the Kamioka Nucleon Decay Experiment, the Kamioka Neutrino Detection Experiment, or both. The original purpose of this facility was to detect the proton decay anticipated in the grand unified theory of elementary particle physics. The construction of Kamiokande was conducted by Koshiba Masatoshi (小柴昌俊) of the University of Tokyo, and Kamiokande was update to Kamiokande-II in the fall of 1985. Since the facility was able to detect cosmic neutrinos, Kamiokande also played a role as a neutrino observatory.

When the supernova SN 1987A erupted, the neutrino burst was detected at three sites: Kamiokande-II (Japan, 25 neutrinos), Irvine-Michigan-Brookhaven (IMB) (USA, 8 neutrinos), and Baksan (Russia, 5 neutrinos). Figure 6.31 illustrates the events of neutrinos observed with Kimiokande-II and IMB (Koshiba 1987). The events are plotted on a time (sec)–energy (MeV) diagram, and the time of 7:35:40 on Feb 23, 1987, given in the figure denotes an estimated initial time of supernova explosion. The levels of detection threshold of Kamiokande-II and IMB are 5 and 20 MeV, respectively. The significance of this figure will be given in Chap. 7.

Owing to his contributions to the construction of Kamiokande and the discovery of neutrinos of supernova origin, Koshiba was awarded the Order of Culture (文化勲章) from the Japanese government in 1997 and the Nobel Prize in Physics in 2002.

6.7 Development of Computer

The age of computer science was initiated in 1946 by the first electronic computer, the Electronic Numerical Integrator and Computer (ENIAC), made in the USA. Afterwards, advances in computer technology were remarkable. Computer speed was measured in floating-point number operations per second (FLOPS). In the early days, computers processed one task at a time, called scalar processors, and computer speed increased from several hundred FLOPS on ENIAC to around 1 megaflops (MFLOPS) in the late 1960s. In 1976, a new type of "vector processor," the CLAY-1, which processes multiple tasks at a time, was introduced. Vector computers operating at speeds exceeding 100 MFLOPS are generally called supercomputers. The development of computers in the USA and Japan from 1940 to 1990 is illustrated in Fig. 6.32, including both scalar and supercomputers. Among the supercomputers, VP-, S-, and SX-series are Japanese productions (Umemura 1990) (Fig. 6.32).

In 1989, a new type of computer designed specifically for the calculation of the gravitational many-body problem was constructed by Sugimoto Daiichiro and his group. It was named GRAvity PipE (GRAPE) and connected to a host computer through an interface. The host computer left the computation of the gravitational many-body problem to GRAPE, allowing the total computer speed to be markedly higher. While ordinary supercomputers solved the n-body problem for 10,000 bodies, the GRAPE computer showed a capacity to solve a one-million-body problem (Sugimoto et al. 1990; Ebisuzaki et al. 1990).

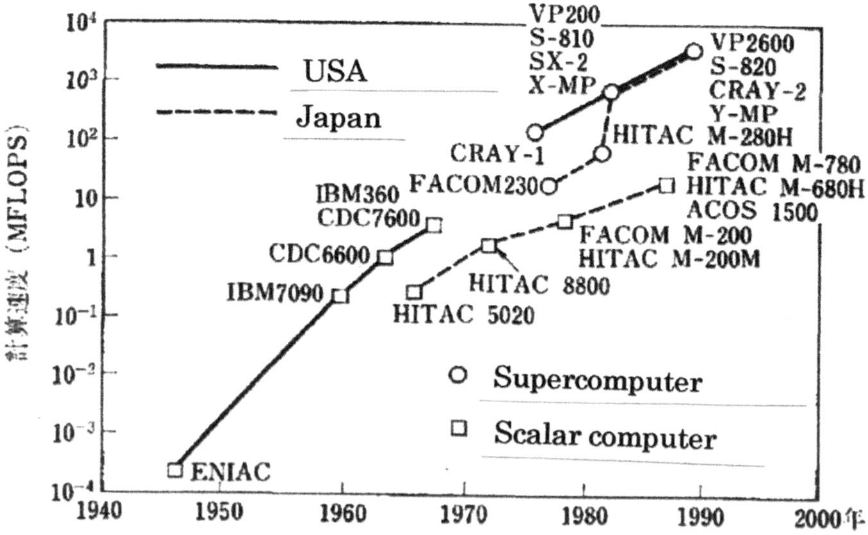

Fig. 6.32 Development of computer. The ordinate is the processing speed in units of MFLOPS. (Umemura 1990)

References

Akabane, K., Morimoto, M., et al. (1974). Detection of interstellar lines of para-formaldehyde and OCS and searches for other molecular lines at 73 GHz. *Publications of Astronomical Society of Japan, 26*, 1–7.

Akabane, T., Saito, Y., et al. (1995). The north polar hood in early autumn on Mars. *Astronomy and Astrophysics, 304*, 595–601.

Asada, K., Kameno, S., et al. (2006). The expanding Radio Lobe of 3C 84 revealed by VSOP observations. *Publications of Astronomical Society of Japan, 58*, 261–270.

Bartholomew, J. R. (1989). *The formation of science in Japan, building a research tradition*. London: Yale University Press.

Clark, G., Tanaka, Y., et al. (2001). Obituaries of Oda Minoru (天文月報, Clark, 田中靖郎: 追悼小田稔氏). *Astronomical Herald, 94*, 308–320.

Ebisuzaki, T., Ito, T., et al. (1990). A special processor for gravitational many-body problems (天文月報. 戎崎俊一: 、重力多体問題専用計算機 GRAPE). *Astronomical Herald, 83*, 234–236.

Forman, W., Jones, C., et al. (1978). Fourth Uhuru Catalogue of X-ray sources. *Astrophysical Journal, Supplement, 38*, 357–412.

Fukui, Y., Abe, R., et al. (1999). A ^{12}CO survey of the LMC with NANTEN. *IAU Symposium, 190*, 61–66.

Gabuzda, D. C. (2009). VSOP's legacy for our understanding of magnetic fields in active galactic nuclei. *ASP Conference Series, 402*, 236–242.

Gordon, A. (2003). *A modern history of Japan* (Table 1.1) (p. 248). Oxford: Oxford University Press.

Hashimoto, J., Maihara, T., et al. (1970). Infrared polarization of the peculiar M-type variable VY CMa. *Publication of Astronomical Society of Japan, 22*, 335–340.

Hayakawa, S., Matsumoto, T., et al. (1968). Infrared observations of the Moon. *Icarus, 9*, 357–359.

Hirabayashi, H. (1998). The VSOP Mission – A New Era for VLBI. *IAU Colloquium, 164*, 11–15.

Hirabayashi, H., & Edwards, P. G. (1999). Recent Results from VLBI Space Observatory Programme (VSOP) Observations. *Astrophysical Letters and Communications, 39*, 141–148.

Honma, M. et al. (2010). Schematic image of the Galaxy (国立天文台ニュース、本間希樹他: 銀河系の模式図). *National Astronomical Observatory of Japan*, News No. 205.

Kaifu, N., Usuda, T., et al. (2000). The first light of the Subaru Telescope: A new Infrared image of the Orion Nebula. *Publications of Astronomical Society of Japan, 92*, 1–8.

Kawabata, K., Ogawa, H., et al. (1984). The 4 m millimeter-wave telescope at Nagoya University (天文月報, 河鰭公昭、小川英夫ほか: 、4-m 短ミリ波望遠鏡). *Astronomical Herald, 77*, 244–247.

Kawaguchi I. (1982). Hida Observatory, Domeless Solar Telescope (天文月報、川口市郎: 飛騨天文台のドームレス太陽望遠鏡). *Astronomical Herald, 75*(7), 188-.

Kawakami, H., Tanaka, W., et al. (1985). Optical imager with a CCD. *Tokyo Astronomical Observatory Report, 20*, 499–517.

Kitamura, M. (2000). Reminiscences of Dodaira (国立天文台ニュース、北村正利: 堂平の思い出). *HAOJ News, 83*, 5–7.

Kogure, T., & Tokunaga, A. T. (Eds.). (1989). *Japanese national large telescope and related engineering developments*. Dordrecht/London: Kluwer.

Kogure, T., Ogura, K., et al. (1992). Spectroscopic observations of emission-line stars in the Orion. *Publications of Astronomical Society of Japan, 44*, 91–99.

Koshiba, M. (1987). Observations of neutrinos from SN 1987A by KAMIOKANDE-II. Proceedings of the ESO Workshop on the SN 1987A (pp. 219–227).

Maddox, J. (2001). Obituary: Minoru Oda (1923–2001). *Nature, 410*, 888–894.

Maehara, H., & Soyano, T. (1987). A search for cool carbon stars. I. Cassiopeia region. *Tokyo Astronomical Observatory Annals, Second Series, 21*, 293–310.

Makita, M. (1979). Determination of the magnetic field from the Zeeman line profile. *Publication of Astronomical Society of Japan, 31*, 575–584.

Morimoto, M., & Kaifu, N. (1972). A search for interstellar para-formaldehyde by its 73 GHz transition. *Memoirs des Societè Royal des Science de Liège III, 467*.

Nakano, M., Wiramihardja, S. D., et al. (1995). Survey observations of emission-line stars in the Orion-region. V. The outer regions. *Publications of Astronomical Society of Japan, 47*, 889–896.

Nishi, K., & Nagasawa, S. (1964). On the observation of the electron corona at the Norikura Corona Station. *Publications of Astronomical Society of Japan, 16*, 285–300.

Noguchi, K., Maihara, T., et al. (1977). Three-micron absorption band of carbon stars. *Publications of Astronomical Society of Japan, 29*, 511–525.

Oda, M. (1965). High-resolution X-ray collimator with broad field of view for astronomical use. *Applied Optics, 4*, 143–144.

Ogawara, Y. (2001). Obituary of Oda Minoru (天文月報, 小河原嘉明: 小田稔先生ご逝去). *Astronomical Herald, 94*, 234–239.

Ogawara, Y., Acton, L. W., et al. (1992). The status of Yohkoh in orbit: An introduction to the initial scientific results. *Publications of Astronomical Society of Japan, 44*, L41–L44.

Okuda, H. (2018). *Draft "A half century of infrared astronomy"*. Private communication.

Orchiston, W., & Ishiguro, M. (2017). The early development of Japanese radio astronomy. In *The emergence of astrophysics in Asia* (Chapter 6, pp. 129–148). Springer.

Sato, S., Kawara, K., et al. (1978). Infrared observations of Nova Vulcullae 1976 (NQ Vul). *Publications of Astronomical Society of Japan, 30*, 419–422.

Sato, S., Nagata, T., et al. (2001). Construction of 1.4 m infrared telescope in South Africa (天文月報、佐藤修二、、長田哲也ほか: 南アフリカ 1.4 m 赤外線掃天望遠鏡の建設の経緯). *Astronomical Herald, 94*(3), 125–129.

Soyano, T., & Maehara, H. (1999). A search for cool carbon stars. IV. Serpentice-Aqulla-Scutum region. *National Astronomical Observatory of Japan, Publications, 1*, 207–215.

Sugimoto, D., Chikada, Y., et al. (1990). A special-purpose computer for gravitational many-body problems. *Nature, 345*, 33–35.

Suzuki, H., & Ohishi, M. (1985). Search for interstellar molecular lines (天文月報, 鈴木博子、大石雅寿: 星間分子スペクトル線の探査). *Astronomical Herald, 78*(8), 209–212.

Takakura, T., & Kai, K. (1961). Spectra of Solar radio Type IV burst. *Publications of Astronomical Society of Japan, 13*, 94–107.

Takase, B. (1980). Counts of ultraviolet-bright galaxies and their distribution in cluster of galaxies. *Publications of Astronomical Society of Japan, 32*, 605–612.

Takase, B., & Miyauchi-Isobe, N. (1993). Kiso survey for ultraviolet-excess galaxies, XVIII. *Publications of the National Astronomical Observatory of Japan, 3*, 100–257.

Takase, B., Kodaira, K., & Okamura, S. (1983). *An atlas of selected galaxies*. Tokyo: Tokyo University Press.

TAO. (1977). *One hundred years of the TAO*. Tokyo University (東京大学、東京天文台百年史).

Umemura, M. (1990). New aspects of astrophysics by supercomputers (天文月報、梅村雅之: スーパーコンピューターが作る新世界). *Astronomical Herald, 83*(1), 4–6.

Wiramihardja, S. D., Kogure, T., et al. (1989). Survey observations of emission-line stars in the Orion region. I. The Kiso Area A-0904. *Publications of the Astronomical Society of Japan, 41*, 155–174.

Yamashita, Y., Nariai, K., & Norimoto, Y. (1976). *An atlas of representative stellar spectra*. Tokyo: University of Tokyo Press.

Yokoyama, K. (2008). History of the Latitude Observatory in Mizusawa for 90 years (日本の天文学の百年、5章、横山紘一: 緯度観測所 90 年の歴史). In *One hundred years of astronomy in Japan,* Kouseisha-Koseikaku (Chapter 5, pp. 43–51).

Chapter 7
Postwar Development of Astrophysics, 1946–2000 (Part II: Astrophysics)

Abstract In parallel with advancements in instrumentation, astrophysical research has progressed to a highly advanced level in the postwar period. Since the appearance of works of Hayashi Chushiro on stellar evolution and of Unno Wasaburo on theoretical astrophysics in the 1950s and 1960s, astrophysics has been taken up by an increasing number of researchers across the country starting in the 1970s.

In this chapter, the development of astrophysics in Japan is traced through notable works related to the Sun, stars, galaxies, and the Universe.

7.1 Development of Astrophysics

The driving force behind astrophysics in the postwar era is clearly the increasing number of researchers in astronomy. Table 7.1 shows the numbers of members of the Astronomical Society of Japan every 10 years from 1950 through 2000. The members are divided into Special members (subscriber to *Publications of Astronomical Society of Japan* and *Astronomical Herald*, mostly researchers) and Normal members (subscriber to *Astronomical Herald*). The increasing rate of research members is remarkable in the second half of the postwar century.

In the postwar period, the fields of research subjects have been markedly expanded with the increase of researchers. Table 7.2 presents the number of papers presented at the Spring Annual Meeting of the Astronomical Society of Japan in the period 1957–1995. In this period, increasing numbers of papers were notable in the fields of stars, galaxies, theory, and instrumentation.

In the following sections, select topics in the aforementioned fields are briefly presented. A general overview on the history of astrophysics and cosmology in this century can be obtained in the works of M. Longair (Longair 2006).

T. Kogure, *The History of Modern Astronomy in Japan*, Historical & Cultural Astronomy, https://doi.org/10.1007/978-3-030-57061-3_7

Table 7.1 Membership of Astronomical Society of Japan in Postwar Period

Year	Special members	Normal members	Total
1950	126	694	820
1960	222	787	1009
1970	360	1427	1787
1980	510	1481	1991
1990	795	1545	2340
2000	1370	1400	2770

After *Astronomical Herald*, Astronomical Society of Japan

7.2 Structure and Evolution of Stars

Theoretical studies of stellar structure, formation, and evolution were initiated by Hayashi Chushiro (林忠四郎, 1920–2010) and advanced by his school in Kyoto University and the University of Tokyo.

7.2.1 Life and Works of Hayashi Chushiro

Hayashi was born in Kyoto the fourth son of Seijiro Hayashi, a manager at a small finance company. In 1940, Chushiro enrolled in the Department of Physics, Tokyo Imperial University, where he was exposed to astrophysics through a paper by G. Gamow and M. Schenberg on the Urca process (1941). Urca was the name of a casino in Rio de Janeiro, called Cassino da Urca, where Gamow and Schenberg first discussed the physical process containing neutrinos. According to the story, Schenberg told Gamow that the energy in the nucleus of a supernova disappeared as quickly as money disappeared at a roulette table (Nadyozhin 1995). The Urca process opened Hayashi's eyes to astrophysics, to which he devoted his life's work.

In 1942, upon graduation, he was conscripted into the navy. After the war ended in 1945, he returned to his hometown of Kyoto and joined the group of Yukawa Hideki at Kyoto University, where he immersed himself in the study of nuclear physics and astrophysics (Fig. 7.1).

In 1957, he was appointed professor of physics at Kyoto University and remained there until his retirement in 1984. His research activity lasted over 60 years, from the 1940s to before his death in 2010. His main works were on the following subjects, which overlapped during his active years:

1. Nuclear physics: 1947–1960
2. Chemical evolution in the early Universe: 1950–1963
3. Stellar evolution, Hayashi track: 1959–1980
4. Formation of planetary system: 1975–1990

In the following section, the main works of Hayashi and his school are presented.

For his achievements in astrophysics, Hayashi received many prizes, including the Order of Culture (1986) (文化勲章) from the Japanese government, the

Table 7.2 Number of papers presented at Annual Meeting of Astronomical Society of Japan

Year	Total number	Positional	Solar system	Sun	Star	Galaxy	Theory	Instrumentation	Other
1957	55	19	6	9	11	5	3	2	0
1960	76	24	4	16	18	8	2	2	2
1965	74	13	3	20	13	18	1	6	0
1970	131	19	14	18	32	26	16	6	0
1975	140	12	18	21	33	40	5	11	0
1980	167	6	22	17	29	65	13	15	0
1985	227	6	14	30	45	76	24	34	0
1990	314	4	16	27	35	124	58	47	3
1995	340	2	10	24	57	133	19	56	9

Notes: The following headings include the respective fields
Position: Positional and practical astronomy
Stars: Normal and peculiar stars, binaries, compact stars
Galaxy: The Galaxy, interstellar matter, galaxies, galaxy clusters
Theory: Celestial mechanics, theoretical astrophysics, cosmology
Other: Astronomical education, history of astronomy
After *Programs of Annual Meetings of Astronomical Society of Japan*

Fig. 7.1 Portrait of Hayashi Chushiro (Sato 2015)

Eddington Medal from the Royal Astronomical Society in 1970, the Kyoto Prize (京都賞) from the Inamori Foundation (1995), and the Bruce Medal of the Astronomical Society of the Pacific (2004). He died from pneumonia at a hospital in 2010 at age 89 (Osaki 2008a, b; Matsuda et al. 2010; Sato 2015).

7.2.2 Chemical Evolution in Early Universe

On the origin of chemical elements in the early phase of Big Bang cosmology, Ralph Alpher, Hans Bethe, and George Gamow (1948) (known as αβγ theory) supposed that primordial matter was composed solely of neutrons under extremely high temperature. As the Universe expanded with rapid cooling, elementary particles such as protons and deuterons were successively produced.

In 1950, Hayashi pointed out the importance of the Urca process at this stage. He showed that electron-positron pair production also occurred at high temperatures in the very early Universe. Taking into account the Urca process and pair production, Hayashi derived a proton-neutron concentration ratio of about 1:4. This value was a marked revision for the neutron model in the αβγ theory (Hayashi 1950).

Based on this proton-neutron ratio, Hayashi and Nishida Minoru (1956) considered the formation of light elements in the expanding universe. In the αβγ theory, under the original assumption of pure neutrons as the constituents of primordial matter, it was difficult to explain the successive building-up process of heavy elements, since there are "crevasses" at mass numbers 5 (boron) and 8 (oxygen). This difficulty could be overcome by the existence of an appreciable amount of original protons, which increased the probabilities of proton capture by light nuclei.

This made it possible to build heavier elements up to neon. That is, the formation of light elements became possible by successive proton capture in Hayashi's theory, against neutron capture in the αβγ theory (Hayashi and Nishida 1956).

7.2.3 *Stellar Evolution*

Hayashi and his school considered the process of stellar evolution from its birth to the final stage of supernova or other collapsing steps.

7.2.3.1 Evolution of Protostars (Hayashi 1966)

Stars are born in dark and cold Molecular clouds that are composed of atoms, molecules, and dust particles. By gravitational instability, some parts of the clouds begin to collapse and form stars. The initial contraction is in a state of free fall and isothermal cloud. The contracting cloud is transparent to interstellar radiation and gradually becomes opaque. The density is low enough so that the cloud remains transparent to the thermal radiation emitted mostly by dust particles.

At some point, the cloud becomes completely opaque, and Hayashi called this stage the initial state of an opaque protostar. The contraction process changes from free fall to adiabatic contraction. Hayashi illustrated the evolutionary process from this point in a Hertzsprung–Russell (HR) diagram ($L/L_{\odot}$– T_{eff}), as shown in Fig. 7.2, where point A denotes the initial protostar. The star evolves by passing points B, C, D, and E and on to point F. The physical processes through these points are as follows.

The protostar, from A through D, is in adiabatic contraction before it reaches gravitational equilibrium. In the adiabatic state, the increasing rate of pressure is much higher in the central core of the protostar. When the contraction reaches a certain point, the central core enters gravitational equilibrium and gives rise to a bounce of the core. The interaction of the bouncing core with its outer contracting regions produces shock waves. When shock waves reach the protostellar surface, the star flares up more than 1000 times the original luminosity.

The time scale of the contraction from A to E is about 20 years, of which the time for the flare-up from D to E is about 100 days. After reaching E, quasi-hydrostatic contraction of the star starts with a wholly convective structure and continues until the star reaches the main sequence at F, where the central temperature becomes high enough for hydrogen to burn.

Hayashi compared the flare-up process from D to E with the flaring-up phenomena of FU Orionis, a T Tauri-type stars. This star showed a remarkable flare-up in 1936 (Herbig 1965). The star increased its brightness over six magnitudes, within about 120 days, and gradually decreased its luminosity. Hayashi argued that this phenomenon could be explained as the last stage of the dynamical evolution of a star.

This evolutionary track has been called the Hayashi track, as opposed to the Henyey track (Henyey et al. 1955), which is derived based on a stellar model in

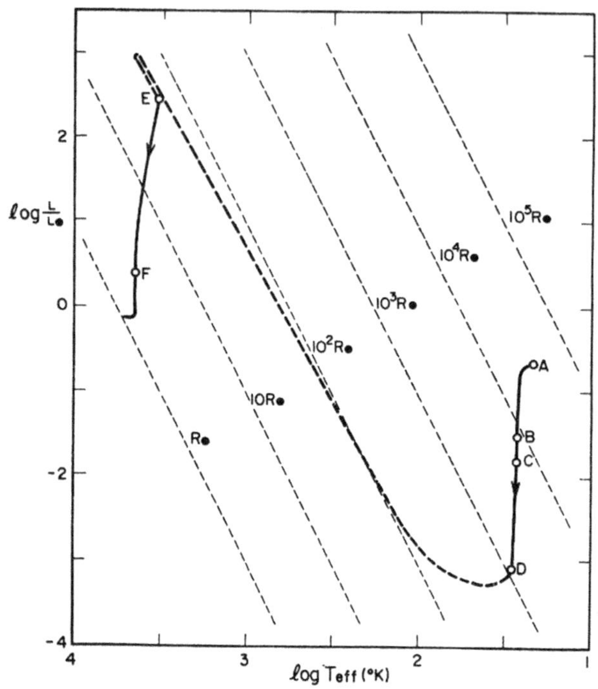

Fig. 7.2 Evolutionary track of protostar of one solar mass from initial state to main sequence in HR diagram (Hayashi 1966)

complete radiative equilibrium. Hayashi supposed that protostars had negligible nuclear energy production due to their low internal temperature, so energy transport takes place predominantly through convection, that is, the stellar interior is in the state of convective equilibrium.

It is noticed that Hertzsprung-Russel (HR) diagram usually expresses the distributions of stars on the spectral-type vs. stellar absolute magnitude diagram for the observational purposes. Instead, Hayashi adopted the stellar-luminosity vs. stellar effective temperature diagram for the purpose of physical understanding of stellar evolution.

7.2.3.2 Stellar Evolution in Advanced Stage

The main-sequence stage, where hydrogen burns and produces a helium core, is the longest period in the life of a star. The helium core develops and gradually contracts due to its strong self-gravity, and when the central temperature exceeds 100 million degrees, helium burning starts, forming a carbon core. The carbon core also shrinks and forms a neon core. The cores of heavy elements are successively formed in this way depending on the stellar mass, and finally an iron core is produced. Since iron is the most stable element, no iron burning will take place, and stars exhaust their energy source.

In Fig. 7.3, evolutionary tracks of stars in the HR diagram are presented for hydrogen, helium, and carbon burnings, indicated by H, He, and C, respectively, in

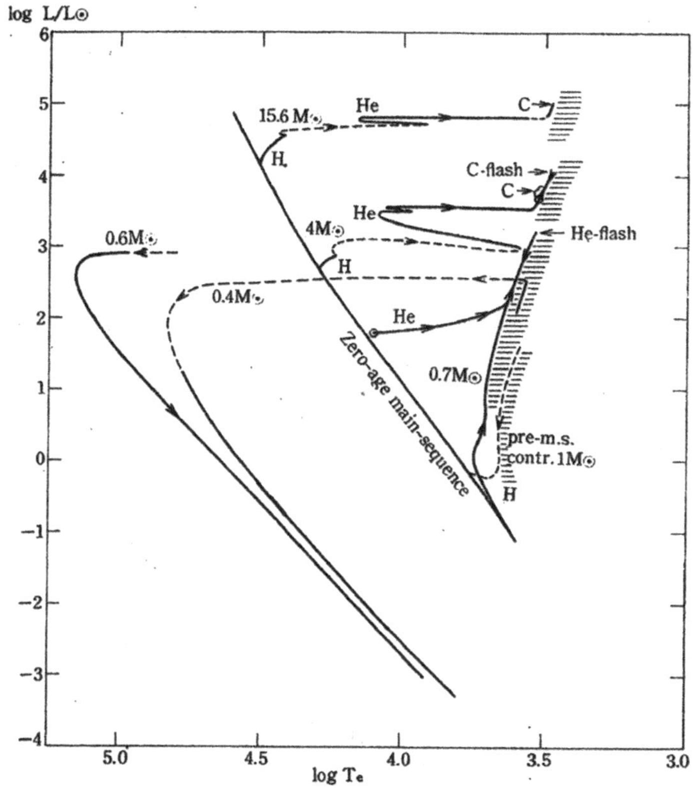

Fig. 7.3 Evolutionary tracks of stars in HR diagram (Hayashi et al. 1962)

the cases of stellar masses of 15.6, 4, and 0.7 $M_\odot$. A phase of pre-main-sequence contraction is shown for the Sun. The dashed areas represent forbidden regions for quasi-static equilibrium, and the dashed curves represent stages with a contracting core (Hayashi et al. 1962).

In addition, a final phase of contraction and cooling is shown in Fig. 7.3 for stars with a mass of 0.6 $M_\odot$ (composed of metal alone) and 0.4 $M_\odot$ (population II star). Both small-mass stars are already on the white dwarf sequence.

7.2.3.3 Formation of Planets

On the origin of the solar system, Hayashi considered the growing process of protoplanets by the successive capture of planetesimals. His work started from the numerical solutions of the restricted three-body problem, of which two major bodies are the Sun and a protoplanet, and the third small body is a planetesimal. The third body can be captured by the protoplanet through Lagrange points under the influence of perturbations from other bodies. Repeating such captures, the protoplanet finally grows up to become a planet (Hayashi et al. 1970).

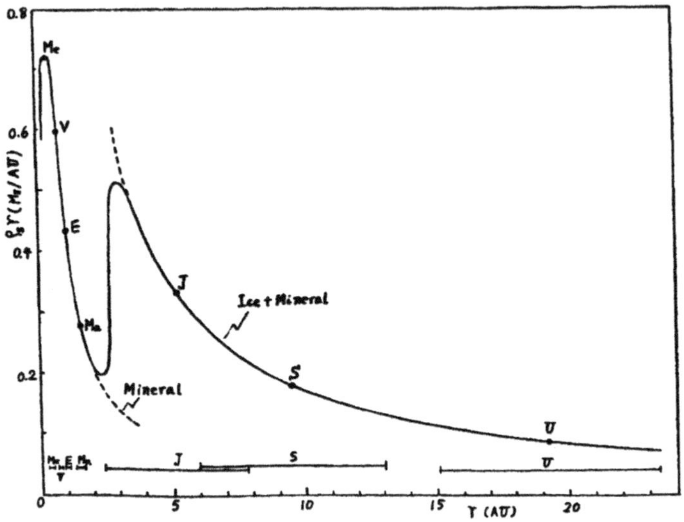

Fig. 7.4 Distribution of dust mass and ranges of planetary perturbation (Hayashi 1981)

On the formation of the planets, Hayashi presented a timetable for an evolutionary sequence of processes, which begins with the formation of the solar nebula and ends in the formation of planets and satellites. In the timetable, four stages are particularly noted: (1) the formation of primordial solar nebula, which is a gas disk rotating around the protosun in a state of thermal and dynamic equilibrium; (2) the fragmentation of the primordial solar nebula due to gravitational instability; (3) the formation of the first giant planet, Jupiter, through the collapse of molecular hydrogen; and (4) the disappearance of the gas component of the disk due to the solar wind and ultraviolet radiation, which were very strong during the T Tauri stage of the Sun's formation (Hayashi 1981).

Hayashi (1981) also demonstrated a relationship between the arrangement of the planets (Bode's low) and the primordial structure of the solar nebula when the nebula reached thermal and dynamic equilibrium. The result of his model calculation is illustrated in Fig. 7.4, where the abscissa denotes the distance from the Sun's center and the ordinate the surface density times distance in units of M_E/AU, with M_E being the mass of the Earth. The curve closer to the Sun indicates the dust distribution composed of minerals, whereas the outer one is the dust distribution composed of ice and minerals. The two curves are connected in between Mars and Jupiter, corresponding to the zone of asteroids. The range of planetary perturbation for each planet is indicated by the segments in the lower region of the diagram, which means the range of primordial nebula condensed to each of the planets. He thus explained the separation of inner planets (earth type) and outer giant planets, along with the existence of the asteroid belt. He also explained the formation of satellites in the same manner as in the case of planetary formation. Hayashi's scheme on the formation of the planetary system was further extended by Sekiya Minoru, Miyama Shoken, and others in the late 1980s (Sekiya and Miyama 1988; Sekiya et al. 1988).

7.3 Variable Stars

In the postwar period, studies on variable stars made progress from both observational and theoretical perspectives with respect to various types of variable stars. Observations were made through an international network VSNET, or the domestic network VWNET-J, initially organized by Nogami Daisaku and Kato Taichi of Kyoto University and others (Nogami et al. 1997). The stars that were observed included various types such as novae, supernovae, cataclysmic variables, symbiotic stars, pulsating stars, and many others. Observed data are collected in VSNET or VSNET-J. Databases were created for network members.

In this section, topics are mostly selected based on theoretical interest in variable stars in connection with certain topics.

7.3.1 Stars of Radial Pulsation

The prototype of radially pulsating stars is δ Cephei found by J. Goodricke in 1786. He observed the period and asymmetric light variation of this star. The binary hypothesis on the origin of stellar variation of Cepheid variables had been adopted for about 130 years. The theory of radial pulsation of single stars was first advocated by H. Shapley in 1914. J. A. Edgar considered the possibility of the excitation of first overtone modes in 1933, referring to the suggestion of Shinjo Shinzo in 1922 (Chap. 2) (Edgar 1933).

In Japan, studies of Cepheid variables had been undertaken by Takeuchi Mine, Aikawa Toshiki, and their group at Tohoku University in the 1980s. Among Cepheid variables some members, such as U Triangulum and TU Cassiopeia, exhibit double periods of pulsation, like beat phenomena. The period ratios are in the range 0.70–0.71. Takeuchi explained these phenomena as a superposition of two periodic pulsations and named these variables double-period Cepheids. He also explained these periods as a combination of the first and second harmonics of pulsations (Takeuchi 1992).

In double-period Cepheids, even after subtracting the harmonic components, some irregular fluctuation remains with small amplitude. For single-period Cepheids also similar, small, chaotic fluctuations are observable. Tanaka Yasuo (Ibaraki University) and Takeuchi (Tohoku University) considered a model oscillator of irregular variation and suggested that such irregular variability was caused by chaotic motions due to nonlinear effects (Tanaka and Takeuchi 1988; Tanaka 1991).

For other Cepheid stars, Takeuchi and Aikawa conducted Fourier analyses of light curves and found the possible formation of double-period Cepheids. Based on a hydrodynamic study of harmonic motion, they found that when the surface gravity was strong, radial pulsation was limited to the fundamental mode, whereas when the surface gravity was weak, multiharmonic pulsations, such as chaotic motions, tended to appear. Actually, they observed chaotic motions in F-type supergiants with low surface gravity (Aikawa 1991).

7.3.2 Nonradial Oscillations of Stars

The existence of nonradial oscillations was first proposed by W. T. Kelvin in 1890, and its theoretical possibility was shown by F. R. Moulton in 1909. O. Struve carried out spectroscopic observations of β Cephei-type stars in the 1950s and found variations of radial velocities in some spectral lines. These stars later became the prototype of nonradial oscillations. Theoretical studies of nonradial oscillations in the postwar period were conducted by P. Ledoux (1974) and J. P. Cox (1976).

In Japan, Unno Wasaburo of the University of Tokyo and his coworkers Osaki Yoji, Ando Hiroyasu, and Shibahashi Hiromoto markedly contributed to the theoretical studies of nonradial oscillations in the 1970s. In their book *Nonradial Oscillations of Stars* (1979), they wrote that stars are generally analogous to musical instruments, which play various modes of oscillation and tones. In fact, nonradial oscillations are observable in stars of a wide range of spectral types, with various modes of oscillations (Unno et al. 1979).

Modes of oscillations are expressed by spherical harmonics l and m, where $l = 0, 1, 2, \ldots$, and m is associated with l as $m = 0, \pm 1, \pm 2, \ldots \pm l$. On the restoring force of oscillations there are two modes: pressure-mode (p-mode) and gravity-mode (g-mode). Analogically speaking, p-mode corresponds to sound waves in the atmosphere and g-mode to the surface waves of a water pond. A sample of a wave pattern in the case of $l = 4$ is shown in Fig. 7.5.

When Osaki began to study β Canis Majoris in β Cephei-type stars in 1971, there were two models, a and b, on beat phenomena and the characteristic variation in line widths. (a) Ledoux (1951) proposed that these stars undergo nonradial oscillations in the presence of rotation. (b) Chandrasekhar and Lebovitz (1962) presented a model for beat phenomena as a coupling of radial and nonradial oscillations. Osaki (1971), based on Ledoux's model, showed that the combinations of stellar rotation and nonradial oscillations were able to explain most of the fundamental characteristics of β Canis Majoris-type stars, as follows (Osaki 1971).

Osaki considered the variations of line profiles and radial velocities of nonradial oscillations corresponding to second harmonics ($l = 2$, $m = 0. \pm 1, \pm 2$) for β Cephei stars. Different modes yield different forms of nonradial oscillations. For example, oscillation waves in the mode $m = -2$ propagate in the same direction as the rotation and take forms symmetric with respect to the stellar equator. He carried out numerical calculations in this case for radial velocity curves and variations in the half-widths. The results of those calculations are illustrated in Fig. 7.6, where the upper panel shows radial velocity curves and the lower panel the variation in line half-

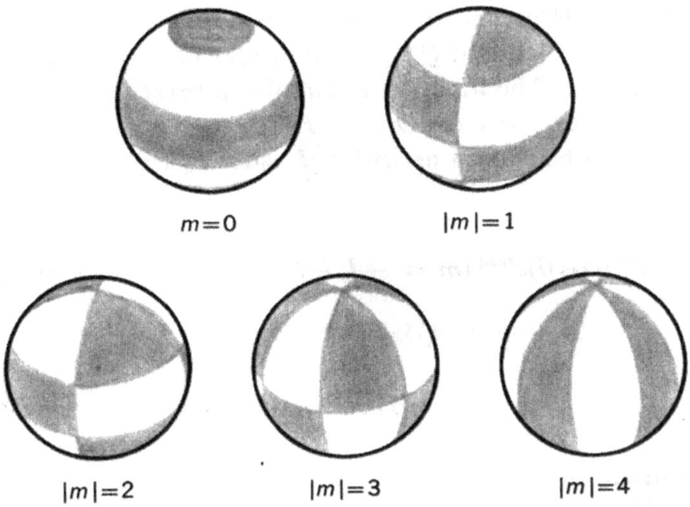

Fig. 7.5 Illustration of spherical surface harmonics for $l = 4$ (Unno et al. 1979)

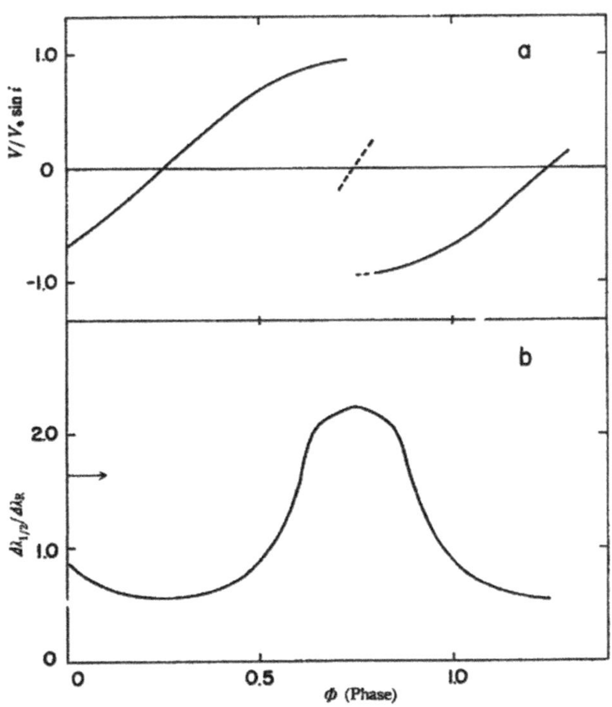

Fig. 7.6 Theoretical curves of radial velocity (**a**) and line half-width (**b**) of nonradial oscillation (Osaki 1971)

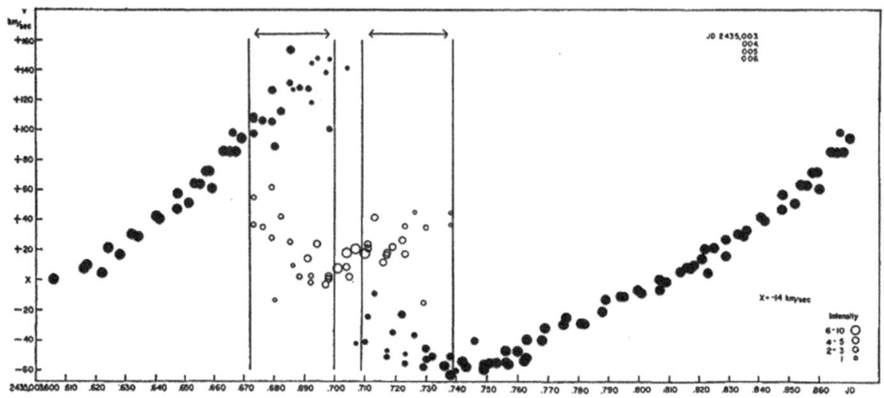

FIG. 1.—Radial velocities of BW Vulpeculae.

Fig. 7.7 Radial velocities in BW Vulpeculae, observed in the lines of H, HeI, OII, and SiIII. The unit of the abscissas is 15 min (Struve 1954)

width. In radial velocity, weak line doublings are seen in the interval of the phase $\Phi = 0.70–0.80$. The arrow in the lower panel indicates the half-width for pure rotation. The radial velocity curve of BW Vulpeculae (β Cep type) observed by O. Struve (1954) is shown in Fig. 7.7 for comparison. Osaki's theoretical study is in good agreement with the observed curve of Struve.

In the 1980s, a new class of pulsating star, called *line-profile variable stars*, was spectroscopically discovered. These stars, mostly B type, manifest their variability in spectral line profiles, with little accompanying change in the central radial velocity. Two subclasses were discovered: slowly rotating B stars (53-Perseus type) (Smith 1977) and rapidly rotating B stars (ζ-Ophiuchus type) (Vogt and Penrod 1983). Vogt and Penrod had shown that observed variations of these stars could be explained by the model profiles of nonradial pulsation. In 1988, Kambe Eiji and Osaki Yoji carried out extensive calculations of the theoretical line-profile variations due to nonradial pulsations in rapidly rotating stars. An example of their calculations is shown in Fig. 7.8 for a harmonic mode of $l = 5$, $m = -4$, and stellar inclination $i = 30°$. A retrograde traveling dump (denoted by r in the figure) is seen together with a prograde traveling dump (p) (Kambe and Osaki 1988).

7.3.3 *Helioseismology and Internal Structure of Sun*

The Sun was once also known as a nonradial pulsator. R, B. Leighton's group first observed the 5-min oscillations on the solar surface in 1992. Since then, it was found that the 5-min oscillations are global p-mode nonradial variations with high spherical harmonics of $l = 200–1000$. Intermediate- and low-mode oscillations ($l = 1–200$)

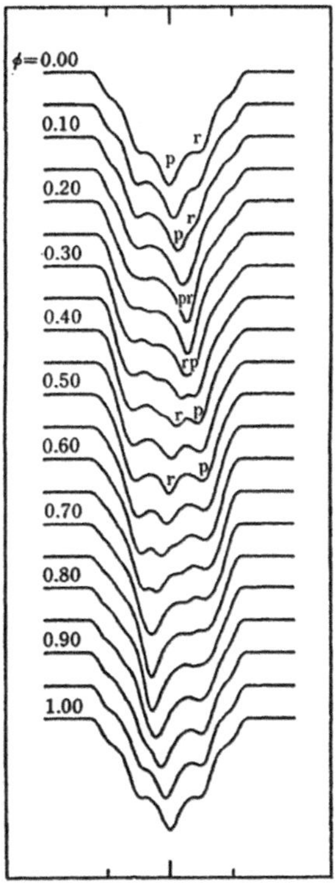

Fig. 7.8 Line-profile variations due to nonradial pulsation with $l = 5$, $m = -4$, and inclination angle 30° (Kambe and Osaki 1988)

were also confirmed. In the 1970s, the existence of g-mode oscillations was also discovered, though not confirmed at the time.

Higher-mode oscillations, such as 5-min oscillations, are confined to the upper layer of the Sun. Since temperature rises toward a deeper layer, and the sound velocity also gets higher, then each mode of oscillation is reflected at some depth and back to the surface with some information on the deep layer. Lower-mode oscillations reach deeper layers of the solar interior. Different modes yield information on different inner layers. Thus, the nonradial oscillation of the Sun may allow for the study of the physical structure of the solar interior. This new branch of science was called *helioseismology*, and it was developed in connection with the deep relationship with the theoretical physics of the internal solar structure starting in the early 1980s (Shibahashi and Sekii 1993).

To observe nonradial oscillations in detail, long-term unbroken and stable observations are essential. For this purpose, two projects were launched in the late 1990s. One was the Global Oscillation Network Group (GONG) project (1995), which was a network of same-size telescopes distributed around the world to avoid discontinuities in observations of the Sun. The second was the SOHO satellite launched by the European Space Agency (ESA) and the National Aeronautics and Space Administration (NASA) in 1995. SOHO was loaded with several instruments for observing the Sun, including telescopes for gravity-mode oscillations. SOHO had an orbit

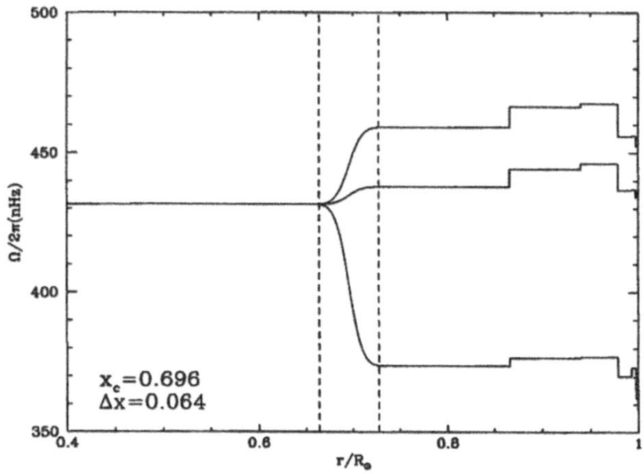

Fig. 7.9 Rotation rate in solar interior. Ordinate: inverse rotation period in nanohertz; abscissa: radial distance from center. In the outer layer, the rotation rate is given at latitudes of 0° (top), 30° (middle), and 60° (bottom) (Sekii 1997)

about the equilibrium point between the Sun and the Earth (Lagrange-1 point). With these projects, observations of the lower p-mode and g-mode oscillations were greatly advanced (Sekii 1998a, b).

Sekii Takashi of the National Astronomical Observatory of Japan (NAOJ) used the data of GONG while at Cambridge, UK, to derive the internal rotation of the Sun (Sekii 1997). He developed a model and compared it with GONG data to derive the rotation rate inside the Sun. The result is illustrated in Fig. 7.9. It is notable that the rotation rate inside the Sun is divided into a rigidly rotating interior and a differentially rotating outer layer, through a transition zone at around 0.7 solar radius (indicated as Xc = 0.698 in the figure). The vertical dashed lines indicate the position of the upper and lower boundaries of the transition zone. The equatorial acceleration of solar rotation has thus been explained as the result of nonuniform rotation of the Sun.

7.3.4 Novae, Dwarf Novae

In modern Japan, a nova was first observed by Yamamoto Issei at Kyoto University (Chap. 4), who rigorously trained amateur astronomers and actively created groups of observers. The tradition of nova observation continued into the postwar period with a notable extension of observation networks for various types of novae and nova-like stars in the network of VSNET.

Novae are known as members of cataclysmic variable stars, which are binary stars that consist of a white dwarf primary star and a mass transferring secondary star (Hoffmeister et al. 1985). Novae are usually classified into three types: classical novae, recurrent novae, and dwarf novae. The first two types are binaries combined with main-sequence stars or subgiants, whereas dwarf novae are binaries connected with evolved red giants. In parallel with active observations, the nature of nova

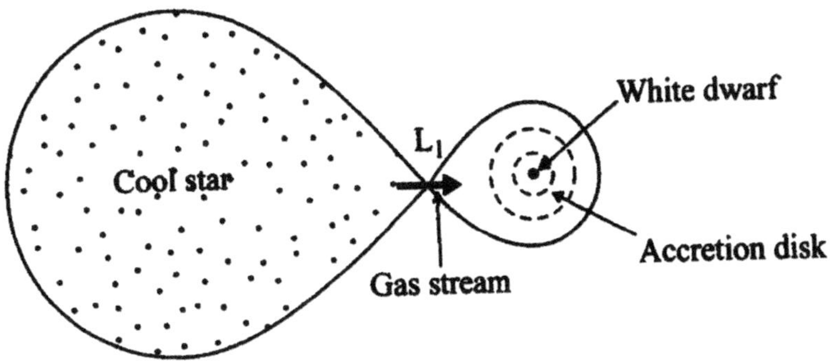

Fig. 7.10 Schematic picture of cataclysmic variable in orbital plane. The cool star expands in the Roche lobe, and a gas stream flows onto the accretion disk of the white dwarf (Hoffmeister et al. 1985)

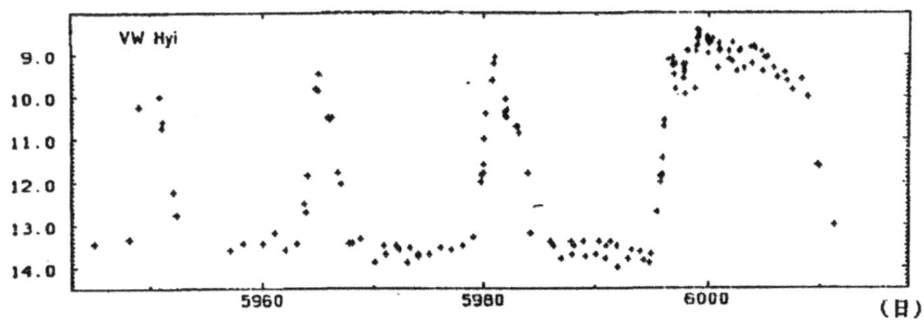

Fig. 7.11 Light curve of VW Hydri. In addition to normal outbursts (first three events), a super-ouburst is seen in the fourth burst (Mineshige 1987)

phenomena has been widely studied in Japan. Particular attention is presented here for the studies of dwarf novae by Osaki Yoji and Mineshige Shin (Fig. 7.10).

As an example of dwarf novae, the light curve of VW Hydri is shown in Fig. 7.11. This nova is of the SU Ursae Majoris type, which is defined by the appearance of super-outbursts in addition to normal outbursts.

On the mechanism of dwarf nova outbursts, two models have so far been proposed: a mass-transfer model by G. T. Bath and a disk-instability model by Osaki Yoji (1974). The mass-transfer model explains the outbursts of dwarf novae by the liberation of energy in the accretion disk through temporal mass-outflow from a red companion (Bath 1973). In contrast, the disk-instability model suggests that the outbursts are caused by intermittent accretion of material onto the surface of a white dwarf from the surrounding disk. The intermittent accretion was supposed to be triggered by some instability within the disk (Osaki 1974, 1983).

A long-lasting debate ensued over the validity of these two models involving many people around the world. In 1986, Mineshige, based on the disk-instability model, reproduced the two types of normal and super-outbursts, as observed in the light curve of VW Hydri. According to this model, a super-outburst occurs in the outer edge of an accretion disk where the disk is cool enough for hydrogen to recombine, whereas normal outbursts appear in the inner hot regions of the disk. Other behaviors in the variations of optical and X-ray spectra have also been interpreted in favor of the disk-instability model (Mineshige 1986).

In dwarf novae there is a rich variety in the light curves of outbursts, such as SU UMa type, U Gem type, and Z Cam type. Osaki presented a unified theory of outbursts within a single frame of the disk-instability model (Osaki 1997).

7.3.5 Supernovae

Supernovae are an order of magnitude brighter than ordinary novae at their maximum light in absolute magnitude. The appearance of supernovae is quite rare in our Galaxy. Only a few cases have been recorded, for example, SN 1054 (Crab Nebula) and SN 1572 (Tycho Brahe's star).

In the latter half of the twentieth century, the discovery of supernovae in other galaxies drastically increased due to the development and popularization of telescopic observations globally, including among Japanese amateur astronomers. This enabled detailed studies of the types and physical properties of supernovae.

Supernovae are basically separated into two types, I and II, depending on whether hydrogen lines are present (Type II) or lacking (Type I) in their spectra. Both types, however, can be classified into several subtypes by their spectral features and different patterns of light curves.

A Type Ia supernova manifests Si absorption at $\lambda 6153$ Å, whereas a Type Ib supernova shows no Si line, but it does show He lines in emissions. Type Ia supernovae are ordinarily attributed to the runaway explosions of white dwarfs in old binary systems composed of a red giant and a white dwarf. When the mass of the white dwarf exceeds the Chandrasekhar limit by the accreting flow from the red giant, a runaway burst occurs in the white dwarf as a supernova explosion. Since both component stars are already exhausted hydrogen, Type Ia supernovae exhibit no hydrogen spectrum. Type Ib is lacking both hydrogen and Si lines. This type of supernova has similarity with Type II as their core collapse type as seen in what follows.

Type II supernovae occur in massive stars when the nuclear energy source has been exhausted and the iron core has collapsed. The collapse causes a violent expulsion of the outer layers of the star as a supernova, while the central part of the star collapses into a neutron star or a black hole according to the original mass of the star.

On February 23, 1987, a supernova appeared in the Large Magellanic Cloud (SN 1987A) (Fig. 7.12). SN 1987A is a Type II supernova characterized by the existence of a hydrogen-line spectrum. The progenitor of this SN was also found to be a blue supergiant with a luminosity of around $10^5 L_\odot$, effective temperature of 15,000 K, and a size of 40 solar radii. Shigeyama Toshikazu, Nomoto Kenichi, and

Fig. 7.12 Appearance of supernova 1987A, near 30 Draco nis Nebura in LMC (Shigeyama et al. 1988)

their group found in 1987 that the early light curve can be accounted for by the diffusive release of energy deposited by shock waves propagating from the star's center. SN 1987A yielded the first observations of the X-rays and gamma-rays emitted from supernovae (Shigeyama et al. 1987, 1988).

Approximately 2 to 3 hours before the optical observations of SN 1987A, a burst of neutrinos was observed as shown in Fig. 6.31. A neutrino burst indicates the time of core collapse, which leads to an outward shock wave. Visible light was emitted only after the shock wave reached the stellar surface.

When a massive star at the end of its life collapses into a neutron star or a black hole, it radiates almost all its binding energy in the form of neutrinos, most of which have energies in the range 10–30 MeV. These neutrinos are emitted over a timescale of several tens of seconds. The neutrino luminosity of a gravitational core collapse is typically 100 times higher than the optical luminosity. These behaviors of supernova neutrinos were confirmed at first by the neutrino observations of SN 1987A (Hirata et al. 1987).

Another supernova, SN 1993 J, appeared in the galaxy M81 and was also studied by Nomoto's group.

SN 1993 J was the second brightest supernova, next to SN 1987A, at the time. This supernova was characterized by the appearance of the secondary light maximum after the primary maximum. Nomoto Ken'ichi, Suzuki Tomoharu, and their group analyzed the time variation of the SN as a combination of Type II and Type Ib

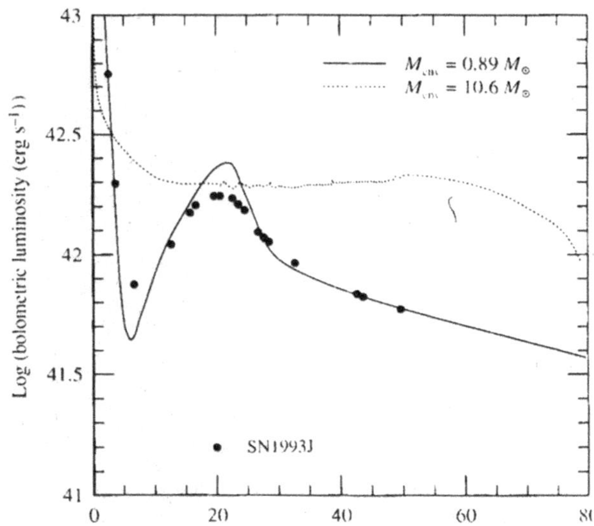

Fig. 7.13 Model calculations of light curves in two cases. The bolometric light variation of SN 1993 J is shown by the filled circles (Nomoto et al. 1993)

(which is also called Type IIb), and they calculated the light curves of this supernova in two models with envelopes of different masses of 0.89 and 10.6 solar masses. The mass of the helium core was the same as four solar masses. The light curve of the small-mass model was well fitted with the observed curve, as shown in Fig. 7.13. According to this model, the spectrum of SN 1993 J has rapidly evolved from Type II to Type I, as observed (Nomoto et al. 1993). The progenitor of this SN was a K-type giant identified on a photographic image taken before the explosion, and the binary component of a B-type supergiant was also confirmed later.

7.4 Solar Physics

7.4.1 Solar Physics in Tokyo

Solar physics in the 1950s and 1960s was mostly promoted by Suemoto Zenzaburo in the Tokyo area and Kawaguchi Ichiro in the Kyoto area.

Suemoto Zenzaburo (末元善三郎, 1920–1991) was born in Toyama Prefecture in August 1920. After graduation from the University of Tokyo in 1943, he started the study of solar physics at the TAO. His first work was to repair the old Solar Tower telescope. After World War II, conditions for the supply of parts and material were dreadful. By cannibalizing parts from disused old instruments, Suemoto finally repaired the telescope. Using this telescope, he advanced spectroscopic observations of solar flares. With these works, he was promoted to associate professor (Fig. 7.14).

He then worked at the Cambridge University Observatory for 2 years. He studied the physics of solar chromosphere under R. O. Redman, who was well known for having determined, in the 1940s, the kinetic temperature of the chromosphere as

Fig. 7.14 Portrait of Suemoto Zenzaburo. (Hiei and Uchida 1993)

being around 30,000 K. Redman and Suemoto analyzed the flash spectra taken at the Khartoum total solar eclipse in 1952 and concluded that the kinetic temperature of the chromosphere should not exceed 10,000 K (Redman and Suemoto 1954). This was confirmation of the low-temperature model of Miyamoto proposed in 1950 (Chap. 4).

After returning to Mitaka, he continued making spectroscopic observations of flares by constructing a new spectrograph attached to the Tower Telescope. Through the analysis of Balmer lines he found that flares consisted of a filamentary structure as thin as around 10 km in diameter for each filament (Suemoto and Hiei 1959).

In 1961, Suemoto was promoted to professor at the TAO. He joined the Department of Astronomy at the University of Tokyo in 1965 and returned to the TAO in 1977 as its director. He also devoted some attention to the development of space observations and served as a professor of the Aeronautical Institute of the University of Tokyo. He contributed to the launch of the solar X-ray satellite HINOTORI in 1981 as general manager of the project.

In 1991, he died at age 71. Suemoto's works have been taken up by Hiei Eijiro, Uchida Yutaka, and others (Hiei and Uchida 1993).

7.4.2 Solar Physics in Kyoto

In the same period as Suemoto, Miyamoto (Chap. 4) and his student Kawaguchi Ichirro (川口市郎, 1924–2019) (Shibata et al. 2019) worked on the theoretical study of solar physics.

Kawaguchi was born in Osaka and entered Matsumoto High School (old), Nagano Prefecture, which was located near high mountains. He loved mountain climbing his entire life. At Kyoto University, Kawaguchi studied astrophysics under Miyamoto. After graduation in 1947 he worked in the same Department of

Fig. 7.15 Portrait of Kawaguchi Ichiro in 1992 (by author)

Astronomy and started theoretical studies of chromosphere and prominence activities. In 1953, he began observational study of the Sun. His first work was the construction of a small coelostat (aperture 14 cm) at the Solar Laboratory of the Kwasan Observatory. For this construction, he designed the optical system and made some parts of the instrument by hand, reflecting the poor state of the early postwar period. The coelostat was reconstructed first with a 30-cm and finally to a 70-cm aperture at the new Solar Laboratory. He was promoted associate professor in 1961 and professor of astrophysics in 1971. Kawaguchi's research interests were active phenomena of the Sun such as sunspots, chromosphere, and active prominences. With these works, he was invited twice by the French government in 1966 for 2 years and in 1975 for 1 year and carried out observations at the Pic du Midi Observatory for the fine structure of the solar photosphere (Fig. 7.15).

In 1979, the Domeless Solar Telescope (DST) was established. As director of the Hida Observatory, Kawaguchi conducted solar observations together with his students, Kitai Reizaburo, Kurokawa Hiroki, Funakoshi Yasuhiro, and others. As stated in Chap. 6, DST has high spatial and temporal resolutions. One of the notable results was the discovery of the close connection between sunspots and magnetic fields. From a morphological study of the evolution of active regions near sunspots, Kurokawa distinguished two types of development of magnetic shear configurations between sunspots: type (a), which is a collision of two sunspots of opposite magnetic polarities, and type (b), representing successive rising of twisted magnetic tubes. Figure 7.16a illustrates two sunspots, A and B, observed in October 11 and 12, 1981. He derived the magnetic lines of force connecting spots A and B and the magnetic tubes rising to the solar surface, as shown in Fig. 7.16b. From these configurations, Kurokawa suggested that type (b) may be the essential form producing major flares (Kurokawa 1987).

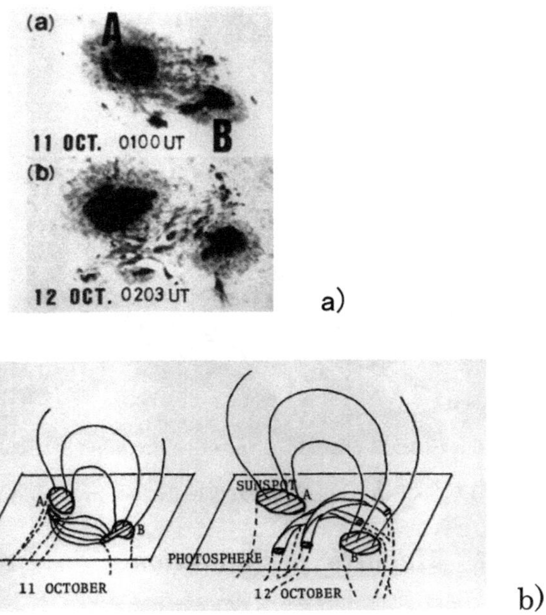

Fig. 7.16 (**a**) Two sunspots A and B observed in October 1981. (**b**) Magnetic field lines of force connecting spots A and B, and rising magnetic tubes of preflare structure (Kurokawa 1987)

7.4.3 Development of Solar Physics

In the 1980s and 1990s, solar observations have been greatly advanced through the construction of the (DST) (1979), the Nobeyama Heliogragh (1992), and the launching of the Yohkoh Solar Satellite (1991–2003).The role of Yohkoh was particularly notable.

Observations of soft-X ray images of Yohkoh and radio images at 17 GHz obtained with the Nobeyama Heliograph for impulsive radio flares that occurred on August 12, 1992, were made simultaneously by Takano Toshiaki, Enome Shinzo, and their group, as shown in Fig. 7.17 (Takano, Enome et al. 1994). The Yohkoh images are shown in (a) through (h) in color in the order of observed time. In images (a) and (g), X-ray images are superimposed by three coronal loops (SXT L − 1, L − 2, and L − 3). The radio flares are illustrated as contours in images (b), (d), and (h), corresponding to the phases of preflare, impulsive flare, and decaying flare, respectively. Image (i) indicates an illustration of the proposed connectivity of the magnetic loops. The radio spectra of flares showed a peak at about 6 GHz with nonthermal feature. Takano's group suggested that the radio emissions in the impulsive phase were of the gyrosynchrotron type emitted from high-energy electrons accelerated in the flare.

Collaborative observations by Yohkoh and DST were carried out in the early 1990s, and a close relationship between X-ray loops and Hα flare kernels was found in some active flares (Kurokawa et al. 1992).

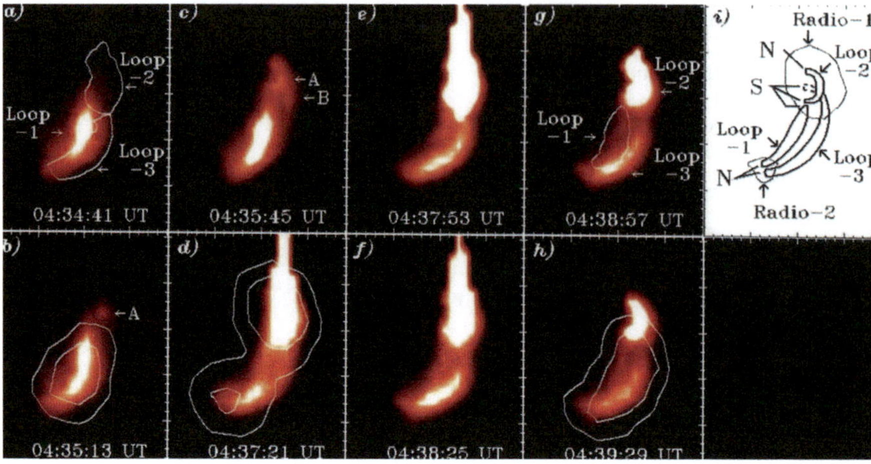

Fig. 7.17 SXT images of a radio flare that occurred on August 12, 1992 (Takano et al. 1994)

7.4.4 Magnetic Reconnection

It is widely accepted that the powerful energy in solar flares is released through magnetic reconnection of antiparallel magnetic fields. It converts magnetic energy into heat and kinetic energy. It is believed that magnetic reconnection is ubiquitous in the universe.

In the Sun, magnetic reconnection is best seen in the case of loop solar flares often observed at the solar limb. Figure 7.18 presents a schematic picture of a loop flare, modeled by Hori Kuniko, Yokoyama Takaaki, and others (Hori et al. 1997). A closed magnetic loop is formed by magnetic tubes rising up from the inner convection layer of the Sun. Outside the closed loop, open magnetic lines extend toward the corona. High above the closed loop, the open magnetic field on both sides of the loop provokes a magnetic reconnection at point X, which produces a large amount of energy and forms upward and downward jet streams. The rising jet flows up to the corona and turns to the coronal streamer, whereas the falling jet collides with the magnetic loop and causes an X-ray loop flare. Theoretical studies were conducted by the group of Shibata Kazunari and Yokoyama Takaaki in the 1990s (Yokoyama and Shibata 1994, 1998).

7.5 Stellar Physics

7.5.1 Physics of Early- and Late-Type Stars

In this section, the extensions of Osawa Kiyoteru's works on early-type stars and those of Fujita Yoshio's works on late-type stars are briefly summarized.

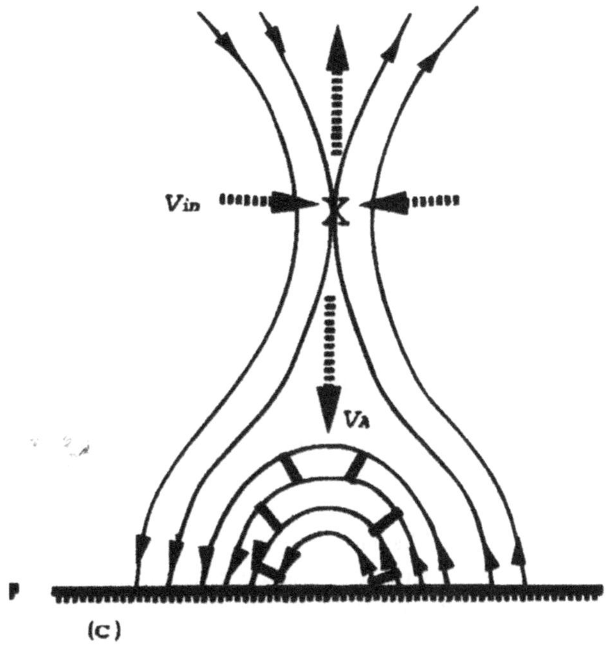

Fig. 7.18 Hydrodynamic modeling of solar loop flare and magnetic reconnection (Hori et al. 1997)

7.5.1.1 Early-Type Stars

In the 1960s, Nariai Kyoji and Nishimura Shiro took up Osawa's work on A and Ap stars (Chap. 3) based on photometric and spectroscopic observations at the Okayama Astrophysical Observatory (OAO). Nariai observed a hydrogen-deficient star, HD 30353, and determined the physical parameters of the atmosphere using model-atmosphere calculations and infrared (IR) observations (Nariai 1967). Nishimura carried out photometric and spectroscopic observations of a metallic-line star, 28 Andromedae, and found that this star was a pulsating variable star with some beat phenomena in the light curve (Nishimura 1969). In the 1970s, Sadakane Kozo and his group, Okyudo Masami, Kato Ken'ichi, and Nishimura Masayoshi started extensive spectroscopic studies of A and Ap stars. Sadakane examined the profile variation of the Ca II K line of the Ap star 73 Draconis and showed that variation of the K-line profile could be explained in terms of a calcium spot (or continent) that appears and disappears as the star rotates. He thus estimated the size of the spot and the concentration of Ca ions in the spot (Sadakane 1974).

In the 1990s, studies of Ap stars were extended by a fine analysis of spectral lines based on the models of nonlocal thermodynamic equilibrium by the group of Sadakane Kozo, Takeda, Yoichi et al. (1993) and Hidai Masahide and Takeda Yoichi (1996) for the upper main-sequence stars and for some early-type supergiants. Takeda Yoichi investigated the stellar magnetic fields in Ap stars based on a fine analysis of a model atmosphere and the analysis of Zeeman-split profiles in

Table 7.3 Strength of magnetic field H in Ap stars (Takeda 1993)

Star	Spectral type	Ion	H (kilo gauss)
52 Her	A2 Vp	Fe II	~4–5
HD 2453	A0 p	Ti II	~3
HD 8441	A2 p	Fe I	~0
HD 110066	Ap	Ti II	~3
HD 118022	A4 p	Ti II	~2–3

certain ions. He thus derived new values of the magnetic field of some Ap stars, as shown in Table 7.3 (Takeda 1993).

7.5.1.2 Late-Type Stars

Fujita Yoshio (Chap. 3) continued spectroscopic observations of carbon stars in the photographic IR region from the1960s up to the 1980s in collaboration with Yamashita Yoshimasa, Tsuji Takashi, Utsumi Kazuhiko, Maehara Hideo, and others. Yamashita and Maehara analyzed Mira-type variables and derived the mass loss rate from Mira Ceti in 1978. From the analysis of line profiles, they found the existence of outwardly progressing shock waves generated in the pulsating atmosphere. Most of the matter returns to the stellar surface, while part of it leaves the star, leading to mass loss. Analyzing the variation of line profiles, they derived a mass-loss rate as high as around 2×10^{-7} solar-mass per year (Yamashita and Maehara 1978a, b).

An extensive study of late-type stars was made by Watanabe Tetsuya in 1988. He measured the widths of the MgII, CaII, and Hα lines and emission-line intensities of MgII for around 70 late-type stars in spectral types F to M and luminosity class V to Ib in a kind of stellar catalogue. Based on a statistical study of physical parameters of these stars, he found that the existence of an extended chromosphere is a universal phenomenon in low-gravity late-type stars (Watanabe 1988).

7.5.2 Close Binaries

In 1955, Z. Kopal of the University of Manchester classified close binary stars into three types: detached, semidetached (Algol type), and contact systems, depending on the distance between the two component stars. In Japan, the study of close binaries was advanced by Kitamura Masatoshi (北村正利, 1926–2012) and his group, which included Yamasaki Atsuma, Okazaki Akira, and Takahashi Chie from the 1950s through the 1990s.

Kitamura was born in Nangoku City, Kochi Prefecture, in 1926. After graduation from the University of Tokyo in 1951, he started his work on the photoelectric photometry of stars at the TAO. He worked at Manchester in 1960 for 2 years and studied close binaries under Kopal. After returning home, Kitamura was promoted to

Fig. 7.19 Photograph of Kitamura Masatoshi in 1985 at XIX General Assembly, New Delhi (by author)

associate professor and then to professor of astronomy in 1972. After retirement in 1986, he continued the study of close binaries and in addition promoted international cooperation through the provision of astronomical facilities to developing countries by 2004. Based on the Official Development Assistance (ODA) project of the Ministry of Foreign Affairs of Japan, Kitamura provided a 45-cm reflector or an intermediate-size planetarium to each of 25 developing countries in Asia, South America, Africa, and other areas (Kitamura 2003). Kitamura died in 2012 at age 86 in Tokyo (Kozai et al. 2012) (Fig. 7.19).

Kitamura's first work was the investigation of the effect of reflection of light in close binary systems. The reflection effect causes the displacement of the light center from the mass center on the apparent disk of each component. By taking into account the effect of reflection, Kitamura came up with a theoretical formulation for the determination of orbital elements of close binaries (Kitamura 1954). In the 1960s, his formulae were widely applied to photoelectric and spectroscopic observations of a number of close binaries by him and his group, especially for Algol-type binaries.

In the 1960s, Kitamura and Takahashi Chie analyzed an eclipsing binary, R Canis Majoris (F1 V type). This star is well known for some unusual properties: first, the irregular change of orbital period, second, overluminosity for its mass, and, third, transient humps just after the primary minima. The change in the orbital period was supposedly caused by gaseous matter outside its Roche lobe and gas streams between two components. Taking into account the effects of gas streams and light reflection, Kitamura derived the orbital elements of this binary system as shown in Fig. 7.20 (Kitamura 1969), where dotted lines indicate the Roche lobes and L1 and L2 are the Lagrange points. Numbers written in the figure are the fractional dimensions (Kitamura and Takahashi 1962; Kitamura 1969).

In the mid-1980s, Kitamura and Nakamura Yasuhisa of Fukushima University conducted extensive investigations on the gravity darkening of highly distorted stars in close binary systems.

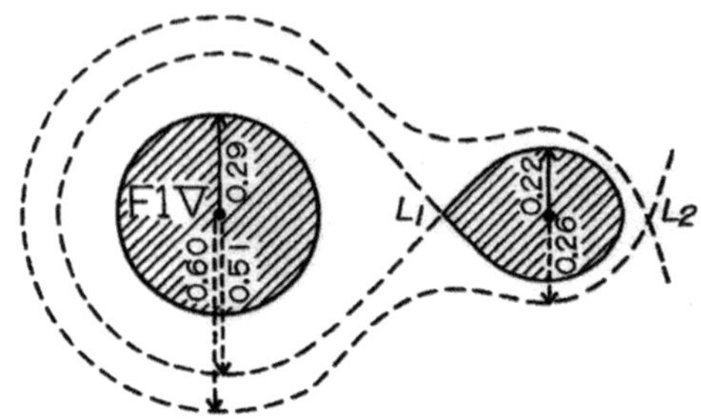

Fig. 7.20 Schematic view of innermost (Roche limit) and outermost contact lobes for R CMa (Kitamura 1969)

Gravity-darkening is a kind of center-limb darkening on a stellar surface that appears in rotationally or tidally distorted stars in radiative equilibrium. In such stars, surface flux at any point of the surface varies proportionally to the local gravity g (von Zeipel's law) as follows:

$$F \propto g. \tag{7.1}$$

According to this law, surface gravity in the equatorial region of a rapidly rotating star is small, so the surface brightness is low compared to the polar region.

In highly distorted stars in close binary systems, gravity darkening deviates from von Zeipel's law in the form

$$F \propto g^{\alpha}. \tag{7.2}$$

where α is the exponent of gravity darkening.

The study of gravity darkening was advanced by E. Budding and Z. Kopal (1970) for the Algol-type binary. They cited, however, the work of Takeda Shin'ichiro (Chap. 4) as the original study in this field. Takeda wrote in his paper that "The effect of this deformation on the light curve is estimated by taking into account both the variation in the surface brightness and darkening to the limb (Takeda 1934).

Kitamura's works could be the modern legacy of Takeda's suggestion. Kitamura made theoretical as well as observational studies in the 1980s and thereafter for every type of detached, semidetached (Algol-type), and contact binary system. Of these, Algol binaries are separated into two types, an ordinary Algol, where the primary system is detached, and the secondary system fills the Roche lobe, as seen in Fig. 7.20. The other type of Algol binary is the reverse Algol, where the primary system fills the Roche lobe and the secondary system is detached.

For detached binary systems, Kitamura and Nakamura found that both components obey von Zeipel's law. Consequently, they adopted the value of $\alpha = 1$ for the detached component in semidetached systems. Under this assumption, they analyzed nine ordinary Algols and found large values of α in a range of 2.25–9.73 for the Roche-lobe filling secondary components. Similarly, they analyzed eight reverse Algol systems and derived the gravity darkening in a range of $\alpha = 1.25$–2.78 for the primary stars. On the difference in the α value in the ordinary and reverse Algols,

Kitamura suggested that it was a result of the difference in mass-loss rate from the Roche-lobe filling component to the detached star in the sense that the mass-loss rate is higher in ordinary Algols, causing higher gravity darkening in the filling component (Kitamura and Nakamura 1987, 1988).

7.5.3 *Emission-Line Stars*

Emission-line stars are stars that show conspicuous emission lines in the optical spectra, and they are widely distributed in HR diagrams (Kogure and Leung 2007).

In this section we consider Be stars and symbiotic stars as examples of well-studied emission-line stars.

7.5.3.1 Be Stars

Be stars are rapidly rotating B-type stars that exhibit emission lines in hydrogen, and sometimes in helium and metallic lines. Emission lines are formed in the nebular disks or rings around a star's equator, as suggested by Otto Struve in 1931. Emission-line profiles are generally classified into three types: single peak, double peak, and double peak with sharp absorption in the line center. Struve explained the different profiles of emission lines by the difference of inclination angles of the rotating disks to the observer, as (a) pole-on stars with small inclination angles where sharp single emission lines are formed, (b) ordinary Be stars with medium inclination angles, where double-peak emissions are formed, and (c) Be shell stars with large inclination angles, where sharp absorption lines are formed in the line centers. These types are illustrated in Fig. 7.21.

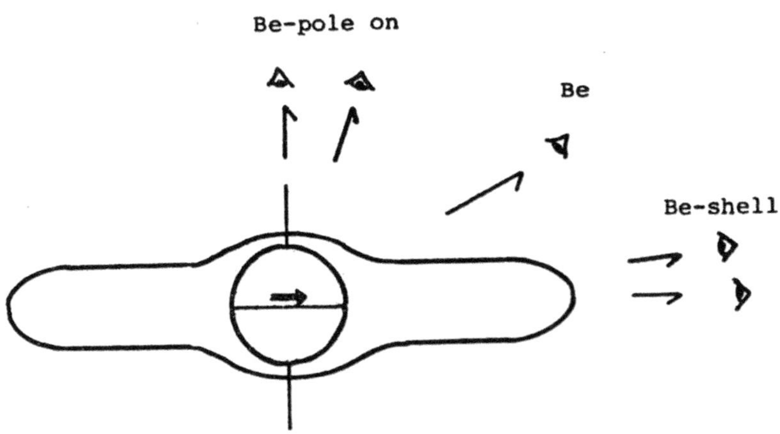

Fig. 7.21 Rotating-disk model of Be stars

The study of Be stars in Japan was initiated by Miyamoto Shotaro in the 1940s (Chap. 4) and was taken over by Kogure Tomokazu on the theoretical radiation field of extended stellar atmospheres in the late 1950s. Spectroscopic observations of Be stars began in 1966 with the Coudé focus of the 188-cm reflector at the OAO. The first study was conducted on some Be pole-on stars, while Kogure remained at the Paris Observatory, Meudon, by making use of spectral plates obtained at the OAO and the Haute Provence Observatory. Based on Miyamoto's revised theory, Kogure analyzed the profile of Hβ emission and derived the disk and ring parameters for some pole-on stars (Kogure 1969).

B and Be stars often show changes in type, such as from B to Be, Be to Be-shell, and vice versa, known as a phase change. Hirata Ryuko and Kogure started observations of Be-shell stars, in particular Pleione (28 Tauri, B8 IVe), which showed remarkable phase changes in the twentieth century as follows: Be (1905–1918) → B (1918–1938) → Be-shell (1938–1956) → Be (1956–1971) → Be-shell (1971–1988) → Be (1988–2005). Hirata and Kogure observed the whole period of the second shell phase in 1871–1988, and they analyzed the formation of shell lines (Hirata and Kogure 1984; Hirata 1995).

The phase change of Pleione was caused by a change in the optical depth of the disk, not by a change in inclination angle, as shown in Fig. 7.21. With the increase in optical depth, shell absorption lines in the Balmer series are produced up to higher members. In strong shell stars, shell lines become visible up to H30 or even H40. Hirata and Kogure analyzed the central depth of shell absorption lines and derived the optical depth at Hα lines, τ (Hα), and the fractional area, β, of a shell-line-forming disk relative to the stellar surface. The variations in τ (Hα) and β in the early period of the shell phase of Pleione are illustrated in Fig. 7.22, where β is composed of two parts: β_1, the fractional area of an optically thicker region (τ (Hα) $\geq$ 2000), and β_2, that of an optically thinner region (τ (Hα) ~ 600). The optical depth τ (Hα) remained at a high level, 10^3, throughout the period (Fig. 7.22).

From this analysis the structure of Pleione's envelope was described as follows. The Be star phase in 1955–1971 was characterized by an optically thin (τ(Hα) < 100) but vertically expanded disk (vertical height $\geq$ stellar radius). The shell phase started with the formation of an envelope, optically thick (τ(Hα) ~4000), but vertically thin near the equatorial plane (the disk area relative to the stellar photosphere was around 0.1 in the initial phase). The vertical height of the absorbing disk gradually increased, as seen in Fig. 7.22, and finally covered the whole photosphere. In contrast, the optical depth τ(Hα) gradually decreased and shell absorption lines disappeared in around 1988, and Pleione changed to the ordinary Be star in the 1990s.

In 1996, Katahira Jun'ichi, Hirata Ryuko, and their group performed a period analysis of the radial velocities of the shell lines of Pleione and found the star to be of a binary nature with a period of 218 days (Katahira et al. 1996). Based on polarimetric observations and considering the effects of binary interaction, Hirata proposed the precession model for Pleione (Hirata 2007). He supposed that the long-term spectroscopic variations could be explained in terms of the disk precession (precession angle 59° with a period of 81 years). He argued that the maximum shell phase occurred at the edge-on view of the disk.

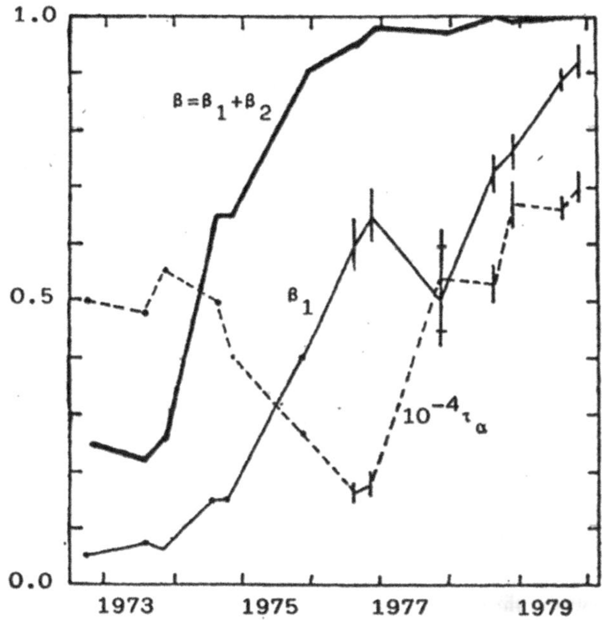

Fig. 7.22 Variations in disk parameters β ($\beta 1$, $\beta 2$) and τ (Hα) of Pleione in its early shell phase (Hirata and Kogure 1984)

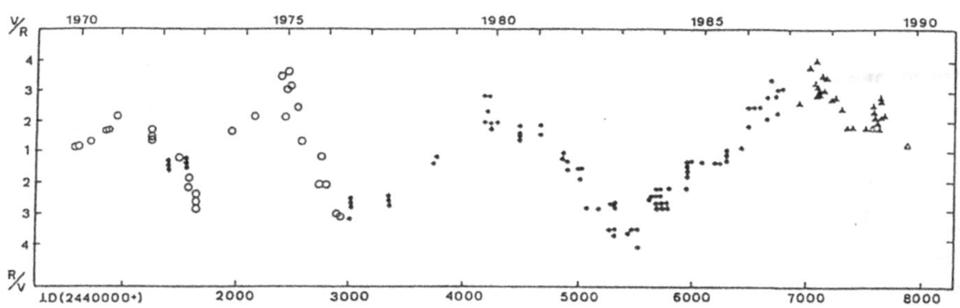

Fig. 7.23 V/R variation in Hβ line of γ Cas (Horaguchi et al. 1994)

Among other spectroscopic variations of Be stars, the V/R variation is notable. When Be stars show double-peaked emission lines, the relative height of violet (V) and red (R) components often exhibits a time variation, known as V/R variation. Figure 7.23 illustrates an example of V/R variation observed in γ Cassiopeiae (γ Cas).

The V/R variations had thus far been spectroscopically interpreted by two models: a symmetric rotating-pulsating model and an elliptical disk or ring model (Ballereau and Chauville 1989). Neither model had any realistic physical reasoning. Instead, from a dynamical point of view, Kato Shoji of Kyoto University first pointed out that one-armed low-frequency waves could exist stably in Keplerian disks. He suggested the applicability of this oscillation to Be stars, and Okazaki Atsuo of Hokkai-Gakuen University developed this theory as the dynamic explanation of the V/R variation. Thereafter, Kato-Okazaki's theory was widely accepted in the study of Be stars (Kato 1983; Okazaki 1991, 1996).

Be stars are generally of the luminosity class, III–V, belonging to some evolved B-type stars. On the other hand, Be and Ae stars, called Herbig Be/Ae stars (Herbig 1960), are pre-main-sequence stars surrounded by MCs. Yoshida Shigeomi, Kogure, and their group observed the Herbig Be/Ae star MWC 1080 in molecular lines with a 45-m radio telescope and in optical spectral lines at the OAO. They found that the CO line revealed the bipolar nature of molecular outflow, whereas the ^{13}CO and CS observations disclosed a disk structure with a cavity of face-on doughnut-like structure. The optical spectra exhibited strong emission lines in the Balmer lines and some metallic lines. The Balmer lines showed prominent P Cygni profiles. From these profiles the mass-loss rate due to the stellar wind was roughly estimated to be $10^{-6}\ M_{\odot}$ year^{-1}, and the maximum wind velocity reaches up to 1000 km s^{-1}. This strong wind was supposed to be responsible for the derivation of molecular outflow and the formation of cavities in molecular disks (Yoshida et al. 1991, 1992).

7.5.3.2 Symbiotic Stars

Symbiotic stars are stars that exhibit combined spectra of early-type and late-type stars. The nomenclature of symbiotic stars is due to P. W. Merrill (1958) of the Mt. Wilson Observatory. Initially, symbiotic stars were assumed to be a type of peculiar star, but it became apparent that they are interacting close binaries composed of an evolved cool giant and an accreting hot component (usually a white dwarf), both components being surrounded by dense nebulae.

Tamura Shin'ichi of Tohoku University made spectroscopic observations of a symbiotic star, HBV 475 (V1329 Cygni), at the OAO (Tamura 1977). This star belongs to a very slow nova (called symbiotic nova) and exhibited a slow explosion in 1964–1967 with maximum light in 1967. In 1977, when Tamura observed this star, the spectrum showed a distinct feature in the coexistence of high-excitation and low-excitation emission lines and their strong variability. The spectra of HBV 475 at three epochs are illustrated in Fig. 7.24, where the most notable feature was the appearance of high-excitation emission lines [FeVII], [FeVI], [CaV] and their strong variations. From the spectral analysis Tamura found that the nebulosity was stratified into two inner-hotter (electron temperature of around 20,000 K) and outer-warmer layer (electron temperature of around 6000 K), with a high electron density (higher than 10^{6} cm^{-3}). The temperature of the central star increased from 60,000 K in 1970 to around 100,000 K in 1977. From these results, Tamura argued that symbiotic stars might be stars in the process of undergoing planetary nebulae formation (Tamura 1977). To test this hypothesis, Tamura (1984) examined the distribution of symbiotic stars and star-like planetary nebulae in the Galaxy. He found that both objects concentrate near the galactic center with somewhat different degrees of concentration. From these distributions Tamura concluded that there certainly exists a close relationship between symbiotic stars and stellar-image planetary nebulae (Tamura 1984, 1997).

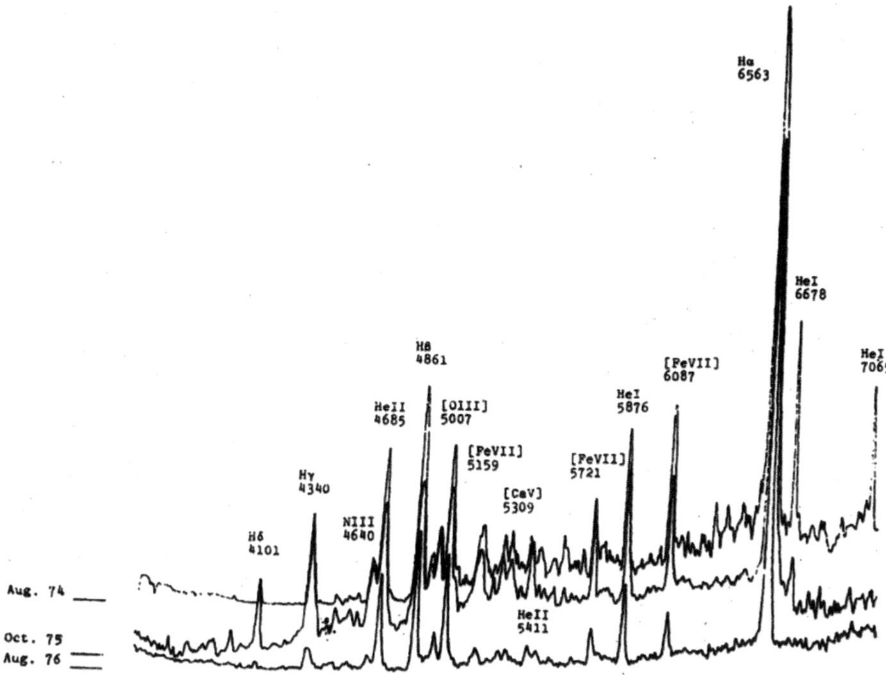

Fig. 7.24 Intensity tracings of HBV 475 at three epochs, 1974, 1975, and 1976 (Tamura 1977)

7.5.4 *Compact Stars*

Compact stars are the final stage of stellar evolution. When their energy source becomes exhausted, stars collapse into compact stars, white dwarfs, neutron stars, or black holes, depending on the original mass of the star.

7.5.4.1 White Dwarf Stars

The first study of white dwarf stars in Japan was made by Araki Toshima in 1934 (Chap. 4). In the postwar period, Suda Kazuo and Hitotuyanagi Zyuiti built a model of white dwarfs composed of an isothermal core and a convection envelope in 1957–1965 (Hitotuyanagi and Suda 1957). Nariai Kyoji analyzed the atmospheres of white dwarfs of hydrogen-poor chemical composition, and he derived the UBV colors of white dwarfs in 1962–1965 (Narial 1962).

Kodaira Keiichi (1967) made spectroscopic observations of the white dwarf Sirius B at the OAO. He analyzed the profile of the Balmer Hγ line and found that Sirius B had a slightly larger radius than that predicted by Chandrasekhar (1939) for white dwarfs consisting of fully degenerate pure helium. To explain this

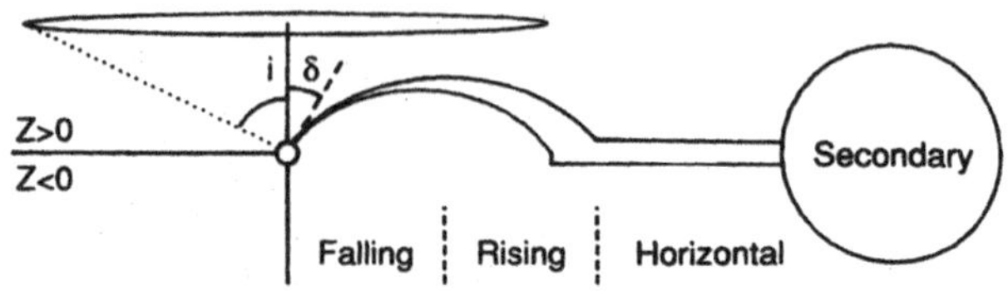

Fig. 7.25 Schematic view of AM Her type along orbital plane (Mukai 1988)

discrepancy, Kodaira proposed the possibilities of some other chemical composition or incomplete degeneracy near the stellar surface (Kodaira 1967).

Since 1970, the theoretical study of white dwarfs has made rapid progress in the following areas: nonradial oscillation (Osaki Yoji and C. J. Hansen 1973), stellar winds (Kato Mariko 1983), internal structure (Hoshi Reiun 1998), and others.

Some white dwarfs are known to have strong magnetic fields, which prevent the formation of accretion disks, and the accreting flows from the second component fall onto the polar region of the white dwarf along the magnetic lines of force. These stars are called polar or intermediate polar according to the strength of the magnetic field.

AM Hercules (AM Her) is a typical polar. Mukai Koji analyzed the accretion streams in AM Herculis-type systems in 1988 and found that the size of the magnetosphere reached nearly to the inner Roche lobe with a strong magnetic field as high as 30 Mega Gauss. A schematic view of the system as seen from the orbital plane is illustrated in Fig. 7.25, where the narrow accretion flow from the secondary system enters the magnetosphere through the inner Lagrange point and falls onto one side of the pole of the white dwarf. A strong supersonic gas flows out from the pole accompanying shock waves, and a postshock region is formed above the polar axis (Mukai 1988). Detailed X-ray spectroscopy of AM Her was carried out with the ASCA satellite by Ishida Manabu, Matsuzaki Keiichi, and others. They found X-ray black-body components and emission lines of iron that form in shocked regions in the poles of white dwarfs and that vary with the rotation phase of the system (Ishida et al. 1997).

7.5.4.2 Neutron Stars

In the postwar period, research on neutron stars started with the theoretical study of star spots on the surface. Fujimoto Mitsuaki and Murai Tadayuki made a model calculation in 1972 on the formation of a star-spot pair with antimagnetic poles on the surface of neutron stars. They explained the mechanism of pulsar emission by assuming that the pair rotates around the axis vertical to the stellar surface (Fujimoto and Murai 1972, 1973).

Starting in the 1980s, theoretical works have been carried out on the following topics:

(a) Spherically symmetric, steady mass-loss process from neuton star surface as a result of nuclear flashes (Kato Mariko 1983)
(b) Surface properties of accreting neutron stars as a trigger of helium flash, and disk structure in weakly magnetized neutron stars (Hoshi Reiun 1981)
(c) X-ray spectra and atmospheric structure of bursting neutron stars (Ebisuzaki Toshikazu and Nomoto Kenichi 1986)
(d) Radial oscillation of extended envelopes of neutron stars (Shibasaki Kiyoto and Ebisuzaki Toshikazu 1989)
(e) Thermal instability in accretion disks of neutron stars (Kakubari Ken'ichi, Hoshi Reiun et al. 1989)

Among these, we make mention the work of Ebisuzaki and Nomoto, item (c), on the model calculation of the stellar atmosphere for X-ray bursting neutron stars. They calculated four cases of hydrogen-rich (HRS), helium-rich (HES), magnesium-rich (MGS), and pure-helium (HED) and constructed an X-ray luminosity (*Lf*) versus X-ray color-temperature (*T*c) diagram. In this diagram, they plotted the data of three outbursts in the X-ray bursting neutron star MXB 1638–536, observed by TENMA, and they found that the helium-rich model was generally in good agreement with observations (Ebisuzaki and Nomoto 1986).

In parallel with theoretical works, X-ray observations have been promoted by satellites, TENMA, GINGA, and ASCA successively.

In the 1980s, Makino's group (1988) and Leahy's group (1985, 1989) at the ISAS observed a massive X-ray binary GX 301–2, using TENMA, for the X-ray intensity and spectral variations. D. A. Leahy, Nakajo Masanori, and their group found an eclipse of a neutron star by a small-scale object in this system (Leahy et al. 1988).

GINGA observed a great number of X-ray pulsars, including some new ones. Koyama Katsuji, Kawada Mitsunobu, and their group discovered four X-ray pulsars and three hard X-ray sources in the 5-Kpc arm of the Galaxy, and they suggested that this arm might be a colony of X-ray pulsars (Koyama et al. 1990). Nagase Fumiaki, Dotani Tadayasu, and their group observed an X-ray pulsar and found two absorption features in the X-ray spectrum, which were identified as cyclotron absorption in strong magnetic fields that are as high as 10^{12} gauss. In such strong fields, electrons were tightly spiraling around the magnetic line of force, and the electron motions were quantized. Thus, the absorption features at 12 and 23 KeV can be interpreted as the first and second transitions from the ground state, respectively (Nagase et al. 1991). With a γ-ray detector on board GINGA, Murakami Toshio and his group observed several γ-ray bursts from highly magnetized neutron stars. In two cases they discovered two absorption features, similarly interpreted as cyclotron scattering in highly magnetized plasma (Murakami et al. 1988; Yoshida and Murakami 1990).

Based on ASCA observations, Asai Kazumi, Dotani Tadayasu, and their group analyzed soft X-ray transients, which are most likely low-mass binary systems containing either neutron stars (NS-LMXB) or black-holes (BH-LMXB) in the quiescent state. The quiescent luminosity of NS-LMXB was estimated to be on the order of 10^{32}–10^{33} erg s^{-1}, which is much lower than the value predicted in the current model of accretion disks. Instead, Asai's group proposed the role of the so-called propeller effect in rapidly rotating magnetic neutron stars. Through the rapidly rotating magnetic field, some parts of the accreting matter are blown out, which leads to the reduction of the X-ray luminosity. They suggested that X-rays are emitted from the magnetic polar regions of neutron stars (Asai et al. 1996, 1998).

7.5.4.3 Stellar Black Holes

In the 1970s, based on data obtained by the UHURU satellite and rocket observations, Oda Minoru and his group (1970–1977) measured the variations of X-ray intensities and spectral features of Cyg X-1, the most like candidate for a black hole. Cyg X-1 (V1357 Cygni) is a spectroscopic binary composed of a supergiant (B0 Ib) and X-ray source, for which Oda's group found several types of X-ray variations, such as (a) the existence of millisecond pulses, (b) variations in the average intensity by a factor of 2 over time on the order of 10^3 seconds, and (c) the oscillatory structure in X-ray counting rate that lasts typically for several seconds and whose frequency drifts repeatedly within a few seconds. Among these, the existence of millisecond pulses implies a small source and, hence, a high specific emissivity at or in the vicinity of the black hole. In addition, the rapid flaring activity indicates a modified accretion structure.

As a result of their X-ray observations, Oda built an accretion-disk model around the black hole of Cyg X-1, as shown in the lower part of Fig. 7.26, where two main sources are indicated: (1) a hot, optically thin, and geometrically thick inner region as the source of high-energy X-rays and (2) a cool, optically thick, and geometrically thin outer region as the source of low-energy X-rays (Oda 1977).

X-ray observations of Cyg X-1 have been carried out extensively in the age of 1980's and 1990's, by TENMA (Kitamoto Shuji's group), by GINGA (Negoro

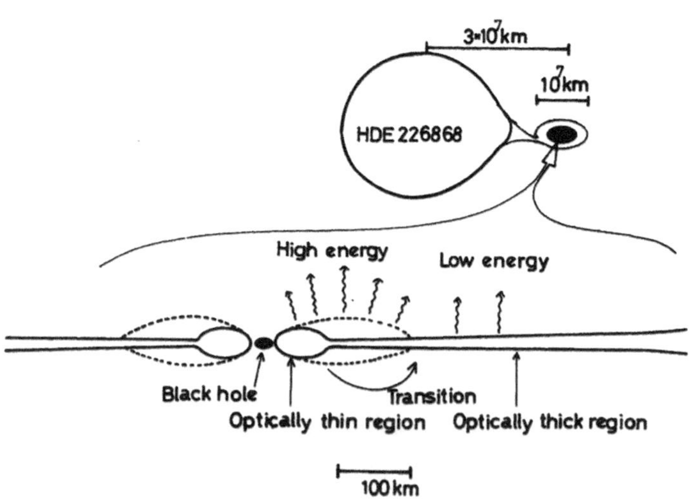

Fig. 7.26 Model of Cyg X-1, a binary black hole with accretion disk (Oda 1977)

Hitoshi' group), and by ASCA (Ebisawa Ken and Dotani Tadayasu' group), successively.

Other black hole candidates observed by these satellites include LMC X-1 (Ebisawa Ken et el. 1989), LMC X-3 (Ebisawa Ken et al. 1993), GX 339–4 (Ueda Yoshihiro et al. 1994), GS 1124–683 and five other candidates (Miyamoto Shigenori 1996), GS 2000 + 25, and others (Ebisawa Ken et al. 1997). Most black hole candidates were found to have two components, hard and soft X-ray spectra with violent time variations. Some of them exhibit quasi-periodic oscillations (QPOs), and some show nova-like variations.

7.6 Celestial Mechanism

In Japan, celestial mechanics was advanced by Hagihara Yusuke, as shown in Chap. 3. His works were taken up by Kozai Yoshihide (古在由秀), Hori Gen-ichiro (堀源一郎), Kinoshita Hiroshi (木下宙), and others in the postwar period (Miyamoto and Kinoshita 2008).

7.6.1 Kozai and Celestial Mechanism

7.6.1.1 Life and Works

Kozai Yoshihide (1928–2018) was an astronomer working in celestial mechanics and discovered the Kozai-Lidov mechanism in the restricted three-body problem, along with M. L. Lidov of the Soviet Union.

He was born in Tokyo into an academic family, where, among others, his father Kozai Yoshimasa (由正) was a historian and his uncle Yoshishige (由重) a philosopher. Yoshihide (由秀) chose astronomy and entered the Department of Astronomy at the University of Tokyo in 1948. He studied celestial mechanics under Hagihara Yushuke (Chap. 3).

Kozai Yoshihide wrote his thesis on the satellite system of Saturn in 1958 at the TAO. In the following year, the first artificial satellite, Sputnik, was launched by the Soviet Union. The celestial mechanics of the orbital motion of an artificial satellite became an important and hot topic of the time. Thus, the motion of artificial satellites became one of Kozai's life works. In honor of his achievements, he received the Imperial Prize of the Japan Academy (日本学士院賞) in 1979 and the Brouwer Award of the American Astronomical Society in 1989; moreover, in 2009, he won the Decoration of Cultural Merit (文化功労章) from the Japanese government. Along with research work, he had an administrative capability and contributed to the construction of the Subaru telescope in the 1980s and served as director of the TAO (1981–1988) and the NAOJ (1988–1994). From 1988 to 1991, Kozai served as president of the IAU. After retirement, he served as director of the Gunma

Fig. 7.27 Portrait of Kozai Yoshihide (Shibata et al. 2018)

Astronomical Observatory (Chap. 8) in 1997–2012). He died in 2018 in Tokyo at age 89 (Osaki 2008a, b; Takahashi 2015; Shibata et al. 2018; Kinoshita 2018) (Fig. 7.27).

7.6.1.2 Kozai's Equations and Kozai-Lidov Mechanism

Following the launch of Sputnik in 1957, Kozai proposed a mechanical theory of artificial satellites (Kozai 1959). He built a theory of perturbations for six orbital elements of satellites moving in the gravitational field of the Earth, which is assumed to be symmetrical with respect to the axis of rotation.

He derived periodic perturbations of the first order of approximation and then secular perturbations up to the second order. His formulae were called *Kozai's equations*, and they were widely applied to the determination of the orbital elements of satellites for practical purposes in space technology of satellite launching.

In 1962, Kozai and M. L. Lidov independently discovered a new type of orbital motion in restricted three-body problems, called the *Kozai-Lidov* or *Lidov-Kozai mechanism*.

1. Kozai considered the secular perturbations of asteroids with high inclination and eccentricity under the attraction of the Sun and Jupiter, assuming that Jupiter's orbit was circular. A secular effect occurs on timescales much longer than the orbital periods of an asteroid or Jupiter. Under these assumptions, he found that the orbit-averaged motion of an asteroid had two conserved quantities, the semimajor axis of the orbit and the angular moment of asteroid motion projected to the angular moment of Jupiter's orbital motion (Fig. 7.28). Then Kozai found

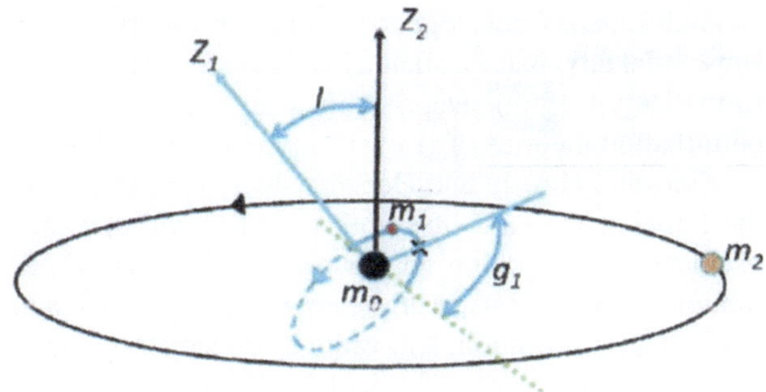

Fig. 7.28 Hierarchical three-body system: m_o (Sun), m_1 (asteroid), and m_2 (Jupiter); Z_1 and Z_2 are the angular momentums of m_1 and m_2 orbital motions, respectively; I denotes the orbital inclination angle; g_1 is the argument of perihelion from the ascending node (Seto 2017)

that asteroids with large inclination angles gave rise to the oscillation of libration at the argument of perihelion g_1 around 90° or 270°, measured from the ascending node (Kozai 1962; Nakai and Kinoshita 1985; Seto 2017).

2. In 1961, M. L. Lidov discovered similar properties in the three-body system consisting of the Earth, Moon, and an artificial satellite orbiting around the Earth with large perigee height. In low-perigee satellites, the effects of the Earth's shape, atmospheric resistance, and other features became important, but Lidov avoided such complications. As a result, he derived similar oscillations of artificial satellites as in the case of Kozai's asteroid. Lidov's paper (1961) was written in Russian and translated into English in 1962 (Lidov 1962).

Since then, the Kozai-Lidov mechanism has been extended into a more accurate theory, such as allowing elliptical orbit of Jupiter, and also applied to many astronomical phenomena, particularly in planetary motions in binary star systems and the dynamic evolution of outer solar planets and of binary black holes (Kinoshita 2018; Seto 2017).

7.6.2 Development of Perturbation Theory

The launch of various types of artificial satellites and space missions gave rise to new problems in celestial mechanics on the treatment of perturbations and on its technological practices in launching programs. In 1962, Hori Gen-ichiro considered the sources of perturbations for artificial asteroids. His main focus was perturbation due to the nonspherical form of the Earth. He showed that, in long-term orbital motions, much longer than the orbital period, there appears a critical orbital inclination at around 68° 28″, and some libration occurs near this critical inclination (Hori 1962).

In 1966, Hori extended the theory of general perturbation based on Lie's theorem in canonical transformation. In this theory he showed that perturbations in any quantity in orbital elements can be presented as a simple formula and in an explicit form. The motion of an artificial satellite is treated as an application of this general theory (Hori 1966).

André Deprit developed perturbation theories based on Lie series that offered substantial advantages, such as the transformation of state variables into an explicit form (Deprit 1969). Deprit's theories were later proved to be equivalent to the perturbation theories of Hori (Campbell and Jefferies 1970).

Kinoshita Hiroshi and Hori considered the stationary motion of satellites around the Earth in the case of elliptical and hyperbolic orbits. They showed that small oscillations of libration appear near the osculating perigee angle of the stationary motion in both orbits in the case of second-order approximations (Kinoshita and Hori 1972). In addition, Kinoshita extended Hori's perturbation method to third- and fourth-order solutions and found third-order periodic perturbations with fourth-order secular perturbations.

A comparison with the results of numerical integration demonstrated that the solution predicts the position of a close-Earth satellite with a small eccentricity with an accuracy greater than 1 cm over 1 month (Kinoshita 1978).

7.7 The Galaxy

7.7.1 Stellar System of the Galaxy

In 1956, Maarten Schmidt of Leiden Observatory constructed a model of the distribution of mass in a flat galactic system based on an equilibrium hypothesis between the centrifugal force of a rotating stellar system and its gravity. This model had been widely accepted. It depends upon the rotation curve of the Galaxy. The observations of rotation curve beyond the solar distance from the center were carried out by Vera C. Rubin (1965) and Maarten Schmidt (1965), and it became apparent that the rotation curve extended with a slow peak at around 9 Kpc from the center. From the analysis of the rotation curves of the Galaxy and other galaxies, Jeremiah Ostriker and James Peebles (1973) proved the existence of large-mass dark halos around the spiral nebulae that control the global structure of galaxies including our own.

In 1975, Miyamoto Masanori and Nagai Ryuzaburo of the TAO constructed a new three-dimensional model for the distribution of mass in the Galax, by taking into account the new galactic rotation curve and the role of the dark halo in the Galaxy (Miyamoto and Nagai 1975). This model, shown in Fig. 7.29, has been accepted widely since then.

The inner region of the Galaxy was also observed by IR observations in the 1970s by Hayakawa Sachio, Okuda Haruyuki, and others of Nagoya University. Hayakawa's group carried out balloon observations and obtained a contour map of the Galaxy between the galactic longitude 290° and 79° at 2.4 μm, as shown in Fig. 7.30 (Hayakawa et al. 1981). They found that the surface intensity in this wavelength was much greater than that expected from the contributions from spiral arms and clusters of young stars.

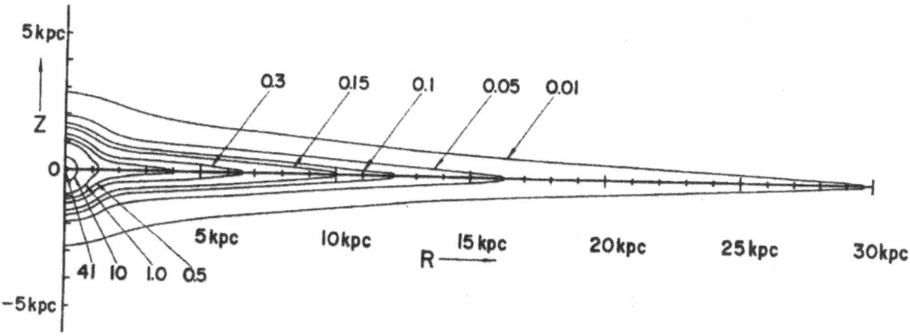

Fig. 7.29 Meridional section of a possible distribution of mass in the Galaxy. The unit of volume density is solar mass per cubic parsec. The volume density at the galactic center and at the Sun are 41 and 0.15 $M_{\odot}\mathrm{pc}^{-3}$, respectively (Miyamoto and Nagai 1975)

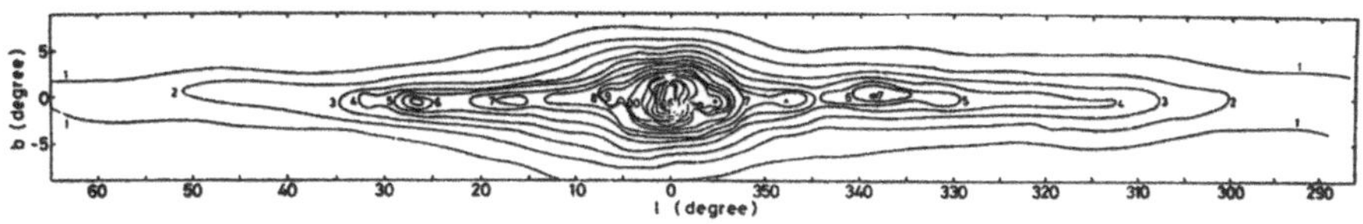

Fig. 7.30 Contour map of 2.4-μm surface brightness in galactic plane (Hayakawa et al. 1981)

On this excess emission, Okuda suggested that near-IR radiation around the galactic center was predominantly emitted from late-type stars, which include a major part of the mass of the Galaxy and, hence, governs its dynamics. According to this view, the contour map of Fig. 7.30 exhibits an edge-on view of the Galaxy, which resembles the dynamical structure of edge-on galaxies (Okuda 1981).

7.7.2 *Spiral Arm and Magnetic Field*

The formation of spiral arms in our Galaxy was studied by a group at Tohoku University from a magneto-hydrodynamic viewpoint in 1965. As stated in Chap. 5 (Sect. 5.3.4), their hydrodynamic models had met with difficulty in the short lives of spiral arms.

To avoid such a difficulty, C. C. Lin and F. H. Shu (1964, 1966) considered the stability problem of the galactic disk, and they showed that the arm structure could be regarded as a density wave in the galactic plane. According to them, the structure of arms depends upon the velocity dispersion of stars. When sufficient velocity dispersion exists among stars, a stable arm can be produced.

In star-predominant density waves, how can star-forming regions (SFRs) be formed along the arm? In 1966, Fujimoto Mitsuaki suggested the presence of large-scale galactic shockwaves along the stellar density waves. Thereafter,

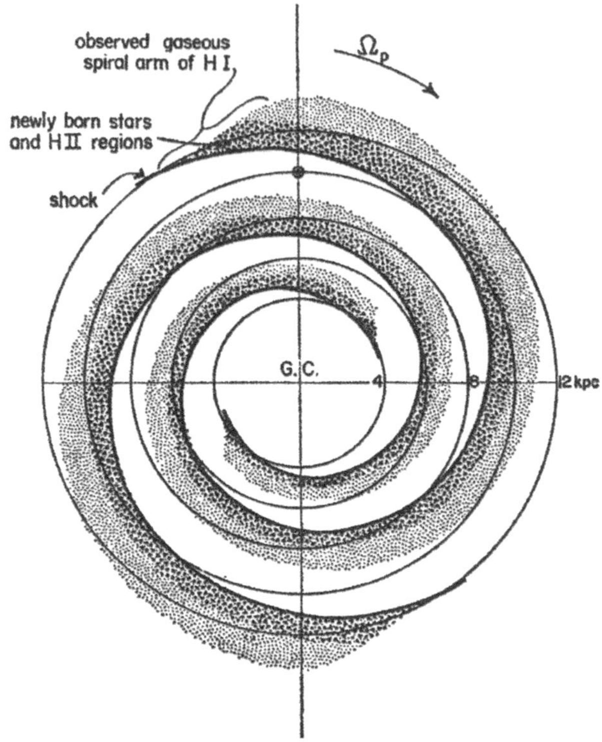

Fig. 7.31 Large-scale shockwave pattern in a two-armed model spiral galaxy. Two shock fronts and the newly formed regions of stars and the HII and HI regions are illustrated (Roberts 1969)

Fujimoto (1968) and William W. Roberts (1969) independently applied the theory of large-scale galactic shockwaves to their two-armed spiral model of the galactic plane (Fujimoto 1968). Figure 7.31 illustrates the spiral-arm model formulated by Roberts for a two-armed galaxy. The locations of the shock front are shown together with the regions of newly born luminous stars and the HII and HI regions on the inner side of the gaseous spiral arm. The propagation of shocks is assumed to be of a trailing type (Fujimoto 1966; Roberts 1969).

7.7.3 Interstellar Medium

The Galaxy contains various kinds of interstellar media, including ionized, atomic, and molecular gas, as well as dust and cosmic rays. The ionized gas around early-type stars is called the HII region and that around supernovae is called a supernova remnant (SNR). Some studies in Japan on HII regions and SNRs are presented briefly in this section.

7.7.3.1 HII Regions

The first radio observations of galactic HII regions were made with 30-m parabola antennae at Kashima in 1967 in a collaborative project that included Green Bank and Nan?ay Radio Observatory in several radio bands in the frequency region 1410–14,500 MHz. Kawajiri Nobuhiro of Radio Research Laboratories, Kashima, and Akabane Kenji of the TAO made observations at 4170 MHz for selected HII regions (Kawajiri and Akabane 1967). They measured the size (ℓ in pc) and mean electron density (Ne in cm^{-3}) and plotted their data in a (ℓ – Ne) diagram. They found the observed points distributed along the line of $Ne\ell$ = constant, as shown in Fig. 7.32. This relation was derived simultaneously at Green Bank and Nan?ay Observatory, and the mean values of $Ne\ell$ are given in Table 7.4. The values of $Ne\ell$ exhibit no systematic difference by radio frequency and roughly indicate that the electron density of HII regions is inversely proportional to the nebular size.

The first photometric observations in the Hα band were made by Ishida Keiichi and Kawajiri Nobuhiro in 1968 using the 50/65-cm Schmidt telescope at the Dodaira Station of the TAO. They derived the distribution of dust in some nebulae by comparing the Hα maps with radio-continuum maps (Ishida and Kawajiri 1968).

In 1982, Mizuno Shun of the Kanazawa Institute of Technology observed HII regions S254–S257 in two bands of E (Hα + [NII]) and V-band using the Schmidt

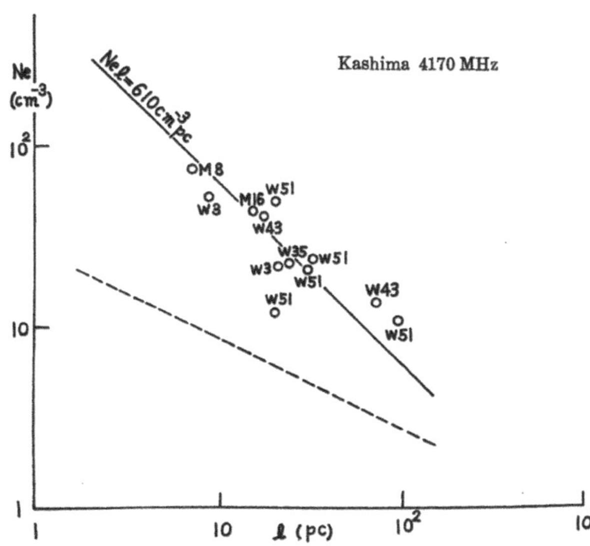

Fig. 7.32 Relation between mean electron density (Ne) and absolute linear size (ℓ pc) (Kawajiri and Akabane 1967)

Table 7.4 Observations of HII regions (Kawajiri and Akabane 1967)

Observatory	Radio frequency	$Ne\ell$
Green Bank	5007 MHz	580 cm^{-3} pc
	14,500	780
Nan?ay	1430	710
Kashima	4170	610

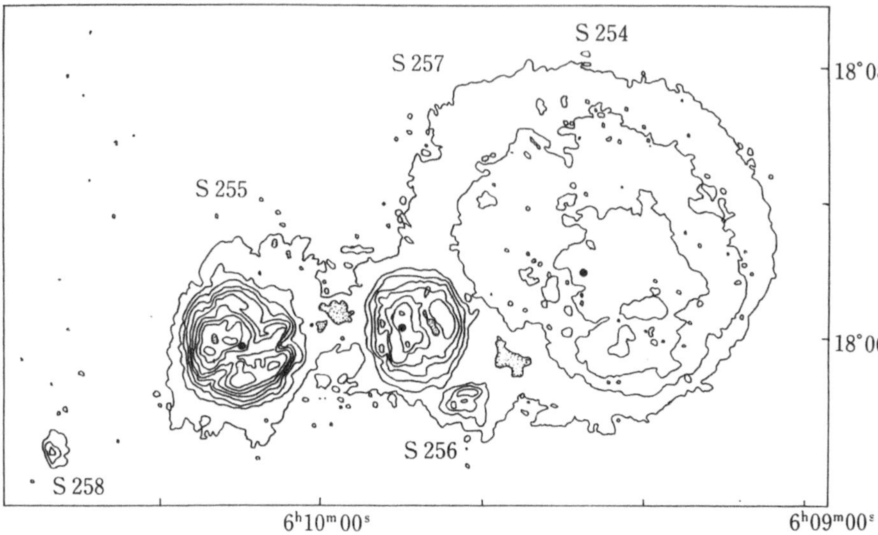

Fig. 7.33 Contour map of HII regions S254–S257 in *E* band (Hα + NII) (Mizuno 1982)

telescope at the Kiso Observatory (Mizuno 1982). Figure 7.33 illustrates the *E*-map of this region. It is apparent that S254 is extended with low Hα intensity, whereas S257 and S255 indicate some shell structure in the Hα band. In contrast, the *V*-map of every region shows a bright center gradually decreasing in luminosity toward the periphery. From these properties Mizuno suggested that the formation age is in the order from S255 to S257 and S254.

In 1989, M. H. Heyer, R. L. Snell, and their group of the Five College Radio Astronomy Observatory obtained a contour map of ^{13}CO (110.2 GHz) in this region, as shown in Fig. 7.34 (Heyer et al. 1989), where one may see that S254 is already almost outside the MC and S257 lies in its edge. In contrast, a high-column-density region of the MC exists between S257 and S255. The filled squares mark the positions of the compact HII regions, suggesting the formation of new stars. The optical HII regions are shown by half-tone circles corresponding to the map in Fig. 7.33. These maps confirm Mizuno's suggestion on the age of HII regions with an additional finding of a new SFR between NGC 257 and NGC255.

7.7.3.2 Supernova Remnants

In the 1960s and 1970s, SNRs were studied mostly from a theoretical point of view in connection with the following topics: evolution of nonthermal radio spectra in expanding phase (Ohtani Hiroshi 1964); formation of X-ray spectra (Itoh Hiroshi 1977, 1979); evolution of SNRs (Ikeuchi Toru 1978); and optical shockwave spectrum (Ohtani Hiroshi 1980).

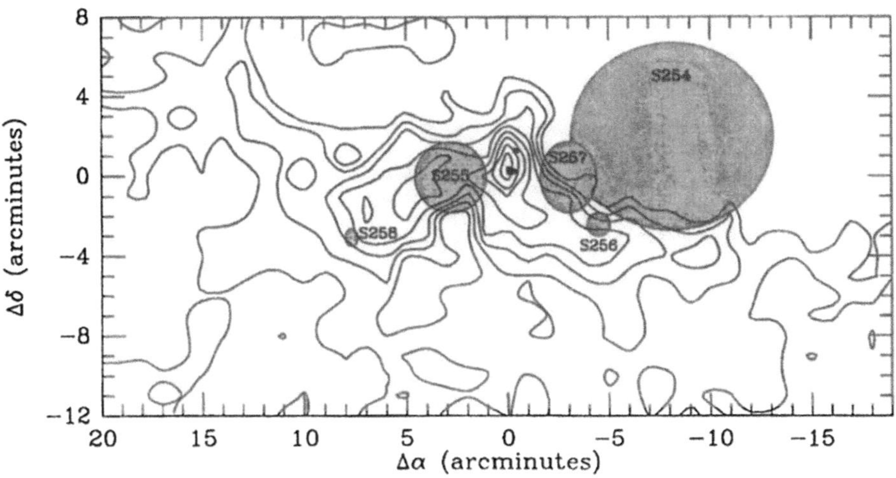

Fig. 7.34 Contour map of S254–S237 region in column density of ^{13}CO molecular line. Gray mark indicates optical HII regions (Heyer et al. 1989)

In the 1980s and 1990s, radio observations at the Nobeyama Radio Observatory were made, including 5- and 10-GHz continuum observations (Sofue Yoshiaki 1980, 1983) and CO line and polarization observations (Tatematsu Ken'ichi 1985, 1987).

X-ray observations were carried out with GINGA and ASCA by Koyama Katsuji, Makino Nobuyoshi, and their group for typical SNRs. They found that, while the types of original supernovae were different, the observed X-ray spectra were wholly thermal in nature. Their spectra exhibited a multitemperature system due to shock-heated gas. An example of ASCA observations made by Fujimoto Ryuichi, Tanaka Yasuo, and others at the Institute of Space and Aeronautical Sciences (ISAS) is illustrated in Fig. 7.35 for SNR W 49B (Fujimoto et al. 1995). Panel (a) illustrates the contour map of W 49B in the range 0.5–1.0 KeV, where the center-filled structure is apparent. Panel (b) shows the X-ray spectrum. Owing to the superior energy resolution of ASCA, the number of emission lines were resolved to the highly ionized ions of Mg, Si, S, Ar, Ca, and Fe. The coexistence of different ionization degrees suggests the multitemperature state of the SNR. It was also noticed that radio emissions are nonthermal in nature, indicating the existence of a sufficient magnetic field inside SNRs regardless of their type.

7.7.4 Star-Forming Regions

A typical SFR is the Orion region where OB stars, HII regions, and dark clouds form a complex structure of ongoing star formation.

Fig. 7.35 (**a**) Contour map of W49B observed with ASCA in 0.5–1.0 KeV energy band. (**b**) X-ray spectrum of entire remnant of W 49B. The individual emission lines of Si, S, Ar, Ca, and Fe were clearly resolved (Fujimoto et al. 1995)

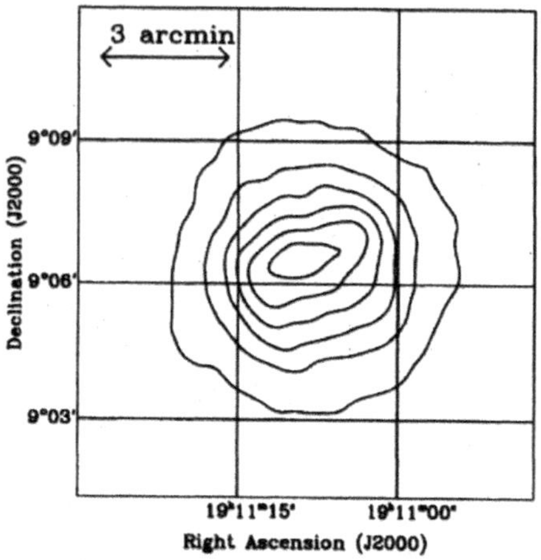

(a) Contour map

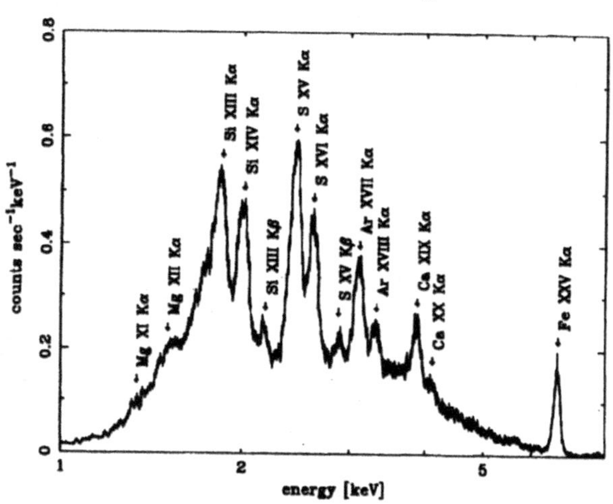

(b) X-ray spectrum

The first postwar observations of the Orion region were made by Isobe Shuzo of the TAO (Isobe 1973). He carried out photographic photometry at the Hα line for the whole region of the Orion OB association (5° × 35°). In parallel, he performed star counting on Palomar Sky Survey prints in the same sky region. Based on these data he found that a considerable part of the interstellar absorption on the line of sight was due to the circumstellar dust grains in the Orion association.

SFRs are roughly divided into two types: giant molecular clouds (GMCs) and ordinary molecular clouds (MCs). GMCs are associated with HII regions and OB associations and are sites of massive star formation, whereas MCs are specified by dark nebulae, which are the formation sites of small- or intermediate-mass stars. In Nobeyama, the first target was the Orion region. Some other GMCs and MCs were subsequently observed, and these observations gradually shed light on the star-forming processes.

7.7.4.1 The Orion Star-Forming Regions

Molecular line observations of the Orion region were made mainly with CO, CS, and HN_3 lines. Among these, a wide field of Orion A was surveyed with CS by Tatematsu Ken'ichi, Umemoto Tomofumi, and others in 1991–1993. The CS (1–0, 149 GHz) molecule is suitable for observations of parts of higher molecular density compared to CO molecules. Orion A is an elongated, narrow radio source centered at the Orion nebula. Tatematsu's group surveyed the main part of Orion A in a $1.5° \times 2.5°$ field extending north and south of the central nebula. The surveyed area is illustrated in Fig. 7.36. In this area they found 125 MC cores as the sites of active star formation (Tatematsu and Umemoto 1991; Tatematsu et al. 1993).

The Orion Kleinmann-low IR (Ori KL) object, discovered in 1967, was first observed by Hasegara Tetsuo, Kaifu Norio, and their group in 1984 with a CS line using the Nobeyama 45-m telescope with an angular resolution of 33″. They found a large rotating disk around Ori KL. The disk is 4×10^4 AU (0.2 pc) in diameter, with a rotating period of 3×10^5 year $\times \sin i$, where i is the inclination angle. The axis of the disk coincides with the direction of the high-velocity outflow of gas from Ori

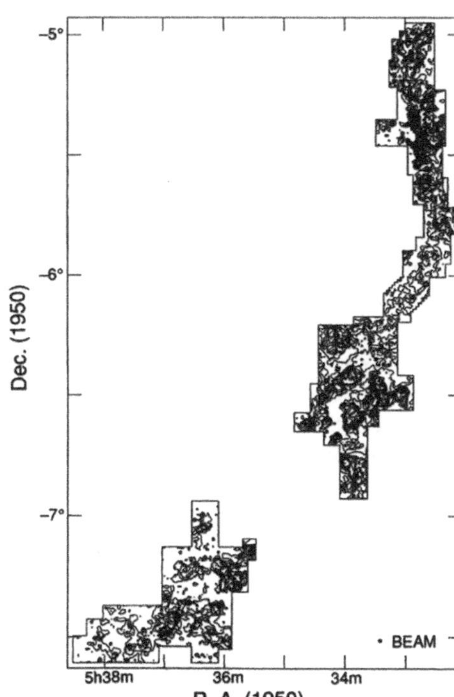

Fig. 7.36 High-resolution CS-line map of Ori A MC, except central region (Tatematsu et al. 1993)

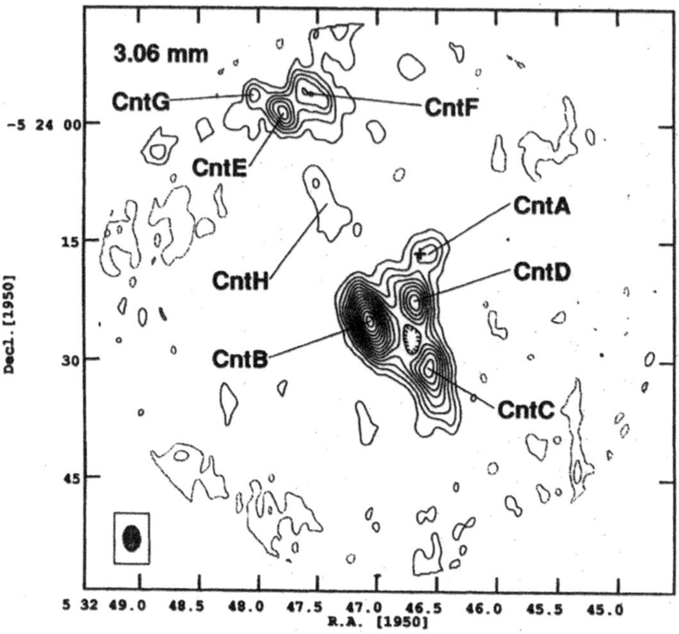

Fig. 7.37 Radio continuum map at 3.01 mm around Orion KL object. Eight compact clumps are identified as Cnt A–Cnt H. These clumps are candidates of protostar condensations (Murata et al. 1992)

KL. They supposed that this disk may be responsible for the bipolar nature of the flow (Hasegawa et al. 1984).

Radio continuum observations of the Ori KL region at λ 3 mm were made using aperture synthesis with the Nobeyama millimeter array by Murata Yasuhiro, Kawabe Ryohei, and their group (1992). With the beam width of around $2'' \times 3''$ for a field of $70'' \times 70''$ around Orion KL, they found eight discrete sources, Cnt A through Cnt H, as shown in Fig. 7.37. Among these, the Cnt A–Con D peaks belong to the Orion KL region, and the Cnt E–Cnt G peaks are called the CS1 cluster. The typical size and mass of these clumps were estimated to be approximately 0.01 pc and a few solar masses, respectively. These clumps are supposed to be candidates for protostar condensations. In parallel with continuum observation, Murata's group compared the KL region with molecular-line maps of NH_3 and HDO, which exhibited rather different pictures compared to the image of the CS line observed by Hasegaws's group (Murata et al.1992). These radio observations indicate a highly complex structure of the KL region.

7.7.4.2 Giant Molecular Clouds and Dark Nebulae

In parallel with the Orion region, molecular-line observations of GMCs associated with HII regions (W 51, NGC 7538, S 140) and MCs associated with dark nebulae (L1551, B335, B1) were widely carried out at Nobeyama.

Among these, NGC 7538 was observed by Kameya Osamu and his group (in 1986–1990) with CS, SO lines and H_2O masers. NGC 7538 is a bright HII region adjoined by a molecular cloud in its southeast side. In the core region of the molecular cloud, Kameya' group found three high-velocity flows, one was associated with compact HII regions and one was associated with a reflection nebula. All of these high-velocity regions are possibly bipolar and associated with infrared sources. They showed that these cores are the candidate sites of the high-velocity molecular-outflows combined with active star formation (Kameya and Takakubo 1988; Kameya et al. 1989).

Molecular line observations of Lynds dark nebula L1551 (one of the MCs) were made by Kaifu Norio's group (1984) and Uchida Yutaka, Kaifu Norio and their group (1987). Kaifu' group used 45-m telescope and found a rotating disk (Kaifu, Suzuki et al. 1984) while Uchida' group, using the Millimeter Array in CO line, observed the detailed structure of rotating disk and bipolar flow. They depicted the overall structure as shown in Fig. 7.38. (Uchida et al. 1987). As the result, they found that both lobes had a clear hollow cylindrical structure and that both lobes were likely spinning with a velocity of $1 \sim 2$ km s^{-1} in the same direction as that of the disk found by Kaifu and his group in 1984. These characteristics were well explained by the magnetohydrodynamic theory proposed by Uchida Yutaka and Shibata Kazunari, who stressed the role of magnetic field in producing such flows (Kaifu et al. 1984; Uchida and Shibata 1985).

7.7.5 *The Galactic Center*

The Galactic Center is located at the radio source Sgr A, as discovered in the radio observations in the mid-1950s. This was confirmed by IR observations in the 1960s.

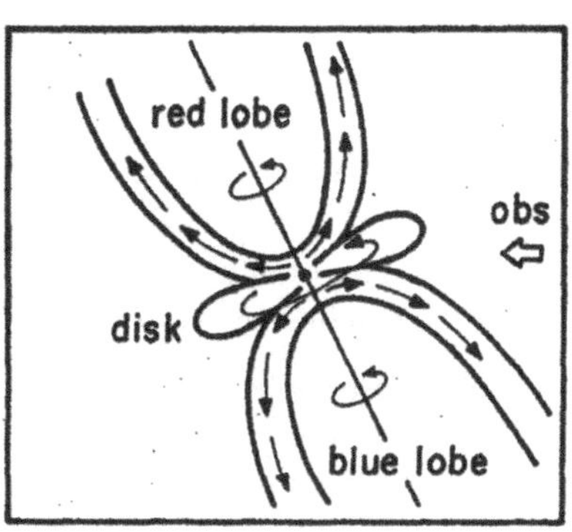

Fig. 7.38 Schematic picture of L1551 (Uchida et al. 1987)

In Japan, first observations were made in the near IR region by Okuda Haruyuki's group in 1971, using the 188 cm and 90 cm telescopes of the OAO (Okuda and Maihara 1971; Maihara et al. 1972). At the Agematsu IR observatory, Kobayashi Yukiyasu and Kawara Kimiaki observed the polarization of IR emission in the Galactic Center (Kobayashi et al. 1980).

In 1990, Okuda and Shibai Hiroshi made oversea IR observations using the IR telescopes at Mauna Kea and found a compact cluster of five luminous infrared sources, which was called the Infrared Quintuplet. These infrared sources were later explained as a cluster of Wolf-Rayet type massive stars (Okuda et al. 1990; Figer et al. 1999).

Radio observations of the Galactic center started in the 1980s in Japan by Kaifu Norio's group, based on the 6-m telescope at the TAO. Subsequently the Nobeyama 45-m telescope was used actively. In the 1990s, X-ray observations with GINGA and ASCA satellites were widely carried out by Koyama Katsuji and his group. Theoretical modellings of activities in the Galactic center were constructed mainly by Uchida Yutaka, Shibata Kazunari, and Sofue Yoshiaki. With these works, the complicated and active nature of the central region has been gradually clarified. Some of remarkable studies include the followings.

7.7.5.1 Expanding Rings

By making use of the 21-cm line data, obtained with the Parkes radio telescope in Australia, Kaifu Norio, Iguchi Tetsuo and Kato Takaji found an expanding ring of neutral hydrogen at 270-pc from the Galactic center in 1974. They supposed that this ring may be an indication of the past explosion of the nucleus of the Galaxy. On the 21-cm data, they also found expanding rings at 2.4 and 3 Kpc from the center as shown in Table 7.5. They noticed that, if the 2.4 Kpc and 3 Kpc rings were the remnants of shock waves caused by past explosions in the Galactic center, the nucleus of our Galaxy might have experienced at least three explosions in these 10^7 years (Kaifu et al. 1974).

7.7.5.2 Radio Lobe and Radio Arc

In the Galactic center there are some active regions, above and below the galactic plane. The typical samples are the radio lobe and the radio arc.

Table 7.5 Comparison of three Galactic neutral hydrogen rings (Kaifu et al. 1974)

Radius	Expanding vel. km s^{-1}	Rotating vel. km s^{-1}	Structure
270-pc	130	50	Ring, tilted
2.4-Kpc	135	151–170	Incomplete ring, tilted
3-Kpc	50	200	Arm-like

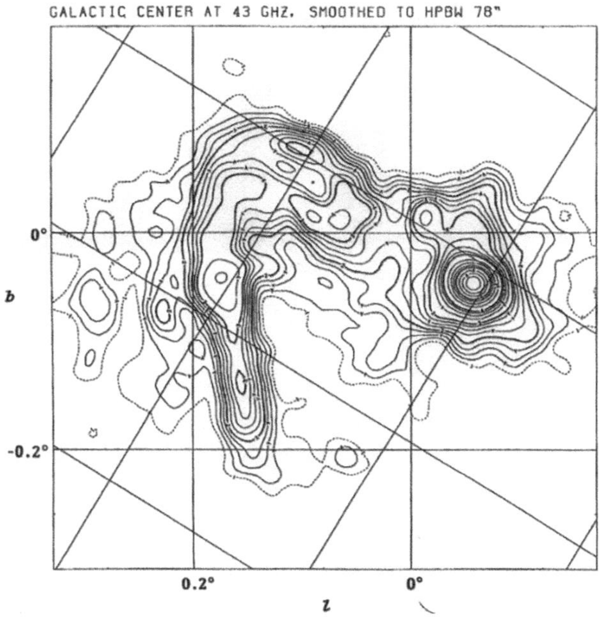

Fig. 7.39 The galactic center and the radio arc observed at 43-GHz continuum of the NRO (Sofue et al. 1986b)

Sofue Yoshiaki and Handa Toshihiro (1984) and their group observed central region of the Galaxy using the Nobeyama 45-m telescope at 10 GHz continuum. They found two spurs forming a giant Ω-shaped radio loop which was called the central galactic lobe (CGL) (Sofue and Handa 1984; Sofue et al. 1986a). The flat spectrum suggests a thermal radiation from ionized gas. A survey of CO line emission showed the high-velocity feature of the lobe expanding vertically from the Galactic plane. Uchida Yutaka and his group (1987) proposed a magneto-dynamic mechanism to produce the CGL by the combined effects of rotating disk and magnetic field near the galactic center (Sofue 1984a, b, 1985; Uchida et al. 1985).

Radio arc was observed by Sofue Yoshiaki, Inoue Makoto and others in 1986 at 43-GHz continuum with the Nobeyama 45-m telescope. The observed map is shown in Fig. 7.39, where the circular source in right side denotes Sgr A (near the galactic center), and the curved source is the radio arc. The arc runs almost perpendicularly to the galactic plane with the nonthermal radio spectra. Sofue's group argued that, if the emission is due to synchrotron radiation, the lifetime of cosmic-ray electrons emitting at 43-GHz is as short as 10^6 years. This infers that the radio arc may be among the youngest active objects in the galactic center (Sofue et al. 1986b; Tsuboi et al. 1995).

7.7.5.3 X-Ray Structure

In the 1980s, when HAKUCHO and TENMA made X-ray survey for a wide field of the Galaxy, no remarkable X-ray source was found in the Galactic center. It was

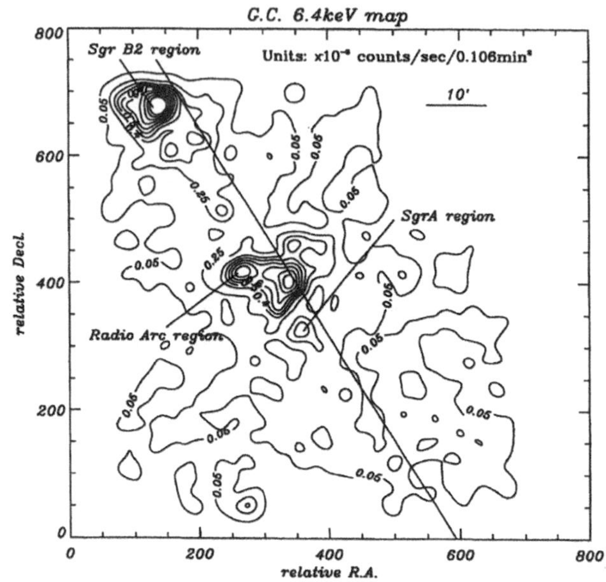

Fig. 7.40 X-ray map of the 6.4-keV line. Some relations with radio sources are indicated. The galactic center is near the Sgr A region (Koyama et al. 1996)

GINGA that found a strong X-ray source in the center from its high spatial resolution. Koyama Katsuji and his group observed the galactic center with a highly ionized iron-emission line at 6.7 keV in 1989 and discovered an optically thin hot plasma cloud, with temperature as high as several 10 to 100 million-degrees. On the origin of such high-temperature plasma, they suggested shock-wave heating caused by a past violent and variable activity of the galactic center (Koyama et al. 1989; Yamaguchi et al. 1990).

The next satellite ASCA has a capability to observe the Galactic center with higher spatial and spectral resolutions in the energy range up to 10 keV. In spectral observations, Koyama Katsuji, Maeda Yoshitomo, and their group found several emission lines of highly ionized metals such as S XV, S XVI, Ca XIX, Fe XXV, Fe XXVI, emitted from diffuse thermal clouds. The mapping observation was made as shown in Fig. 7.40, where there are two strong X-ray sources along the galactic plane. Northern source coincides with radio source Sgr B2, and southern source lies near the galactic center (Sgr A). These sources coincide with the main part of the galactic radio arc (Koyama et al. 1996).

7.7.5.4 Central Black Hole

In 1981, R. H. Sanders of Groningen University of Netherlands proposed a model of recurring activity in normal galactic nuclei based on the assumed presence of massive black holes. J. M. Moran of Harvard-Smithsonian Center for Astrophysics, and his group including Miyoshi Makoto, Nakai Naomasa and Inoue Makoto of the NAOJ performed observations of water masers for about 300 galactic nuclei, and

Table 7.6 Galaxies with possible massive black holes (Moran et al. 1996)

	Hubble	Distance	Massive black hole		
			Mass	Diameter	Density
Galaxy	Type	Mpc	$10^6 M_\odot$	pc	$10^6 M_\odot pc^{-3}$
Our Gal.		0.0055	2	0.3	90
M32	E2	0.7	2	0.7	12
M31	Sb	0.7	30	0.7	160
NGC 4258	Sb	6.4	35	0.26	3800
NGC 3115	S0	8.4	1000	10	2
NGC 4594	Sb	9.2	500	18	0.2
NGC 3377	E5	9.9	80	24	0.01
M87	E6	15	2400	36	0.1
NGC 4261	E2–3	30	1200	14	0.5

they found the presence of highly rotating molecular disks in 16 galaxies. Among them, they suggested the existence of massive black holes in nine galactic nuclei, including our Galaxy and the Andromeda galaxy (M31).Table 7.6 gives the list of galaxies with possible massive black hole candidates (Moran et al. 1996).

As seen in Table 7.6, massive black holes widely exist in the centers of various types of galaxies including normal and active galaxies.

7.8 Galaxies

7.8.1 Observations in Early Postwar Phase

The 188-cm telescope of the OAO, particularly by the installation of the image intensifier at the Cassegrain focus, opened a new window to the physics of galaxies in Japan in 1969.

In the early 1970s, Karoji Hiroshi and Kodaira Keiichi of the University of Tokyo observed an extremely compact galaxy CGCG 1622–4112 (Catalogue of Galaxies and Clusters of Galaxies, Zwicky and Herzog 1966). This galaxy appears almost stellar, surrounded by a faint halo. Emission lines in Balmer series, HeII, and forbidden lines of [OIII] and [NIII] were measurable. From the analysis of these lines, they found that this galaxy is closely related to Seyfert galaxies in its activity, while the absolute magnitude is low in the range of normal galaxies (Karoji and Kodaira 1972).

In Kyoto, Sakka Kazushi, Wakamatsu Ken'ichi and Oka Shozaburo carried out imaging, photometric and spectroscopic observations of a hot-spot-nucleus galaxy NGC 2782, which was once classified as a member of Seyfert galaxies, though questioned thereafter. NGC 2782 has 4 hot spots near the center, and one of them coincides with its galactic center. These spots were large scale ionized clouds excited by a lot of OB stars. The central spot was furthermore a source of strong infrared and nonthermal radio emission. From these behaviors, Sakka's group suggested that, while the energy scale is comparable with Seyfert galaxies, the form of activity is rather different (Sakka et al. 1973). Kyoto group continued the observations of peculiar galaxies such as Haro galaxies and polar-ring galaxies up to the 1980s at the OAO. The polar-ring galaxy is a type of active galaxy characterized by an outer ring of stars and gas. The ring is vertical to the galactic plane and rotating around the center of the galaxy.

7.8.2 Normal Galaxies

7.8.2.1 Andromeda Galaxy (M31)

M31 is the nearest giant spiral galaxy, composed of central spheroidal bulge and outer disk component.

In Japan, first observations of the central bulge of M31 were made by Matsumoto Toshio, Murakami Hiroshi and Hamajima Kiyotoshi (Nagoya University and TAO) in 1977. They observed at 7000–9000 Å and 2.2 μ m, using the 100-cm Agematsu Infrared Telescope. They measured the position angle (PA) and axial ratio (a/b) of the spheroid, and found that both PA and a/b exhibited larger values in 2.2 μ m than in λ7000–9000 Å band. Since infrared light comes from late-type giants, they suggested that the larger values of these parameters infer the distribution of evolved giant stars which form a more elongated prolate bar-like structure in the bulge of M31 (Matsumoto et al. 1977).

For the outer disk component, Sawa Takeyasu and Sofue Yoshiaki calculated the spatial distribution of neutral hydrogen gas in M31, based on the HI data obtained by the 100-m radio telescope at the Max Planck Institute for Radio Astronomy. In this analysis, they applied the density-wave theory to determine the positions of spiral arms. The result is shown in Fig. 7.41, where several spiral arms are visible, making a two-armed trailing spiral structure especially at radii greater than 15 kpc from the center (Sawa et al. 1981). In 1985, Ichikawa Takashi, Nakano Makoto and their group observed the southern arm of M31 with CO line using the 45-m telescope of the NRO. They found conspicuous dark clouds and a complex of HII regions located in the arm, where CO emitting regions are likely avoiding HII regions (Ichikawa et al. 1985).

Makishima Kazuo, Ohashi Takaya and their group observed a wide field of M31 with GINGA in 2–20 keV range of X ray, and they detected a weak X-ray source at the center of M31. The observed X-ray spectrum showed the spectra of low-mass X-ray binaries. From these results they suggested that the low-mass X-ray binaries dominate in the central part of M31 (Makishima et al. 1989).

Fig. 7.41 Photograph of Spiral-arm structure of M31. A two-armed spiral pattern is recognized in the outer part (Sawa et al. 1981)

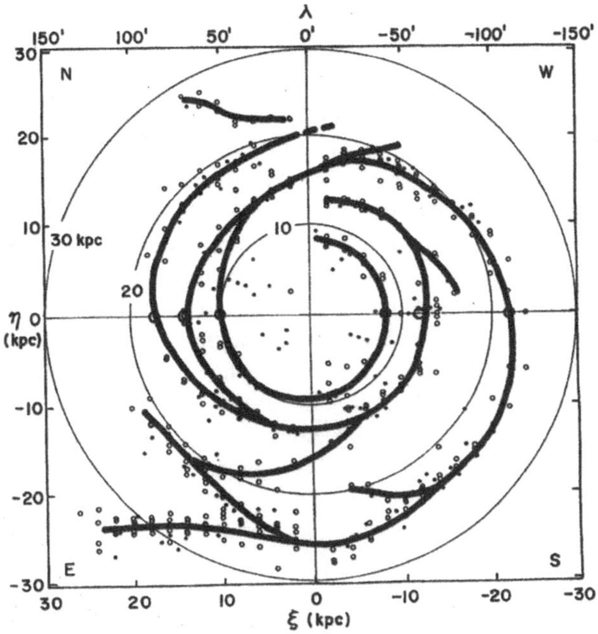

7.8.2.2 Edge-on Galaxies

Edge-on galaxies are the spiral galaxies seen from the plane of spiral arms. These galaxies provide the information on the vertical structure of spiral galaxies. Observations had been made mainly by Sofue Yoshiaki, Handa Toshihiro and their group in the early 1990s with the 45-m telescope using the CO lines.

In the edge-on galaxy NGC 891, Sofue's group found a prominent spur of molecular gas emerging vertically from the disk, at 5.5 kpc from the center of the galaxy. The height of the spur from the galactic disk was 520 pc, and the velocity width was less than 20 km s^{-1}. On the origin of spur ejection, they supposed the possibilities of a supper-bubble in the disk, or of some instability in the gas-disk, though not conclusive (Handa et al. 1992). On the structure of gas disk, they found the presence of a molecular gas ring at 3.5 kpc and a rigidly rotating inner-disk or a bar of radius of 3 kpc. From these properties they concluded that NGC 891 is a SBb galaxy in the Hubble type. In addition, they found that the thickness of the CO disk is larger than that of our Galaxy by factor 2. They suggested that the geometrically thick CO disk of NGC 891 might be due to active star formation in the disk (Sofue and Nakai 1993; Handa et al. 1994).

Other edge-on galaxies, NGC 3079, NGC4565, NGC 4631 and NGC 5907, were similarly observed by the same group in the CO molecular lines in the early 1990s.

7.8.3 Seyfert Galaxies

Seyfert galaxies are one of the most active spiral galaxies. They are characterized by the existence of very luminous nuclei with strong emission lines. Their active behaviors were first remarked by C. Seyfert of the Mt. Wilson Observatory in 1943.

In Japan, Seyfert galaxies were observed widely in optical and infrared regions in the 1970 - 1980s. Kaneko Noboru, Ohtani Hiroshi and their group (in 1968–1972) made spectroscopic analysis of emission lines and infrared continuum from the central region of NGC 1068. They found a star-like nucleus emitting a hot black-body radiation. Kodaira Keiichi and Iye Masanori and others (1979) observed a cluster of galaxies, VV 150, and Seyfert galaxy NGC 3227. The brightest member of VV 150 (VV150 A) exhibited Seyfert-type spectrum, resembling to NGC 3227 in its spectral feature.

Taniguchi Yoshiaki, Kameya Osamu and Nakai Naomasa (in 1990–1992) carried out CO molecular-line observations of nuclear regions of starburst galaxies and Seyfert galaxies using the Nobeyama 45-m telescope. They showed that the CO gas made a ring structure both in starburst galaxies and Seyfert galaxies. Based on these observations, Taniguchi proposed a starburst-implosion model, that is, molecular gas rings are active SFRs where the high velocity gas is formed by successive supernova-explosions inside these rings. The ejected gas partly falls into the galactic center as the steady energy source of active galactic nuclei (Taniguchi 1992).

Seyfert galaxies were classified into two types, Seyfert 1 and 2, based on the presence or absence of broad emission lines in the optical spectra. This classification, however, soon becomes questionable. In 1985, J. R. Antonucci (National Radio Astronomical Observatory) and J. S. Miller (Lick Observatory) carried out optical polarimetric observations of an archetypical Seyfert-2 galaxy NGC 1068, and found broad emission lines in the wing parts of Hα and SII lines. These wings are highly polarized, as compared to the central narrow parts in low polarization which is a typical feature of Seyfert 2 galaxy. This discovery gave a motivation to the proposal of the unified model of Seyfert galaxies with a single parameter, that is, the viewing angle of the active galactic nucleus hidden within a torus of narrow-line emission region. In 1991, Awaki Hisamitsu, Koyama Katsuji and their group observed 16 Seyfert galaxies, including 9 Seyfert-2 galaxies, with GINGA satellite. They showed that the X-ray properties can be explained by the unified model of accretion-torus as shown in Fig. 7.42. In this model strong X-ray radiation is beamed within a small solid angle toward the rotation axis of the torus, and weak soft X-ray is emitted from the dotted region in the Figure. The dark region represents the torus emitting narrow optical lines. The variety of Seyfert galaxies is possibly explained by the different view angle. The size and density parameters can take different values from galaxy to galaxy (Awaki et al. 1991).

In 1992, Mouri Hideaki and his group conducted observations of infrared [Fe II] 1.644 μ m line emission in Seyfert galaxies, NGC 1068 and NGC 4151, and they found the stratification of emission regions from inner ionized region to outer neutral

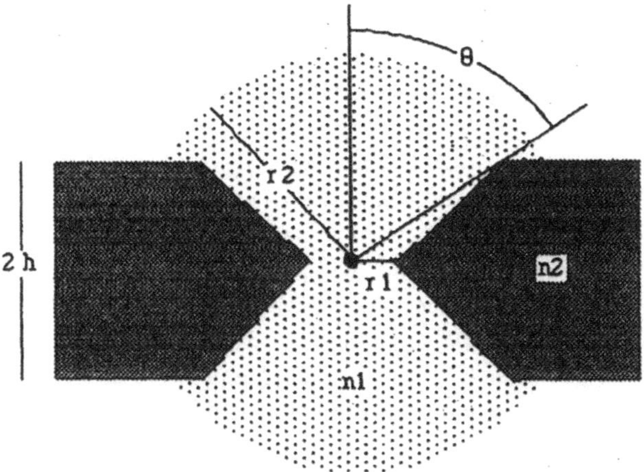

Fig. 7.42 Cross section of the torus model of Seyfert galaxies (Awaki et al. 1991)

region via transition region. The [Fe II] emission together with [O I] line are formed in the transition region excited by electron collisions, whereas Hα and [N II] lines are produced in the ionized region (Mouri and Taniguchi 1993).

In 1994, Aoki Kentaro, Ohtani Hiroshi and their group performed spectroscopic observations of the extended emission-line region of the Seyfert 1 galaxy NGC 3516 with the Spectronebulargragh of the OAO, which is an area spectroscopic-system to sweep images in the H$\boldsymbol{\alpha}$, [OIII] and other emission lines. They thus confirmed the existence of both the narrow and broad emission-line regions. With their results they posed a question to the simple unified scheme of Seyfert nuclei (Aoki et al. 1994).

Nakai Naomasa and his group (1995) searched for water-maser emission with extremely high radial velocity, using the 45 m Nobeyama telescope and the 64 m Parkes telescope in Australia.

They surveyed 18 Seyfert galaxies, among which maser emission was detected in seven Seyfert galaxies with a velocity width of range around ± 1000 km s^{-1}. They thus found that the maser features are consistent with a compact disk which is rotating around a massive black hole at the galactic center (Nakai et al. 1995).

Taniguchi Yoshiaki and Murayama Takashi also considered the physical properties of dusty tori of Seyfert nuclei probed by the H_2O maser emission at 22 GHz observed with the US Very Large Allay (VLA) and US Very Long Baseline Array (VLBA). Assuming a simple cylinder model for the dusty torus, they derived that the tori structure as vertically thin, and its outer radius around several to ten parsecs. However, the observed maser regions are concentrated in the inner 1 pc regions of the torus. For this property, they argued that only the inner region has necessary physical conditions enough to cause the maser emission, where the temperature is as high as several hundred K, and the density is as high as ~10^{10} cm^{-3} (Taniguchi and Murayama 1998).

7.9 Cosmology

7.9.1 Nariai and Cosmology

In the early postwar period, the modern cosmology was based on Einstein's general relativity as mostly promoted by Nariai Hidekazu (成相秀一, 1923–1990) (Tomita 1991).

Nariai was born in Daisha (Izumo City, Shimane Prefecture), and studied astrophysics under Matukuma Takehiko of Tohoku University (See Chap. 5). After graduation in 1947, he started cosmological study almost by himself to derive a new solution of Einstein's field equation. In 1950, he found a solution for the homogeneous and static universe with spherical symmetry, including the cosmological constant. His solution was eventually in parallel with de Sitter's model of an empty and expanding universe with positive cosmological constant. Nariai showed in 1952 that, while both were mutually independent having different metrics in space-time, both solutions take the same form only when the propagation of light is treated in the radial direction. It was in 1984 that Nariai unified both solutions in higher dimensional space-time and he considered its characteristics in detail (Fig. 7.43).

In 1953, he moved to the Research Institute for Theoretical Physics (理論物理学研究所), Hiroshima University, at Takehara City, as a research assistant and promoted to an associate professor in 1962 and a professor of cosmology in 1972.

In 1955–1956, Nariai considered the homogeneous cosmic turbulence in an expanding medium. For this purpose, he derived the generalized Navier-Stokes equations based on the general relativity. He thus built a general theory of homogeneous turbulence for the quasi-incompressible fluid. Main features of his dynamical equations were the appearance of a new dissipation term and the preservation of kinematic viscosity both due to the cosmic expansion. The effects of dissipation and viscosity were different for different cosmic scales of turbulence. Nariai compared

Fig. 7.43 Nariai Hidekazu at his retirement lecture in 1987 (by author)

three types of cosmic scale of turbulence: large, intermediate, and small, and showed that the intermediate scale was the only case to be taken as the trigger of the galaxies-formation. In the 1960s, Nariai and his colleagues including Tomita Kenji and Kimura Toshiei developed the theory of the formation of proto-galaxies by the gravitational instability in the cosmic turbulence in the hot big-bang universe.

In the 1960s and thereafter, Nariai immersed himself in the studies of theoretical physics and cosmology (Nariai 1984). His main subject areas include the followings.

1. Quantization of gravitational field in the expanding universe (in 1962–1964).
2. Creation of particles in Hoyle's cosmology and in other several types of universe (in 1964–1982)
3. Gravitational instability in an expanding universe (in 1965–1973)
4. Generation of sound from the primordial turbulence in an expanding universe (in 1973–1974)

Quantization of gravitational field is important in the stage of very-early universe. Nariai's works had the basic significance in the theory of inflation universe. Other subjects were also basic ones and have been carried out under collaboration with his colleagues and students in the respective ages (Nariai 1984). After some illness, he died at age 64 in 1990 at Takehara City.

7.9.2 Development of Cosmology

Nariai's study of cosmology in Japan has been succeeded and extended by the next generation since the 1970s. In Kyoto, the works of Tomimatsu Akira, Sato Humitaka and Sato Katsuhiko are to be noticed, all of them were the disciples of Hayashi Chushiro in Kyoto University.

The exact solution of Einstein's field equations in general relativity was first discovered for a point mass by Karl Schwarzschild in 1916. In the next year, Hermann Weyl extended the exact solution to the field around a deformed mass (axially symmetric mass). In 1963, Roy P. Kerr generalized Schwarzschild's solution to a rotating mass. Kerr's solution just describes the space-time around a rotating black-hole.

In succession, Tomimatsu Akira and Sato Humitaka derived the fourth solution by adding a rotation term to Weyl's solution known as T-S solution (Tomimatsu and Sato 1973). That is,

Schwarzschild solution → rotation→ Kerr solution
↓
deformation
↓
Weyl solution → rotation→ T-S solution

In this scheme, Tomimatsu and Sato found that the space-time in their solution has ring singularities outside the event horizon, that is, there exist naked singularities. They thus suggested that the naked singularity may act as an active energy source of gravitational waves.

Sato Katsuhiko, moved from Kyoto University to the University of Tokyo, was one of the pioneers of the theory of an inflation universe in the early 1980s (Sato 1981).

The big-bang theory, posited by G. Gamow, was widely accepted and its standard theory on the origin of our universe was constructed in the 1970s. In those years, elementary-particle physics was also markedly advanced, and this new physics pointed out several difficulties lying in the standard big-bang theory, particularly on the origin of the large-scale homogeneity, and on the small-scale inhomogeneous structure of the universe. In order to overcome these difficulties, Sato Katsuhiko proposed a model of the universe exponentially expanding at its very-early phase (Sato 1981). In the same year, Alan Guth of the Stanford University presented the similar model and coined the name *the inflation universe*. This naming has been widely accepted.

According to the inflationary universe model, the universe underwent extreme supercooling. As the result, the universe approaches to a false vacuum, which is a very peculiar state of matter, the energy density of the vacuum remains constant even as the universe expands, and gravity works as the repulsion. The gravitational repulsion would produce a very rapid expansion as compared to the expansion of the standard model. This is the inflation period, which occurred in approximately between 10^{-34} to 10^{-32} s after the big-bang explosion. The scale factor of expansion reached as high as 3×10^{43} in R ($\propto e^{t/T}$), accompanied by strong cooling and reheating. Thereafter, the false vacuum changed to a stable vacuum. Comparison of the evolution of standard big-bang and inflation models are schematically shown in Fig. 7.44.

The inflation theory became confirmed by the observations of cosmic background radiation by COBE satellite launched in 1989 by NASA. In the inflationary epoch, the state of matter is supposed to be in quantum fluctuation. After the inflationary period, the fluctuations are imprinted in the cosmic microwave background. The whole sky map obtained by COBE indicated general homogeneity of the universe and its small-scale fluctuation. The amplitude of fluctuation is of the order of 10^{-5} and this can be supposed to be the remnant of quantum fluctuation. These features are well explained by the inflation cosmology.

7.10 Theoretical Astrophysics

7.10.1 Age of Cosmic Gas Dynamics

In August 1948, a symposium on cosmical aerodynamics was held in Paris under the co-sponsors of the IAU and IUTAM (International Union of Theoretical and Applied Mechanics). The purpose of the Symposium was to introduce to each other the gas

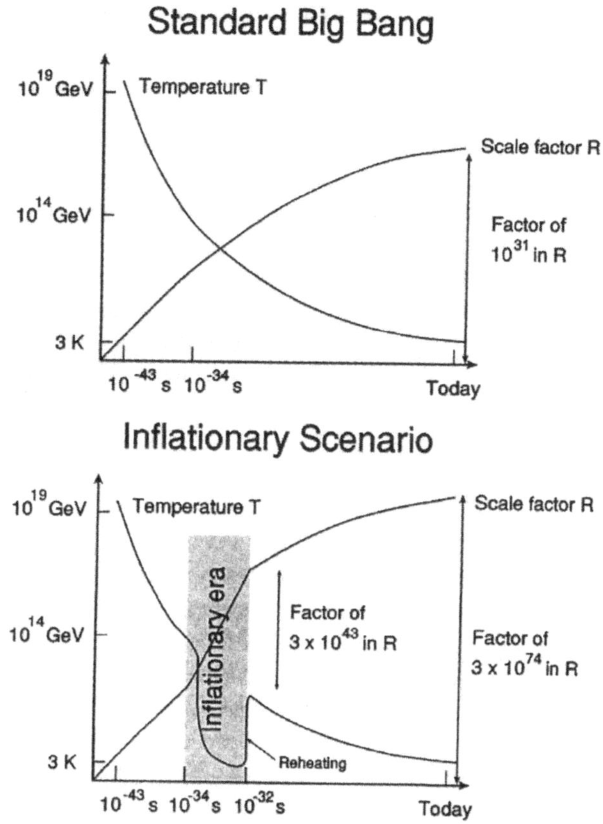

Fig. 7.44 Comparison of the evolution of the scale factor and temperature in the Standard Big-Bang and Inflationary cosmology (Longair 2006, Fig. 16.2)

dynamics in the cosmic scale, that had been unfamiliar to both astronomers and physicists. Main topics were basic theories of shock waves, turbulence, and dynamics of expanding shells of novae and supernovae. Since the 1920s, astrophysicists had been worked in the spectroscopy and stellar structure based on quantum mechanics and nuclear physics, and few people interested on the gas-dynamical phenomena of the universe. In this sense, this symposium was to declare the opening of the age of new cosmic gas dynamics in astrophysics.

In Japan, Ono Yoro and Sakashita Shiro and their group of Hokkaido University developed the theory of shock waves under the title of "Propagation of shock waves in inhomogeneous gases ", (in 1960–1969), mainly on the shock waves in stellar interior. Kogure and Osaki Toru of Kyoto University worked on the structure and propagation of shock waves in stellar atmospheres in 1961–1962. Unno Wasaburo and his group of the University of Tokyo also contributed markedly to the development of cosmic gas dynamics.

The age of cosmic gas dynamics tapered off at the end of the 1960s with the development of computer technique. Since then, the present style of astrophysics based on numerical analysis and modelling has taken over the classical cosmic gas dynamics.

7.10.2 Unno Wasaburo and Theoretical Astrophysics

Theoretical astrophysics in postwar period was actually initiated by Unno Wasaburo (海野和三郎, b. 1925) of the University of Tokyo. He was born in Urawa City (present Saitama City) and entered the University of Tokyo where he studied astrophysics under Hagihara Yusuke (Chap. 3). After graduation, he continued studies of astrophysics, and was promoted to an associate professor in 1952 and to a professor of astronomy in 1963 at the University of Tokyo. He often stayed at foreign institutes, such as Princeton University Observatory, New Jersey; University of Michigan Observatory, Ann arbor; Max Planck Institute for Physics, Munich; and Commonwealth Scientific and Industrial Research Organization, Sydney, in the 1950s and 1960s. After retirement from the University of Tokyo in 1986, he was appointed as a professor of Kinki University in Osaka, where he continued his work for many years (Fig. 7.45).

Unno's work on theoretical astrophysics over his lifetime covered the following three main areas with some overlapping of time:

(a) Planetary nebulae (in 1950–1961)
(b) Solar physics (in 1956–1961, 1974–1991)
(c) Stellar physics (in 1961–1998).

Unno conducted these works mainly by his own as well as in collaboration with his young colleges, Kato Shoji, Uchida Yutaka, Shibahashi Hiromoto, Osaki Yoji, and others.

Fig. 7.45 Photograph of Unno Wasaburo (ca 1995) (by author)

7.10.2.1 Planetary Nebulae

Initially, V. A. Ambarzumian of Byurakan Observatory derived the radiation pressure by assuming a rectangular contour of the $L\alpha$ flux in 1932. According to his calculation, the radiation pressure is quite strong and blows up the whole nebular matter. Unno started his work from the theoretical derivation of radiation pressure due to hydrogen Lyman-alpha line ($L\alpha$). In 1950, Unno solved the radiative transfer of $L\alpha$ line, by taking the Zanstra effect of redistribution in frequency in the line contour into account, and obtained significantly reduced radiation pressure, enough to avoid the blow-up catastrophe. In highly excited planetary nebulae, hydrogen is almost ionized, and the $L\alpha$ line of ionized helium takes place the main role of nebular motion. Unno (1952) solved the radiative transfer of this line in the same way with hydrogen and again derived the reduced radiation pressure to give the expansion velocity of nebulae in the order of 10 km s^{-1}. Based on these results, Unno and Takakubo Keiya examined the internal motions and stratifications of planetary nebulae in 1952. They showed that high-excitation nebulae are stratified in the order of HeIII, HeII, HeI and HI from inner to outer layers, and expanding velocity increases from 5 to 15 km s^{-1} toward outer layers, which well coincided with observations (Unno 1950, 1952; Unno and Takakubo 1952).

7.10.2.2 Solar Physics

In the 1950s, Unno treated two problems in Solar physics: convection layer and radiative transfer. On the convection layer, he derived the production rate of acoustic waves by turbulence and showed that the acoustic energy thus produced may be enough for the heating of the solar corona (Unno and Kawabata 1955). On the radiative transfer, Unno (1956) considered the line formation in magnetic field such as in sunspots, where Zeeman effects are important in the line formation. He derived and solved the equations of radiative transfer for each Zeeman component and calculated the strength of polarized lines (Unno 1956).

In 1974, Unno' work shifted to the dynamic process of the solar atmosphere: the structure and motion of spicules (1974); active regions of the photosphere (1981), turbulent convection (1981); and nonlocal convection zones (1989). Investigations of the stability of the solar core and the dynamics of nonradial oscillations were also carried out in 1975.

7.10.2.3 Stellar Physics

1. Dynamical process in stellar atmospheres

Under collaboration with Kato Shoji and Makita Mitugu, Unno considered the condition of the onset of thermal convection in polytropic atmosphere. This problem had a long history since John W. Strutt (Lord Layleigh) (1916) and widely treated by many authors. One of the difficulties is laid on the large density gradient in the undisturbed atmosphere. Unno et al. (1960) proposed a variation method for computing the condition of the onset of thermal convection. With this method they derived a nondimensional parameter which determines the condition of marginal stability for polytropic atmospheres.

2. Radial and nonradial oscillations

In the 1960s, Unno mostly worked on the stabilities of stars for radial and nonradial pulsation under collaboration with Kamijo Fumio and Kato Shoji. These works became the basis of later development in this field, as seen in Sect. 7.3.

3. Supermassive objects.

In the late 1960s, Unno, along with Osaki Yoji and Kato Shoji, paid attention to the super-massive objects as a possible origin of extremely energetic objects. In 1963, Fred Hoyle and William A. Fowler suggested the possibility of super-massive stars as the origin of strong radio sources. Since then, several investigations have been made on this subject. Among them, Osaki (1966) studied the evolution and pulsation of super-massive stars and found that the masses larger than $10^{5.5}$ solar mass eventually continue to collapse towards the Schwarzschild limit, whereas masses smaller than this level underwent pulsation of large amplitude. In 1971, Unno considered the physical state and stability of super-massive objects which were defined as a homogeneous mixture of gas and stars. He supposed that the evolution of a super-massive object takes three steps: (1) contraction of gas, (2) dissipation of stellar kinetic energy, and (3) formation of dense stellar system. The structure and stability at each step were analyzed in detail. Unno suggested that when mass is higher than 10^6 solar mass, this object possibly evolves to a quasar (Osaki 1966; Unno 1971).

7.10.3 *Development of Theoretical Astrophysics*

7.10.3.1 Accretion Disks and Jets

Disk phenomena are ubiquitous in stars and galaxies. Disks are formed mostly by accretion flows from the accompanying objects. The theoretical study of accretion disks was first conducted by Nikolai Shakura and Rashid Sunyaev in USSR in 1973. They showed that, although the nature of viscosity was not well understood, basic properties of thin accretion disks were explained independent of the specific form of the viscosity, and their model was called the standard model, and this model was successfully applied to many active objects. A general relativistic version was also described in 1973 by I. D. Novikov and K. S. Thorne.

In Japan, theoretical studies on accretion disks began soon after. Osaki Yoji proposed an accretion model for the outbursts of dwarf novae (Osaki 1974), Kato Shoji and Fukue Jun examined trapped radial oscillations of accretion disks around a black hole (Kato and Fukue 1980). Kyoto group, including Matsumoto Ryoji, Okazaki Atsuo, Fukue Jun, and Kato Shoji, examined the characteristics of topological structure and stability of the transonic flows towards black holes, and further the wave trapping nature of relativistic disks.

Standard disks were found to be unable to describe all types of accretion phenomena around compact objects. Models of slim disks were thus developed in the end of 1980s by Abramowicz group in Europe (Abramowicz et al. 1988). Mineshige and Ohsuga group in Kyoto took much attention to this model in relation to ultra-luminous X-ray sources (ULXs). Super-Eddington accretion flows and outflows (including jets) were extensively studied by their group (Ohsuga et al. 2005). As another type of accretion flows, the advection-dominated accretion flows (ADAFs) were extensively developed in 1990s by Narayan group in Harvard, Abramowicz group in Europe and by Kato's group in Japan (Mineshige and Kato 1996; Kato 2000). ADAFs were cooled by advection (heat captured by matter rather than radiation), so that ADAFs were much less luminous than the standard models. ADAFs were first proposed in 1977 as a model of Cyg X-1 by Ichimaru Setsuo (Ichimaru 1977), but his priority had been forgotten in the major groups of accretion disk community till a meeting at Lund in Sweden (or at Reykjavik in Iceland) in the end of 1990s, where Roland Svensson pointed out priority of Ichimaru's work.

Disks around the young proto-stars and galactic nuclei were also studies by Uchida Yutaka and Shibata Kazunari in 1989–1990. They showed that, in every case, magnetic field plays essential role in the formation of disks and jets (Uchida et al. 1990; Shibata and Uchida 1989).

Hydrodynamical and radiation-hydrodynamical models of bipoloar jets, disk winds and shocks have been examined in various aspects by Fukue Jun of Osaka Kyoiku University (Fukue 1987, 1991) and his group in the 1990s.

7.10.3.2 Molecular Clouds and Star Formation

The average stellar mass is around one solar mass, whereas the mass of original molecular cloud reaches as high as several thousand solar mass. In the formation of stars, the molecular clouds must be fragmented by the effects of self-gravity in some stage of evolution. Miyama Shoken and Eriguchi Yoshiharu accomplished a simulation for the fragmentation of isothermal sheet-like molecular clouds. The result of simulation is shown in Fig. 7.46, where one may see that the initial perturbations grow to form filamentary structure and stars are born from these dense filaments (Miyama et al. 1989; Miyama and Eriguchi 1990).

7.10.3.3 Globular Clusters and Galaxies

In the context of N-body simulation, merging of binary globular clusters and colliding spherical galaxies are the same problem.

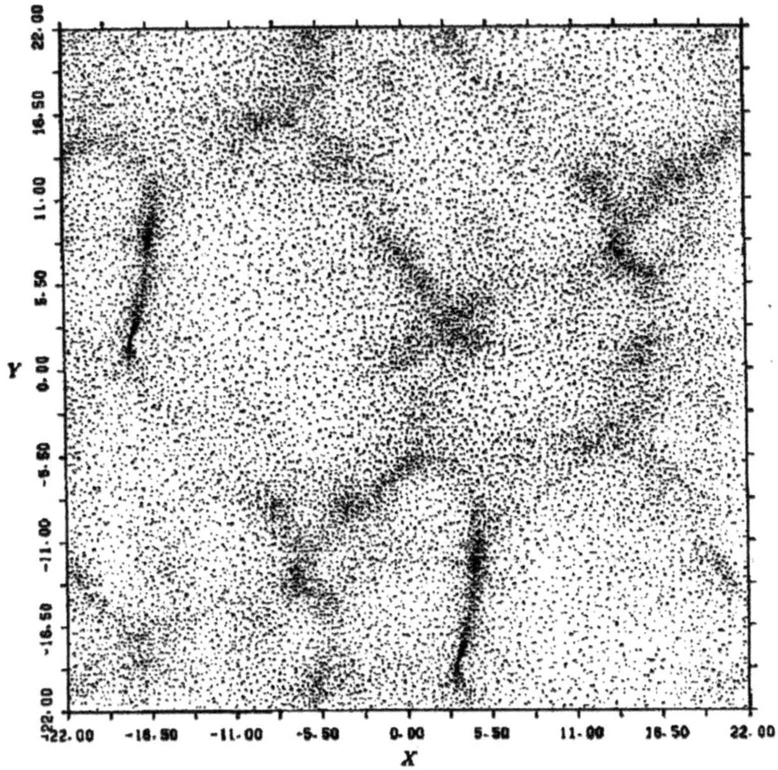

Fig. 7.46 Distribution of particles projected onto a plane. The original sheet-like cloud is fragmented into the filamentary structure (Miyama and Eriguchi 1990)

The simulations of a colliding process of galaxies have been promoted in some cases with the computer GRAPE. An example of simulations for colliding galaxies is illustrated in Fig. 7.47. The upper panel denotes the initial state of two spherical galaxies, both composed of 8000 particles. The middle panel shows the two galaxies approaching through the rejection of particles between galaxies, and the lower panel denotes the final merged galaxy, which resembles to an elliptical galaxy in its elliptical form and distribution of stars in the halo. It is suggested that some elliptical galaxies have their origin in this type of galaxy collision (Ebisuzaki et al. 1990; Sugimoto et al. 1991).

7.10.3.4 Disk Galaxies

The structure and stability of disk stellar systems had been usually studied by making use of the N-body simulations. A different approach based on collisionless Boltzmann simulation (originally called Vlasov equation) was promoted by Kato Shoji's group, including Inagaki Shogo, Watanabe Yoshio, Nishida Mitsugu and

Fig. 7.47 Numerical experiment of the merging of binary galaxies (Ebisuzaki et al. 1990)

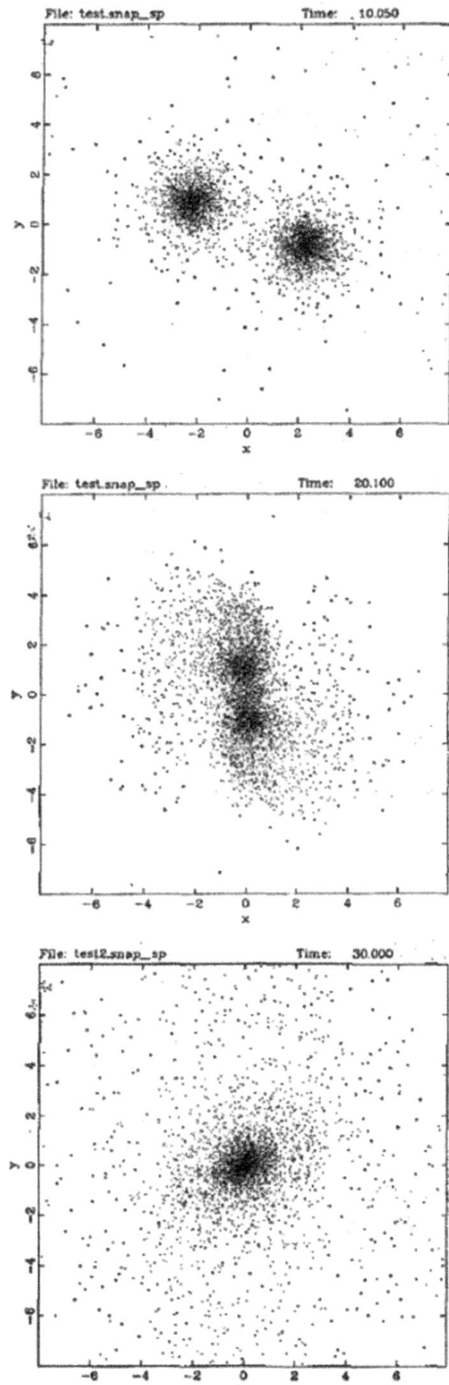

others in Kyoto University in around 1980 (Watanabe et al. 1981). Watanabe's group numerically solved the collisionless Boltzmann equation describing axisymmetric disk galaxies, combined with the Poisson equation for gravitational potential field. They found two different solutions, one was the perturbations which oscillate with time, and the other was the ones which grow up exponentially and saturate at some stage due to increase of velocity dispersion. Nishida's group extended the above axisymmetric simulation to a nonaxisymmetric one (Nishida et al. 1981). They found that a fan-shaped density perturbation grows up pertinently to some saturated state. Studies of disk galaxies by collisionless Boltzmann equation have not been much developed in the 1990s, but some attempts were promoted to study the time evolution of neutrino gas by making use of the collisionless Boltzmann equation.

References

Aikawa, T. (1991). Hydrodynamic pulsation models of yellow supergiant stars in the post-asymptotic giant branch. *Astrophysical Journal, 374*, 700–706.

Aoki, K., Ohtani, H., et al. (1994). Area spectroscopy of the extended emission-line region in the Seyfert galaxy NGC 3516. *Publications of the Astronomical Society of Japan, 46*, 539–552.

Asai, K., Dotani, T., et al. (1996). ASCA observations of soft X-ray transients in quiescence: X1608-52 and Cen X-4. *Publications of the Astronomical Society of Japan, 48*, 257–263.

Asai, K., Dotani, T., et al. (1998). ASCA observations of transient X-ray sources in quiescence. *Publications of the Astronomical Society of Japan, 50*, 611–619.

Awaki, H., Koyama, K., et al. (1991). X-ray implications of a unified model of Seyfert galaxies. *Publications of the Astronomical Society of Japan, 43*, 195–212.

Ballereau, D., & Chauville, J. (1989). Long-term and mid-term spectroscopic variations of the Be-shell star HD 184279 (V1294 Aql). *Astronomy and Astrophysics, 214*, 285–294.

Bath, G. T. (1973). Periodicities and disks in dwarf novae. *Nature, 246*, 84–87.

Campbell, J. A., & Jefferies, W. H. (1970). Equivalence of the perturbation theories of Hori and Deprit. *Report from Applied Mechanics Research Laboratory*, 1–12.

Deprit, A. (1969). Canonical trans formation depending on a small parameter. *Celestial Mechanics, 1*, 12–30.

Ebisuzaki, T., & Nomoto, K. (1986). X-ray spectra of bursting neutron stars and the color-temperature versus luminosity diagram. *Astrophysical Journal, 385*, L67–L70.

Ebisuzaki, T., Ito, T., et al. (1990). A special processes for gravitational many-body problems. GRAPE, *Astronomical Herald, 83*, 224–227. (天文月報、戎崎俊一: コンピュータシミュレーションが切り拓く天文学).

Ebisuzaki, T., Makino, J., & Okumura, S. K. (1991). Merging of two galaxies with central black holes. *Nature, 354*, 212–214.

Edgar, J. A. (1933). The pulsation theory of Cepheid variables. *Monthly Notices of the Royal Astronomical Society, 93*, 422–441.

Figer, D. F., McLean, I. S., et al. (1999). Massive stars in the Quintuplet near the Galactic center. *Astrophysical Journal, 514*, 202–220.

Fujimoto, M. (1966). Gas flows through a model spiral arm. *Astronomical Journal, 71*, 856.

Fujimoto, M. (1968). Modeling of gas flow through a spiral sleeve. *IAUS, 29*, 453–463. (in Russian).

Fujimoto, M., & Murai, T. (1972). Formation of star spots and its connection with pulsars. I. *Publications of the Astronomical Society of Japan, 24*, 269–279.

Fujimoto, M., & Murai, T. (1973). Formation of star spots and its connection with pulsars. II. *Publications of the Astronomical Society of Japan, 25*, 75–90.
Fujimoto, R., Tanaka, Y., et al. (1995). ASCA observations of the supernova remnant W 49B. *Publications of the Astronomical Society of Japan, 47*, L31–L35.
Fukue, J. (1987). A hydrid jet of beam-type and wind-type models. *Publications of the Astronomical Society of Japan, 39*, 895–905.
Fukue, J. (1991). Magnetic pressure-driven jets from a torus. *Publications of the Astronomical Society of Japan, 43*, 131–145.
Handa, T., Sofue, Y., et al. (1992). A molecular spur in the edge-on galaxy NGC 891. *Publications of the Astronomical Society of Japan, 44*, 1227–1230.
Handa, T., Sofue, Y., et al. (1994). A vertical structure of the edge-on galaxy, NGC 891. *Astronomical Society of the Pacific, Conference Series, 59*, 361–362.
Hasegawa, T., Kaifu, N., et al. (1984). CS around Orion KL: A large rotating disk. *Astrophysical Journal, 283*, 117–122.
Hayakawa, S., Matsumoto, T., et al. (1981). Distribution of near infrared sources in the Galactic disk. *Astronomy and Astrophysics, 100*, 116–123.
Hayashi, C. (1950). Proton-neutron concentration ratio in the expanding universe at the stages preceding the formation of the elements. *Progress of Theoretical Physics, 5*, 224–235.
Hayashi, C. (1966). Evolution of proto stars. *Annual Review of Astronomy and Astrophysics, 4*, 171–192.
Hayashi, C. (1981). Formation of the planets. *IAU Symposium, 93*, 113–128.
Hayashi, C., & Nishida, M. (1956). Formation of light nuclei in the expanding universe. *Progress of Theoretical Physics, 16*, 613–624.
Hayashi, C., Hoshi, R., & Sugimoto, D. (1962). Evolution of stars. *Progress of Thoretical Physics, Supplement. No. 22*, 1–183.
Hayashi, C., Nakagawa, K., & Adachi, I. (1970). Long-term behavior of planetesimals and the formation of the planets. *Publications of the Astronomical Society of Japan, 29*, 163–196.
Henyey, L. G., Lelevier, R., & Levée, R. D. (1955). The early phase of stellar evolution. *Publications of Astronomical Society of the Pacific, 67*, 154–160.
Herbig, G. H. (1965). On the interpretation of FU Orionis. *Vistas in Astronomy, 2*, 109–125.
Heyer, M. H., Snell, R. L., et al. (1989). A CO and far-infrared study of the S 254 – S 258 regions. *Astrophysical Journal, 346*, 220–230.
Hiei, E., & Uchida, Y. (1993). In memorium – Zenzaburo Suemoto, 1920–1991. *Solar Physics, 146*(203), 2–5.
Hirata, R. (1995). Interpretation of the long-term variation in late-type active Be stars. *Publications of the Astronomical Society of Japan, 47*, 195–218.
Hirata, R. (2007). Disk precession in Pleione. *Astronomical Society of the Pacific, Conference Series, 361*, 267–273.
Hirata, R., & Kogure, T. (1984). The Be star phenomena II. Spectral formation and structure of envelopes. *Bulletin of Astronomical Society of India, 12*, 109–151.
Hirata, K., et al. (1987). Observations of a neutrino burst from the supernova SN 1987A. *Physical Review Letters, 58*, 1490–1493.
Hitotuyanagi, Z., & Suda, K. (1957). Stellar models with isothermal cores and intermediate convective zones, I. *Publications of Astronomical Society, 9*, 25.
Hoffmeister, C., Richter, G., & Wenzel, W. (1985). Chapter 2: Pulsating stars; Chapter 3: Eruptive stars. In *Variable stars*. Springer.
Horaguchi, T., Kogure, T., et al. (1994). The Be star Gamma Cassiopeiae: X-ray, Far UV, and optical observations in early 1989. *Publications of the Astronomical Society of Japan, 46*, 9–26.
Hori, G. (1962). *Astronomical Herald, 55*(3), 52–55. Motions of artificial satellites (天文月報, 堀源一郎: 人工衛星の運動).
Hori, G. (1966). Theory of general perturbations with unspecified canonical variables. *PASJ, 18*(4), 287–296.

Hori, K., Yokoyama, T., et al. (1997). Pseudo-two-dimensional hydrodynamic modeling of solar flare loops. *Astrophysical Journal, 489*, 426–441.

Hoshi, R. (1981). Theory of steady accretion disks. *Progress of Theoretical Physics Supplement, 70*, 181–201.

Hoshi, R. (1998). Hydrogen Shell burning models for RX J0925.7 – 4758. *Publications of the Astronomical Society of Japan, 50*, 501–507.

Ichikawa, T., Nakano, M., et al. (1985). CO observations of a spiral arm in M 31. *Publications of the Astronomical Society of Japan, 37*, 439–450.

Ichimaru, S. (1977). Bimodal behavior of accretion disks: Theory and application to Cygnus X-1 transitions. *Astrophysical Journal, 214*, 840–853.

Ishida, K., & Kawajiri, N. (1968). Absolute iso-photometry of Galactic H II regions in Hα light. *Publications of the Astronomical Society of Japan, 20*, 95–120.

Ishida, M., Matsuzaki, K., et al. (1997). Detailed X-ray spectroscopy of AM Herculis with ASCA. *Monthly Notices of the Royal Astronomical Society, 287*, 651–662.

Isobe, S. (1973). Dust, stars and hydrogen distribution in the Orion Association. *IAU Symposium, 52*, 433–445.

Kaifu, N., Iguchi, T., & Kato, T. (1974). Gaseous structure in the Galactic center region. *Publications of the Astronomical Society of Japan, 26*, 117–128.

Kaifu, N., Suzuki, S., et al. (1984). Rotating gas disk around L 1551 IRS-5. *Astronomy and Astrophysics, 134*, 7–12.

Kakubari, K., Hoshi, R., et al. (1989). Thermal instability in optically-thick accretion disks and emergent X-ray spectra. *Publications of the Astronomical Society of Japan, 41*, 1035–1048.

Kambe, E., & Osaki, Y. (1988). Line profile variation due to nonradial oscillations in rapidly rotating stars. *Publications of the Astronomical Society of Japan, 40*, 313–329.

Kameya, O., & Takakubo, K. (1988). The interface between the NGC 7538 H II region and its molecular cloud core. *Publications of the Astronomical Society of Japan, 40*, 413–420.

Kameya, O., Hasegawa, T., et al. (1989). High-velocity flows in the NGC 7538 molecular cloud. *Astrophysical Journal, 339*, 222–230.

Karoji, H., & Kodaira, K. (1972). Extremely compact galaxy CGCG 1622+4112. *Publications of the Astronomical Society of Japan, 24*, 239–246.

Katahira, J., Hirata, R., et al. (1996). Period analysis of the radial velocity in Pleione. *Publications of the Astronomical Society of Japan, 48*, 317–334.

Kato, M. (1983). Neutron star wind. *Publications of the Astronomical Society of Japan, 35*, 33–48.

Kato, S. (2000). Transition from standard disk to advection-dominated accretion flow. *Astrophysical Journal, 538*, 295–308.

Kato, S., & Fukue, J. (1980). Trapped radial oscillations of gaseous disks around a black hole. *Publications of the Astronomical Society of Japan, 32*, 377–388.

Kawajiri, N., & Akabane, K. (1967). A morphological study of galactic H II regions. *Publications of the Astronomical Society of Japan, 19*, 476–487.

Kinoshita, H. (1978). Dynamics of planets and satellites and theories of their motion. In *Proceedings of the forty-first Colloquium, Cambridge, England, August 17–19, 1976* (pp. 241–257). Dordrecht: D. Reidel Publishing Co.. Third-order solution of an artificial-satellite theory.

Kinoshita, H. (2018). Obituary, Professor Kozai and celestial mechanics. *Astronomical Herald, 111*(7) (天文月報, 木下宙: 古在先生と天体力学).

Kinoshita, H., & Hori, G. (1972). On the stationary solutions for the problem of artificial satellites. *Tokyo Astronomical Observatory Report, 16*(60), 230–240. (木下宙, 堀源一郎: 人工衛星の運動の定常解について).

Kitamura, M. (1954). The effect of reflection upon the reduction of mass from the observed velocity-curved of close binary systems. *Publications of the Astronomical Society of Japan, 6*, 217–228.

Kitamura, M. (1969). Spectroscopic study of the eclipsing system R CMa. *Astrophysics and Space Science, 3*, 163–174.

Kitamura, M. (2003). Instrumental aid by Japanese ODA for astronomy in developing countries. *Astronomical Herald, 96*(8), 428–435. (天文月報, 北村正利: 日本のODA による発展途上国への天文機器援助).

Kitamura, M., & Nakamura, Y. (1987). The gravity-darkening of highly distorted stars in close binary systems. *Annals of the Tokyo Astronomical Observatory, Second Series, 21*(4), 387–397.

Kitamura, M., & Nakamura, Y. (1988). The gravity-darkening of highly distorted stars in close binary systems. V. The contact components of W UMa-type systems. *Astrophysics and Space Science, 145*, 117–156.

Kitamura, M., & Takahashi, C. (1962). A photoelectric study of the eclipsing variable R Canis Majoris. *Publications of the Astronomical Society of Japan, 14*, 44–55.

Kobayashi, Y., Kawara, K., et al. (1980). 2.2-m Polarization mapping of the galactic center. *Publications of the Astronomical Society of Japan, 32*, 291–294.

Kodaira, K. (1967). A spectrum of Sirius B. *Publications of the Astronomical Society of Japan, 19*, 175–179.

Kogure. (1969). Contri bution à l'ètude des profils d'èmission d'ètoiles Be dites "pole-on". *Astronomy and Astrophysics, 1*, 253–269.

Kogure, T., & Leung, K. C. (2007). *The astrophysics of emission-line stars*. Springer.

Koyama, K., Awaki, H., et al. (1989). Intense 6.7-KeV iron line emission from the Galactic centre. *Nature, 339*, 603–605.

Koyama, K., Kawada, M., et al. (1990). Is the 5-Kpc Galactic arm a colony of X-ray pulsar? *Nature, 343*, 148–149.

Koyama, K., Maeda, Y., et al. (1996). ASCA view of the Galactic center: Remains of past activities in X-rays? *Publications of the Astronomical Society of Japan, 48*, 249–255.

Kozai, Y. (1959). The motion of a close earth satellite. *Astronomical Journal, 64*, 367–377.

Kozai, Y. (1962). Secular perturbations of asteroids with high inclination and eccentricity. *Astronomical Journal, 67*, 591–598.

Kozai, Y., Kogure, T., et al. (2012). Obituaries of Kitamura Masatoshi. *Astronomical Herald, 105*, 720–726. ⟨天文月報, 古在ほか: 北村正利追悼文集⟩.

Kurokawa, H. (1987). Two distinct morphological types of magnetic shear development and their relation to flares. *Solar Physics, 113*, 259–265.

Kurokawa, H., Kawai, G., et al. (1992). Detailed comparison between H-alpha and Yokoh soft X-ray images of a confirmed two-ribbon flare. *Publications of the Astronomical Society of Japan, 44*, L129–L133.

Leahy, D. A., Nakajo, M., et al. (1988). Detection of an X-ray intensity dip from GX 301–2. *Publications of the Astronomical Society of Japan, 40*, 197–208.

Lidov, M. L. (1962). The evolution of orbits of artificial satellites of planets under the action of gravitational perturbations of external bodies. *Planetary and Space Science, 9*(10), 719–759.

Longair, M. (2006). *The cosmic century*. Cambridge University Press.

Maihara, T., Okuda, H., & Sato, S. (1972). Polarizations of the infrared stars and the Galactic Center in near infrared. *Memoirs de Socitè Royale des Sciences de Liége, 3*, 417–424.

Makishima, K., Ohashi, T., et al. (1989). X-ray spectral study of M 31 with Ginga. *Publications of the Astronomical Society of Japan, 41*, 697–708.

Matsuda, T., Sugimoto, D., et al. (2010). Recollections of Hayashi Chushiro. *Astronomical Herald, 103*, 394–403. (天文月報, 松田卓也, 杉本大一郎ほか:, 林忠四郎先生追悼文集).

Matsumoto, T., Murakami, M., & Hamajima, K. (1977). Near-infrared surface photometry of the central region of M 31. *Publications of the Astronomical Society of Japan, 29*, 583–591.

Mineshige, S. (1986). Disk-instability model for outbursts of dwarf novae. III. *Publications of the Astronomical Society of Japan, 38*, 831–852.

Mineshige, S. (1987). Outbursts of dwarf novae. *Astronomical Herald, 80*(10), 284–289. (天文月報, 嶺重慎: 矮新星のアウトバースト).

Mineshige, S., & Kato, S. (1996). Advection-dominant disks: Another standard disk model. *Astronomical Herald, 89*(9), 388–392. (天文月報, 嶺重慎, 加藤正二: アドベクション優勢ディスク, もう1つの標準ディスクモデル).
Miyama, S., & Eriguchi, Y. (1990). Formation and equilibrium of stars. *Astronomical Herald, 83*(3), 60–63. (天文月報, 観山正見, 江里口良治: 星の形成と平衡形状).
Miyama, S., Narita, S., & Hayashi, C. (1989). Fragmentation of isothermal sheet-like clouds. II. *Progress in Theoretical Physics, 78*, 1273–1287.
Miyamoto, M., & Kinoshita, H. (2008). Chapter 9: Position and motion of celestial bodies, Section 2, Celestial mechanics. In *One hundred years of astronomy in Japan* (pp. 146–147) (*日本の天文学の百年*, 第9章, 2節, 宮元昌典および木下宙: 天体力学).
Miyamoto, M., & Nagai, R. (1975). Three-dimensional models for the distribution of mass in galaxies. *Publications of the Astronomical Society of Japan, 27*, 533–543.
Mizuno, S. (1982). Surface photometry of the HII regions S254–S257. *Astrophysics and Space Science, 87*, 121–148.
Moran, J. M., Greenhill, L. J., et al. (1996). Probing with water masers. *IAU Symposium, 175*, 211–215.
Mouri, H., & Taniguchi, Y. (1993). The excitation mechanism of forbidden FeII 1.664-micron emission in Seyfert and starburst galaxies. *Astrophysical Journal, 406*, 52–59.
Mukai, K. (1988). Accretion streams in AM Her-type systems. *Monthly Notices of the Royal Astronomical Society, 232*, 175–197.
Murakami, T., Fujii, M., et al. (1988). Evidence for cyclotron absorption spectral features in gamma-ray bursts seen with GINGA. *Nature, 335*, 234–235.
Murata, Y., Kawabe, R., et al. (1992). Aperture synthesis observations of small clumps in Orion KL. *Publications of the Astronomical Society of Japan, 44*, 381–389.
Nadyozhin, D. K. (1995). Gamow and the physics and evolution of stars. *Space Science Review*, 455–461.
Nagase, F., Dotani, T., et al. (1991). Cyclotron line features in the spectrum of the transient-ray pulsar X0115+634. *Astrophysical Journal, 375*, L49–L52.
Nakai, H., & Kinoshita, H. (1985). Secular perturbations of asteroids in secular resonance. *Celestial Mechanics, 36*, 391–407.
Nakai, N., Inoue, M., et al. (1995). Search for extremely-high-velocity H_2O maser emission in Seyfert galaxies. *Publications of the Astronomical Society of Japan, 47*, 771–799.
Nariai, K. (1967). Analysis of the hydrogen deficient star HD 30353. II. *Publications of the Astronomical Society of Japan, 19*, 63–87.
Nariai, H. (1984). *Gravitation and universe*. Scientific Reports of the Research Institute for Theoretical Physics, Hiroshima University.
Narial, K. (1962). Atmospheres of white dwarfs. *Publication of Astronomical Society of Japan, 14*, 199–207.
Nishida, M. T., Yoshizawa, M., et al. (1981). Vlasov simulations of stellar disks. II. Nonaxisymmetric case. *Publications of the Astronomical Society of Japan, 33*, 567–582.
Nishimura, S. (1969). Variability of the metallic-line star 28 Andromedae. *Astrophysic and Space Science, 3*, 77–79.
Nogami, D., Kato, T., et al. (1997). Joint discussion 12, twenty-third general meeting of the IAU, Kyoto, Japan. Meeting abstract id.2. Electron publishing, now and future.
Nomoto, K., Suzuki, T., et al. (1993). A type IIb model for supernova 1993J. *Nature, 364*, 507–509.
Oda, M. (1977). Cyg X-1/A candidate of the black hole. *Space Science Reviews, 20*, 757–813.
Ohsuga, K., Mori, M., et al. (2005). Supercritical accretion flows around black holes: Two dimensional, radiation pressure-dominated disks with photon trapping. *ApJ, 628*, 368–381.
Okazaki, A. T. (1991). Long-term V/R variations of Be stars due to global one-armed oscillations of equatorial disks. *Publications of the Astronomical Society of Japan, 43*, 75–94.
Okazaki, A. T. (1996). On the confinement of one-armed oscillations in discs of Be stars. *Astronomy and Astrophysics, 318*, 548–560.

Okuda, H. (1981). The large-scale infrared emission in the Galactic plane – Observations. *IAU Symposium, 96*, 247–260.
Okuda, H., & Maihara, T. (1971). The nucleus of our galaxy. *Astronomical Herald, 64*, 184–189. (天文月報, 奥田治之, 舞原俊憲: 銀河系の中心部).
Okuda, H., Shibai, H., et al. (1990). An infrared quintuplet near the galactic center. *Astrophysical Journal, 351*, 89–97.
Osaki, Y. (1966). The pulsation and evolution of super-massive stars. *Publications of the Astronomical Society of Japan, 18*, 384–420.
Osaki, Y. (1971). Non-radial oscillations and the Beta Canis Majoris phenomenon. *Publications of the Astronomical Society of Japan, 23*, 485–502.
Osaki, Y. (1974). An accretion model for the outbursts of U Geminorum stars. *Publications of the Astronomical Society of Japan, 26*, 429–436.
Osaki, Y. (1983). *Proceedings of Japan-France Seminar* (pp. 156–161) Disk instability model for outburst of dwarf novae.
Osaki, Y. (1997). Dwarf nova outbursts: A unification theory. *Lecture Note in Physics, 487*, 92–102.
Osaki, Y. (2008a). Chapter 9: An interview with Professor Hayashi Chushiro (尾崎洋二: 林忠四郎先生へのインタビュー). In *One hundred years of astronomy in Japan* (日本の天文学の百年) (pp. 271–277).
Osaki, Y. (2008b). Chapter 9: An interview with Professor Kozai Yoshihide (尾崎洋二:古在由秀先生へのインタビュー). In *One hundred years of astronomy in Japan* (日本の天文学の百年).
Osaki, Y., & Hansen, C. J. (1973). Nonradial oscillations of cooling white dwarfs. *Astrophysical Journal, 185*, 277–292.
Redman, R. O., & Suemoto, Z. (1954). Temperature and turbulence in the chromosphere. *Monthly Notices of the Royal Astronomical Society, 114*, 524.
Roberts, W. W. (1969). Large-scale shock formation in spiral galaxies and the implication on star formation. *Astrophysical Journal, 158*, 123–143.
Sadakane, K. (1974). Variable K line profile of 73 Draconis. *Publications of the Astronomical Society of Japan, 26*, 93–102.
Sakka, K., Wakamatsu, K., & Oka, S. (1973). Hot-spot galaxy NGC 2782. *Publications of the Astronomical Society of Japan, 25*, 158–173.
Sato, K. (1981). First-order phase transition of a vacuum and the expansion of the Universe. *Monthly Notices of the Royal Astronomical Society, 195*, 467–479.
Sato, H. (2015). *All the works of Hayashi Chushiro*. Kyoto University Press. (佐藤文隆: "林忠四郎の全仕事", 京都大学学術出版会).
Sawa, T., Sofue, Y., et al. (1981). Neutral hydrogen in M31. II. Kinematical properties and density waves. *Publications of the Astronomical Society of Japan, 33*, 665–677.
Sekii, T. (1997). Internal solar rotation. *IAU Symposium, 181*, 189–202.
Sekii, T. (1998a). Recent topics in helioseismology. *Astronomical Herald, 91*, 92–98. (天文月報, 関井隆: 日震学の最近の話題から).
Sekii, T. (1998b). Internal solar rotation. *IAU Symposium, 181*, 189–202.
Sekiya, M., & Miyama, S. (1988). Stability of the primordial Solar Nebula. *Progress of Theoretical Physics Supplement, 96*, 85–94.
Sekiya, M., Miyama, S., & Hayashi, C. (1988). Gas capture by proto-Jupiter and proto-Saturn. *Progress of Theoretical Physics Supplement, 96*, 274–280.
Seto, N. (2017). Kozai-Lidov mechanism and its recent development. *Butsuri (Physics), 73*(4), 282–283. (日本物理学会誌, 物理, 瀬戸直樹: 古在—Lidov 機構とその最近の進展).
Shibasaki, K., & Ebisuzaki, T. (1989). Radial oscillations of extended envelopes of neutron stars. *Publications of the Astronomical Society of Japan, 41*, 641–650.
Shibahashi, H., & Sekii, T. (1993). Invitation to the Helioseismology, I, II. *Astronomical Herald, 86*, 416–422; 461–468. (天文月報, 柴橋博資, 関井隆: 日震学への招待).

Shibata, K., & Uchida, Y. (1989). Formation of astrophysical jets by a contracting accretion disk. *Astrophysics and Space Science, 118*, 443.

Shibata, K., et al. (2018). Obituary of Professor Kozai Yoshihide. *Astronomical Herald, 111*(7), 483. (柴田一成ほか, 天文月報, 古在先生追悼文集).

Shibata, K., Hiei, E., et al. (2019). Memorials of Prof. Kawaguchi Ichiro. *Astronomical Herald, 112*, 406–416. (天文月報, 柴田一成ほか: 追悼 川口市郎先生).

Shigeyama, T., Nomoto, K., et al. (1987). Light-curve models for supernova SN1987A in the Large Magellanic Cloud. *Nature, 328*, 320–323.

Shigeyama, T., Kumagai, S., et al. (1988). Suprnova SN 1987A in Large Magellanic Nebula. *Astronomical Herald, 81*(2), 36–40. (天文月報, 茂山俊和, 熊谷紫麻見: 大マゼラン雲の超新星 SN 1987A).

Sofue, Y. (1984a). The radio lobe above the Galactic plane. *Astronomical Herald, 77*, 188–190. (天文月報, 祖父江義明: 銀河中心 の電波ローブ).

Sofue, Y. (1984b). Shock wave from a galactic nucleus into a halo and intergalactic space. *Publications of the Astronomical Society of Japan, 36*, 539.

Sofue, Y. (1985). The galactic centre lobe. *Publications of the Astronomical Society of Japan, 37*, 697.

Sofue, Y., & Handa, T. (1984). A radio lobe over the Galactic center. *Nature, 310*, 568–569.

Sofue, Y., & Nakai, N. (1993). CO observations of edge-on galaxies. III. NGC 891: Threshold radius for a star-formation disk. *Publications of the Astronomical Society of Japan, 45*, 139.

Sofue, Y., Handa, T., et al. (1986a). The galactic center radio lobe – A cosmic jet in our galaxy? *Astrophysics and Space Science, 118*, 371.

Sofue, Y., Inoue, M., et al. (1986b). 43-GHz continuum observations of the Galactic Center. *Publications of the Astronomical Society of Japan, 38*, 475–483.

Struve, O. (1954). The problem of time resolution in stellar spectroscopy: BW Vulpeculae. *Publications of the Astronomical Society of the Pacific, 66*, 329–334.

Suemoto, Z., & Hiei, E. (1959). Balmer series lines of the flare and its structure. *Publications of the Astronomical Society of Japan, 11*, 185–194.

Sugimoto, D., Ebisuzaki, T., & Makino, J. (1991). Dynamical evolution of binary clusters and colliding galaxies. *IAU Symposium, 148*, 217–218.

Takahashi, K. (2015). A long interview with Professor Kozai Yoshihide. *Astromical Herald, 108*, 4–8. (天文月報, 高橋慶太郎: 古在由秀氏ロングインタビュー).

Takano, T., Enome, S., et al. (1994). Behavior of Accelerated electrons in a small impulsive solar flare on 1992 August 12. *Publications of the Astronomical Society of Japan, 46*, L21–L25.

Takeda, S. (1934). Distorted outer layers of the stars. *Memoirs of College Science, Kyoto Imperial University, 17*, 197–217.

Takeda, Y. (1993). Investigation of stellar magnetic fields based on the strengths of spectral lines. Application to Omicron Pegasi and other cool Ap stars. *Publications of the Astronomical Society of Japan, 45*, 453–469.

Takeuchi, M. (1992). Order in irregular stellar variability. *Astronomical Herald, 85*, 496–601. (天文月報, 竹内峯: 不規則変光星のなかの秩序).

Tamura, S. (1977). High excitation emission lines of HBV475. *Astrophysical Letters, 19*, 57–60.

Tamura, S. (1984). A case study of the planetary nebula formation. *Sendai Astronomiaj Reportoj, 263*, 59–67.

Tamura, S. (1997). A review on the planetary nebula research during the last one decade and a perspective. *Astronomical Herald, 90*(11), 512–517. (天文月報, 田村真一: 惑星状星雲研究この十年と今後の展望).

Tanaka, Y. (1991). Variable stars and nonlinear dynamics. *Astronomical Herald, 84*(12), 397–400. (天文月報, 田中靖夫: 変光星と非線形力学).

Tanaka, Y., & Takeuchi, M. (1988). A model oscillator of irregular stellar variability. *Astrophysics and Space Science, 148*, 229–237.

Taniguchi, Y. (1992). Circumstellar molecular gas in Seyfert and starburst galaxies and starburst-implosion fueling of AGNs. *Astronomical Society of the Pacific, Conference Series, 31*, 357.

Taniguchi, Y., & Murayama, T. (1998). Dusty tori of Seyfert nuclei probed by the water vapor maser emission. How large are the dusty tori? *Astrophysical Journal, 501*, L25–L28.

Tatematsu, K., & Umemoto, T. (1991). Molecular cloud cores in the Orion A cloud. *Societá Astronomica Italiana, Memorie, 62*(4), 813–821.

Tatematsu, K., Umemoto, T., et al. (1993). Molecular cloud cores in the Orion A Cloud. I. Nobeyama CS (1-0). *Astrophysical Journal, 404*, 643–662.

Tomimatsu, A., & Sato, H. (1973). New metric of exact solution for gravitational fields of spinning masses. *Progress of Theoretical Physics, 50*(1), 95–110.

Tomita, K. (1991). Reminis cences of Prof. Nariai Hidekazu. *Astronomical Herald, 84*, 93–96. (天文月報, 冨田憲二: 成相秀一先生を偲ぶ).

Tsuboi, M., Kawabata, T., et al. (1995). Galactic center arc -polarized plumes complex at 43 GHz. *Publications of the Astronomical Society of Japan, 47*, 829–836.

Uchida, Y., & Shibata, K. (1985). Magneto dynamical acceleration of CO and optical bipolar flows from the region of star formation. *Publications of the Astronomical Society of Japan, 37*, 515–535.

Uchida, Y., Sofue, Y., & Shibata, K. (1985). Origin of the galactic center lobes. *Nature, 317*, 699.

Uchida, Y., Kaifu, N., et al. (1987). Observations of the detailed structure and velocity field in the CO bipolar flows associated with L1551 IRS-5. *Publications of the Astronomical Society of Japan, 39*, 907–924.

Uchida, Y., Shibata, K., & Matsumoto, R. (1990). Magnetohydrodynamical simulation in astrophysics – Jets, loops, and flares. *Astronomical Herald, 83*(6), 181–184. (天文月報, 内田豊ほか: 天体の電磁流体現象のシミュレーション).

Unno, W. (1950). On the radiation pressure in a planetary nebula. I. *Publications of the Astronomical Society of Japan, 2*, 53–55.

Unno, W. (1952). On the radiation pressure in a planetary nebula. II. The problem of HeII. *Publications of the Astronomical Society of Japan, 3*, 178–183.

Unno, W. (1956). Line formation of a normal Zeeman triplet. *Publications of the Astronomical Society of Japan, 8*, 108–125.

Unno, W. (1971). Physical state and stability of super-massive objects. *Publications of the Astronomical Society of Japan, 23*, 123–147.

Unno, W., & Kawabata, K. (1955). On the acoustic noise in the solar atmosphere. *Publications of the Astronomical Society of Japan, 7*, 21–26.

Unno, W., & Takakubo, K. (1952). On the internal motions in planetary nebulae. *Publications of the Astronomical Society of Japan, 3*, 195–208.

Unno, W., Osaki, Y., et al. (1979). *Nonradial oscillations of stars*. University of Tokyo Press.

Watanabe, T. (1988). An indication of extended chromosphere in line width – Luminosity relations. *Publications of the Astronomical Society of Japan, 40*, 475–490.

Watanabe, Y., Inagaki, S., et al. (1981). Vlasov simulations of stellar disks. I. Axisymmetric case. *Publications of the Astronomical Society of Japan, 33*, 541–565.

Yamaguchi, S., Kawada, M., et al. (1990). Optically thin hot plasma near the galactic center: Mapping observations of the 6.7 KeV iron line. *Astrophysical Journal, 365*, 532–538.

Yamashita, Y., & Maehara, H. (1978a). The spectrum of the companion of Mira Ceti. *Publications of the Astronomical Society of Japan, 29*, 319–329.

Yamashita, Y., & Maehara, H. (1978b). Mass loss from Mira Ceti. *Publications of the Astronomical Society of Japan, 30*, 409–417.

Yokoyama, T., & Shibata, K. (1994). What is the condition for fast magnetic reconnection? *Astrophysical Journal, 436*, L194–L200.

Yokoyama, T., & Shibata, K. (1998). A two-dimensional MHD simulation of chromospheric evaporation in a solar flare based on a magnetic reconnection model. *Astrophysical Journal, 494*, L113–L116.

Yoshida, A., & Murakami, T. (1990). Cyclotron absorption features in γ-ray bursts and related magnetic fields of neutron stars. *Astronomical Herald, 83*, 38–42. (天文月報, 吉田篤正, 村上敏夫: γ線バーストのサイクロトロン吸収線と中性子星磁場).

Yoshida, S., Kogure, T., et al. (1991). The Herbig Be/Ae stars MWC 1080 and its associated molecular cloud. I. Molecular line observations. *Publications of the Astronomical Society of Japan, 43*, 363–379.

Yoshida, S., Kogure, T., et al. (1992). The Herbig Be/Ae stars MWC 1080 and its associated molecular cloud. II. Optical spectroscopic observations. *Publications of the Astronomical Society of Japan, 44*, 77–89.

Chapter 8
Education and Popularization of Astronomy, 1946–2010

Abstract Education and popularization are basic requirements for the development of astronomy and astrophysics. Education in elementary, secondary, and high schools has been based on *The Course of Study* issued by the Ministry of Education and Science. Problems of astronomy education in this system are briefly reviewed here. In universities, astronomy is separated into general education and technical education; of these two, technical education in graduate schools has shown remarkable progress in the postwar period.

The popularization of astronomy has been promoted through the construction of large planetariums and public observatories equipped with large telescopes. These facilities have contributed markedly to progress in astronomical education.

8.1 Astronomy Education

8.1.1 Course of Study

8.1.1.1 Education Based on *Course of Study*

With the defeat of Japan in World War II in 1945, the education system completely changed from a militaristic to a democratic governing system. In the postwar period, all education in elementary schools (6 years), lower secondary schools (3 years, secondary schools in short) and higher secondary schools (3 years, high schools in short) was provided under the *Course of Study* (学習指導要領) issued by the Ministry of Education and Science. Astronomy was taught in the science course (理科) in the elementary and secondary schools, and in Earth science (地学) in the high schools (Manual of the Course of Study 2007–2009).

The first edition of *Course of Study* was issued in 1947 and updated in 1951, during the Occupation by Allied Forces (October 1945–April 1952). Its basic concept was to emphasize the importance of social studies for democracy, including morals and civics. With the end of the Occupation, it was generally accepted that school curricula should be thoroughly revised to make them more suitable for the

T. Kogure, *The History of Modern Astronomy in Japan*, Historical & Cultural Astronomy, https://doi.org/10.1007/978-3-030-57061-3_8

Table 8.1 Astronomical subjects in science course in elementary schools (*Course of Study*1989)

Class year	Children's age	Subjects in astronomy
3rd	8	Sunny side and shadows Solar position and motions in sky Atmospheric temperature and humidity
4th	9	(Meteorology)
5th	10	Earth and Universe Sunrise and sunset Lunar phase and solar position Lunar surface
6th	11	Brightness and colors of stars Stellar constellations

Japanese educational system. Thus, the *Course of Study* was largely revised in 1955, but the basic aim of social studies remained as it was in this case.

This attitude toward education underwent a significant overhaul in the 1958 edition. In that edition, systematic understanding of science was emphasized in every school, rather than morals and civics, to expand students' scientific and technological understanding.

The decade of the 1960s was a period of rapid economic growth and technological progress. To keep pace with the times, the *Course of Study* were revised for the elementary schools in1968, the secondary schools in 1969, and the high schools in 1972. The basic aims of this revision were twofold: first, to reorient the curricula to reflect the rapidly changing and increasingly complex social structure and, second, to redesign the curricula to foster the acquisition of basic knowledge and technical skills and to foster the development of creativity, emotional intelligence, and will.

In 1977, the *Course of Study* was again largely changed, with the adoption of Yutori education, where Yutori (ゆとり) means an attitude of giving more free time designed to take pressure off children. Class hours were noticeably reduced. Thus, the level of achievement declined in every subject, particularly in mathematics and science.

In 1989, the *Course of Study* was revised again, and Yutori education was abandoned, so that the systematic study of school subjects was revived. This basic principle was followed in the revision of 2003 and thereafter.

Science in schools is generally divided into three divisions: A (life and the environment), B (materials and energy), and C (Earth and space). Astronomical topics fall into division C. Table 8.1 shows the subjects and contents of astronomy in the *Course of Study* in the case of elementary schools. The contents have remained almost unchanged since 1989.

In the secondary schools, division C (Earth and Space) is given in the third year under the following subjects: (1) motions of stars, including diurnal motion and Earth's rotation, annual motion, and revolution of the Earth; (2) Solar System, including solar surface and lunar motion and eclipses; and (3) Universe, including distribution of stars and the Galaxy.

As seen in Table 8.1, astronomical subjects in the elementary schools are mostly limited to the Solar System and the visible sky. In the secondary schools, astronomical subjects expand to stellar systems and the Universe.

8.1.1.2 Problems in the *Course of Study*

In the elementary and secondary schools, many teachers and students experience difficulties in the study of astronomy based on the *Course of Study*. In 1988, Owaki Naoaki and Isobe Shuzo conducted a survey of teachers that contained two questions: First, according to teachers, how do children feel about their difficulties with respect to studying astronomy? Second, what kind of difficulties do teachers have in teaching astronomy? Owaki and Isobe prepared several selective answers to each question. They collected responses from 471 teachers around the country (Owaki and Isobe 1988).

The main points of the answers are summarized as follows, with the response percentages.

Answers to Question 1 for the children's feelings in the teachers's perspective were the difficulties on the following points:

(a) Three-dimensional aspect of space (54%)
(b) Very large scales of space and time (47%)
(c) Shortage of basic mathematics and physics (60%)
(d) Separation of daily life from astronomical phenomena (33%)
(e) Understanding celestial phenomena by students' observational experiments (34%).

Answers to Question 2 on teachers' feelings were as follows:

(a) Astronomy does not receive sufficient hours in the classroom (47%).
(b) Only indirect observations are feasible (66%).
(c) Astronomical phenomena cannot be explained with laboratory experiments (33%).
(d) Teachers themselves have a hard time understanding astronomy (33%).
(e) It is not easy to select and manage astronomical material and instruments (22%).

Similar surveys of teachers of astronomy were also conducted by Miyawaki Ryosuke in 1990 (Miyawaki 1990). These surveys revealed the following problems in astronomy education in schools:

(a) Astronomical phenomena are explained merely indirectly during school hours.
(b) There is a gap between students' interests and the curriculum.
(c) Not enough observations are made.
(d) The number of teachers who have never studied astronomy at schools of education is gradually increasing. This reflects the insufficient state of general education in astronomy at the universities (Sect. 8.1.2).

Table 8.2 Teaching of Earth science in high schools (Agata 1992)

Earth science in high schools rate	Number of prefectures
0.4%–10%	29
10%–20%	14
20%–24%	4

In high schools, the science program is divided into four subjects: physics, chemistry, biology, and Earth science, where Earth science includes meteorology, geology, and astronomy. Among these, the astronomy curriculum consists of the following subjects (*Course of Study*, 2009):

(1) solar System

  (a) Rotation and revolution of Earth
  (b) Moon and planets, their structure and orbital motions
  (c) Solar activity, observations of solar surface.

(2) Stars and the Galaxy

  (a) Properties of stars and their evolution
  (b) Structure of the Galaxy, stellar and nonstellar objects, and their distributions

(3) Galaxies and the Universe

  (a) Variety of galaxies based on structure, distribution, and recession velocity
  (b) Expanding Universe; outline of modern cosmology

Clearly, astronomy is sufficiently treated in Earth science. The problem, however, is that a small number of students select Earth science in high school. Students must select three subjects from four electives over the course of 3 years, and the number of students selecting Earth science is generally low. In many high schools, Earth science is not even offered due to the small number of students selecting this subject. In 1992, Agata Hidehiko of the NAOJ conducted a survey among all 47 Prefectures on the numbers of high schools where Earth science was offered in science courses, and he obtained the results shown in Table 8.2 (Agata 1992), where original results are summarized in three percentage levels.

The national average is around 10%, that is, one student out of ten studies Earth science in high school. With this survey, Agata expressed his concern about a further decrease in school hours spent on Earth science in high school.

8.1.1.3 Progress in Astronomy Education

In the 1990s, astronomy education made definitive progress due to the following background developments:

(a) Sonstruction of planetariums and public astronomical observatories (Sect. 8.2)
(b) Foundation of Japanese Society for Education and Popularization of Astronomy (天文教育普及研究会) in 1989. The main subjects of the society are school education in elementary, secondary, and high schools and widespread popularization of astronomy

Table 8.3 Junior Session at Annual Meeting of the Astronomical Society of Japan (ASJ home page)

Order of session	Year	Number of presentations		
		Total	Oral	Poster
1	2000	20	17	3
2	2001	13	10	3
3	2002	7	6	1
4	2003	23	20	4
5	2004	17	14	3
6	2005	34	26	8

(c) New activity of Astronomical Society of Japan for astronomy education in the 1990s and establishment of the Public Relations Center (天文情報センター) in NAOJ in 1998.

These activities have greatly supplemented the insufficient state of astronomy education in lower schools. The numbers of students interested in astronomy have gradually increase across the country, and students' understanding of astronomical phenomena has shown steady progress in astronomy classes. Some student groups have taken up astronomical observations and their scientific reductions.

Reflecting these activities, the Astronomical Society of Japan prepared a new Junior Session at the Annual Meeting of the ASJ for the presentations of papers submitted by students. The numbers of presentation from the first to sixth Junior Sessions are given in Table 8.3, which shows the steady progress in astronomy activities in secondary and high schools from the 2000s.

8.1.2 Astronomy Education in Universities

In universities, there were no Courses of Study specifically stated by the Ministry of Education and Science. The curriculum of education was under self-regulation of the faculty or department in each university, with some flexibility. This suggests that the curriculum is different from university to university. The curriculum of astronomy education is not an exception. In this section, astronomy education is considered in two categories: general education and technical education.

8.1.2.1 General Education

When the education system was revived in 1948, a college of liberal arts (教養部) was established in all 4-year universities. A faculty of education (教育学部) was also established in the main national universities and the universities of education. These colleges and faculties offered general astronomy education, but its actual status had long been far from sufficient.

A survey was conducted on the state of astronomy education in universities in 1974 by Kiyonaga Kaichi of Osaka Gakugei University in collaboration with the Astronomical Society of Japan. He surveyed 54 colleges of liberal arts (responses 35; 65%), and 56 faculties of education (responses 48; 86%) (Kiyonaga 1974).

The results of the survey for each case are as follows:

(1) College of Liberal Arts

Among the 35 surveyed colleges, 13 (37%) had registered teachers of astronomy, and the remaining 22 (63%) had none. In addition, astronomy experiments, such as telescope handling and observations of solar and planetary surfaces, were carried out at only three (8.6%) colleges. In every college, the number of astronomy courses was quite low compared to other fields of science, like physics, chemistry, and biology.

(2) Faculty of Education

In the faculties of education, astronomy was a part of Earth science, which was one of the compulsory courses (the others being physics, chemistry, and biology). In Earth science (astronomy, meteorology, mineralogy, geology), however, astronomy was a minor branch in the teaching staff and lecture hours. This became evident through the survey of Sawa Takeyasu in 1983. He counted the number of teachers in each branch of Earth science for all faculty of education in Japan (Sawa 1983). The result is shown in Table 8.4, which shows the percentage of teachers is given in each branch of Earth science for the whole faculty of education. More than 60 percent of teachers were in the geological sciences, while teachers of astronomy made up a minority.

These relative numbers of teachers remained almost unchanged from Sawa's survey in the 1990s.

8.1.2.2 Technical Education

In the 1950s, technical education and research in astronomy were limited to the department of astronomy in three National Universities in Tokyo, Kyoto, and Sendai. Starting in the 1960s, astrophysics became a widely studied subject in departments of physics throughout the country, mainly in the fields of cosmology, nuclear astrophysics, and stellar photometry. In the ensuing time period, the number of universities teaching astrophysics has gradually increased in all fields of astrophysics.

In 1993 Sawa Takeyasu and his group at Aichi Kyoiku University started a survey of universities that already had programs for teaching astronomy at a technical level. This survey was conducted once every few years. Table 8.5 presents the results of the

Table 8.4 Percentage of teachers in four branches of Earth science in faculties of education in Japan (Sawa 1983)

Branch	Astronomy	Meteorology	Mineralogy	Geology
Number (%)	11.4	23.3	23.5	41.5

Table 8.5 Number of universities with graduate programs in astronomy or space science in 2001 (Sawa 2005)

Type of University	National	Public	Private
Doctorate	27	3	20
Master's	12	0	3
Total	39	3	23

survey from 2001, and the number of universities offering graduate courses in astronomy and space science is shown. The universities are separated into three types: national, public (established by local government), and private (Sawa 2005).

This table indicates that every year more than 60 new astronomers entered the Japanese astronomical community in the fields of research and education. This reflects the background of the development of astrophysics in the last quarter of the twentieth century in Japan.

8.2 Popularization of Astronomy

8.2.1 Planetariums and Science Museums

Planetariums are theaters that bring the awe and wonder of the Universe to children and citizens. Planetariums also act as the centers of science education.

In Japan, the first planetariums were built in Osaka (Osaka Electric Science Museum, 大阪電気科学館) in 1937 and in Tokyo (Tonichi Astronomical Museum 東日天文館) in 1938. In the postwar era, the first planetarium was Goto Planetarium (五島プラネタリウム), constructed in 1956 in Tokyo, and subsequent construction has taken place since the 1960s. The number of planetarium facilities increased almost in direct proportion to the country's population growth. The diameters of planetariums have also increased. The number of operating planetariums, categorized by dome size, are illustrated in Fig. 8.1.

As shown in Fig. 8.1, the construction of planetariums was concentrated in the 1970s and 1980s at every scale of dome size. Starting in the 1990s, construction gradually declined, mainly due to the closedown of old planetariums after 30 or 40 years, the average lifetime of planetariums. The rate of new construction reduced with the closedown of planetariums, reflecting the state of the Japan economy at the time.

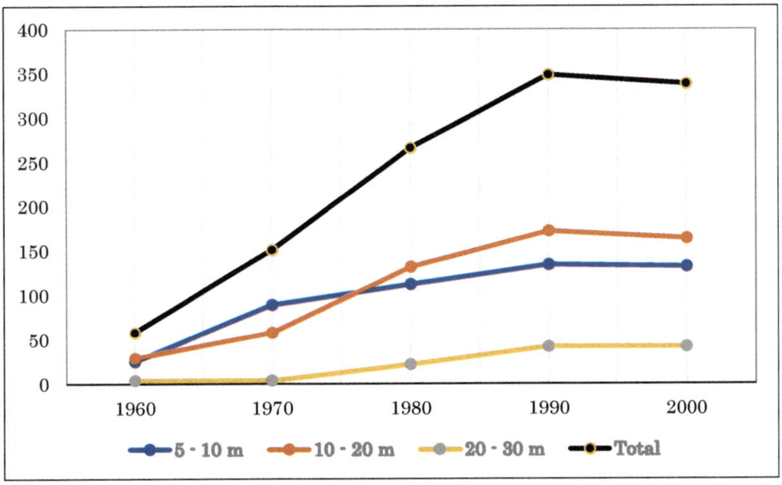

Fig. 8.1 Number of operating planetariums by dome diameter (based on Watanabe 2001)

Table 8.6 Top five countries in terms of number of planetariums in 2001 (Watanabe 2001)

Country	Number of planetariums
USA	802
Japan	347
China	90
Germany	60
Russia	50

Table 8.7 Large planetariums constructed in 1980–2012 in Japan

Year	Dome size (m)	Name of planetarium
1985	25.6	Tsukuba Expo Center (つくばエキスポセンター)
1987	27.0	Miyazaki Science Museum (宮崎科学技術館)
1989	26.5	Osaka Science Museum (大阪市立科学館)
1993	27.5	Tama Rokuto Science Museum (多摩六都科学館)
1993	27.0	Himeji Science Museum (姫路科学館)
2008	27.5	Sendai City Observatory (仙台市天文台)
2011	35.0	Nagoya Science Museum (名古屋市科学館)

Despite this situation, Japan ranks second among countries with large planetariums having dome diameters larger than 5 m, as shown in Table 8.6, which shows the five countries with the largest numbers of planetariums in 2001.

Planetariums may serve two main roles: as an institute of astronomy education for school children and citizens and as a venue for enjoying sky watching. These roles may in part depend on a planetarium's dome size.

Small-scale planetariums with domes less than 10 m in diameter tend to focus on astronomy education, particularly for schools and kindergartens. Small planetariums are fully equipped to function as a means of introducing fundamental concepts of astronomy. In addition, mobile planetariums of this size are widely distributed around the country. This type of planetarium is built from parts brought in and assembled in schools or kindergartens to show the night sky to children and students.

Planetariums with domes larger than 10 m can act as astronomical information centers for education and enjoyment for local visitors. Most large planetariums are places to show off the beauty of the natural sky and to provide the space exhibits based on modern astrophysical research.

The largest type of planetariums, which are larger than 25 m, have been particularly concentrated in Japan since the 1980s, as shown in Table 8.7. These planetariums usually have tilted floors and various types of projectors, such as an all-sky-panorama projector and video projector, in addition to the main planetarium projectors.

For example, the Osaka Science Museum is shown in Figs. 8.2 and 8.3. This museum functions largely as an interactive educational facility whose target audience is children and students. In its role as a science museum it showcases thematic exhibitions on astronomy, physics, chemistry, and modern technology over four floors. The planetarium is a separate facility and holds regular shows throughout the day. Every year the museum organizes more than 50 programs, such as science lectures, workshops, and star parties for young people.

Fig. 8.2 Exterior of Osaka Science Museum (courtesy Osaka Science Museum)

Fig. 8.3 Interior of Planetarium in Osaka Science Museum. Dome size 26.5 m with tilted floor. (Courtesy Osaka Science Museum)

8.2.2 Public Astronomical Observatories

A public astronomical observatory (PAO) is a facility with a dome, a telescope, and some observational devices that is open to the public for star gazing. In addition PAO often provides the observational devices such as CCD camera, photometer and spectrograph for astrophysical observations by advanced visitors.

In Japan, the first PAO was the Kurashiki Observatory (倉敷天文台) with a 32-cm reflector in the city of Kurashiki, which opened in 1926. The second was the Ikoma Astronomical Museum (生駒天文博物館) with a 60-cm reflector, which opened in 1936. Both facilities were established under the guidance of Yamamoto Issei of Kyoto University (Chap. 4). In 1941, Yamaguchi Museum (山口博物館), equipped with a 10-cm refractor, was established by the local government of Yamaguchi Prefecture.

Since the postwar period various types of public observatories have been established. According to a survey conducted by the Japan Public Observatory Society (JAPOS), the types can be classified as given in Table 8.8 (White Paper 2006). In this table, the PAOs are included in the public astronomical facilities (PAFs). PAOs, however, are distinguished by certain advanced astronomical equipment and relatively large telescopes.

Table 8.8 Classification of PAFs (White Paper 2006)

Type	Number in 2005
Public Astronomical Observatory (PAO) (公共天文台)	62
Science museum (科学博物館)	57
Outdoor activity center (野外活動センター)	47
Public education facility (社会教育施設)	32
Facility associated with schools (学校教育付属施設)	13
Welfare facility for children (児童福祉施設)	10
Facility associated with accommodation (宿泊施設)	18
Total	252

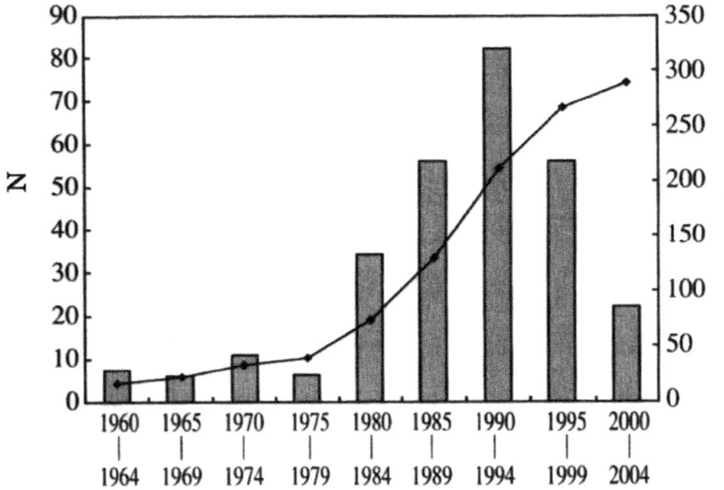

Fig. 8.4 Number of PAFs in Japan every 5 years (dark column), denoted by N in left ordinate, and the total number is given by the line graph, denoted by A in the right ordinate (Kuroda 2008)

Table 8.9 Construction of large telescopes (White Paper 2006)

Telescope aperture range	Years of construction	Number of telescopes
60–80 cm	1980–2000	41
80–100 cm	1989–1997	8
100–200 cm	1993–2006	10

The total number of PAFs has gradually increased since the 1960s, as shown in Fig. 8.4, where one may see that the construction rate suddenly increased in the 1980s and the peak appeared in the first half of the 1990s.

In parallel with planetariums, the construction of PAOs has been promoted mainly by local governments, particularly in the 1980s and early 1990s. In this period, the size of telescopes increased, as shown in Table 8.9, and PAO sites have also spread over the country, as illustrated in Fig. 8.5.

The main role of PAFs is the popularization of astronomy, independent of the size of telescopes. Star gazing through an eyepiece allows visitors to observe various

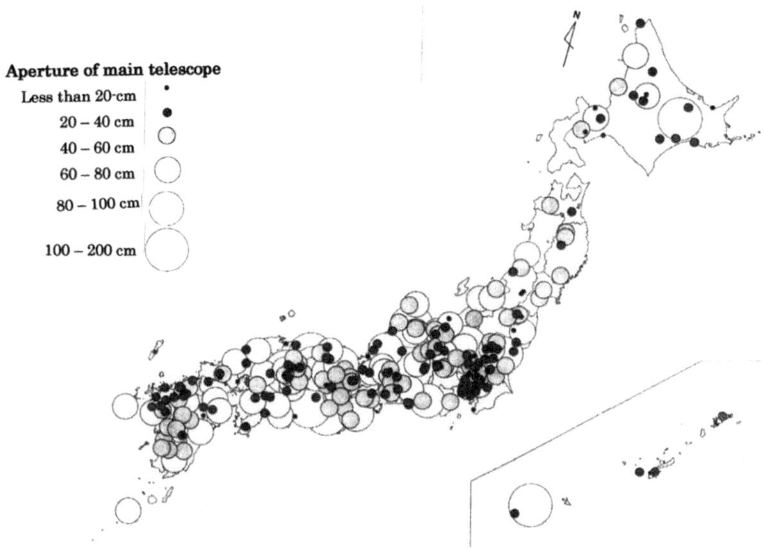

Fig. 8.5 Distributions of PAOs in Japan, by telescope aperture (White Paper 2006)

Table 8.10 PAOs with main telescopes having apertures larger than 100 cm, constructed in 1990–2006 (White Paper 2006)

Name (AO = Astronomical Observatory)	Aperture (cm)	Year of establishment
Bisei AO (美星天文台)	101	1993
Saji Astropark (さじアストロパーク)	103	1995
Misato AO (みさと天文台)	105	1995
Kawabe Ast. Park (かわべ天文公園)	100	1996
Anan City Science Center (阿南市科学センター)	113	1997
Toyama City AO (富山市天文台)	100	1997
Rikubetsu AO (りくべつ銀河の森天文台)	115	1998
Gunma AO (群馬天文台)	150	1999
Nishiharima AO(西はりま天文台)	200	2004
Ishigaki-shima AO (石垣島天文台)	115	2006

astronomical phenomena. Larger telescopes can reach deeper skies. The telescope size has gradually increased, as seen earlier.

Most of the telescopes in Table 8.9 are equipped with observational devices such as charge-coupled device (CCD) cameras, video cameras, photoelectric photometers, and spectrographs with cooled CCD detectors. They allow for astrophysical observations of various celestial objects such as comets, planetary surfaces, variable stars, novae, supernovae, and galaxies, often in collaboration with universities, NAOJ, and the Astronomical Society of Japan. These observations take place in parallel with ordinary popularization activities.

Astronomical observatories equipped with telescopes having apertures than 100 cm are listed in Table 8.10. These telescopes are of the largest class in the world for public observatories.

Fig. 8.6 Bisei Astronomical Observatory, opened in 1993. The dome is equipped with a 101-cm reflecting telescope. (By author)

Fig. 8.7 Nishi-Harima Astronomical Observatory, opened in 1989. A new 200-cm telescope was constructed in 2004. The dome of this telescope is shown (Annual Report of Nishi – Harima Astronomical Observatory, No. 14, 2004)

Fig. 8.8 200-cm telescope of Nishi-Harima Astronomical Observatory, constructed in 2004. (Courtesy Nishiharima Astronomical Observatory)

Among these, two examples of the PAO are exhibited in Fig. 8.6 (Bisei Astronomical Observatory) and in Figs. 8.7 and 8.8 (Nishi-Harima Astronomical Observatory). In both observatories, the telescopes are equipped with all devices for astrophysical observations and are open to the public.

References

Agata, H. (1992). Reduction of astronomical subjects from the course of study and from school textbooks (天文月報、縣秀彦:指導要領から削減・縮小が進む天文分野). *Astronomical Herald, 85*(11), 521–524.

Kiyonaga, K. (1974). On the results of questionnaire on astronomical education (天文月報、清永嘉一: 天文教育に関するアンケート調査の結果について). *Astronomical Herald, 67*(12), 382–385.

Kuroda, T. (2008). *One hundred years of astronomy in Japan* (日本の天文学の百年), Koseisha-Koseikaku (恒星社厚生閣), pp. 245–248, Public observatories (黒田武彦: 公開天文台).

Manual of the Courses of Study. (2007–2009). Ministry of Education and Science (ed.), Manuals of the Course of Study in Elementary Schools (2007), in Secondary Schools (2008), and in High Schools (2009) (文部科学省篇、学習指導要領解説).

Miyawaki, R. (1990). Teaching of astronomy in Asian-Pacific region. *Bulletin of Teaching of Astronomy, 2*, 9–17.

Owaki, N., & Isobe, S. (1988). A note on astronomy education in Japan. *Vistas in Astronomy, 31*, 863–867.

Sawa, T. (1983). Present State of Astronomical Education in the Universities. *Report of Education Center, 7*, 181–189 (愛知教育大学教育センター報告、沢武文:教育系大学における天文教育の現状).

Sawa, T. (2005). Universities where astronomy can be studied, and universities where astronomers work (天文教育, 沢武文:宇宙を学べる大学, 天文学者のいる大学). *Astronomical Education, 17*(6), 8–38.

Watanabe, Y. (2001). On the national census of planetariums (大阪市立科学館研究報告、渡部義弥:プラネタリウムの国勢調査について). *Publications of the Osaka Science Museum, 11*, 31–38.

White Paper. 2006.*White paper of public observatories in Japan*, Japan Public Observatory Society (日本公開天文台協会編: 公開 天文台白書).

Appendix: The Rise of Astrophysics in Western Countries

Astrophysics was born in the eighteenth and nineteenth centuries and developed in Western countries in the broad fields of stellar spectroscopy, stellar structure, and galactic astronomy (Kogure 2015). The rise of astrophysics had no impact on Japan in the Tokugawa period, as seen in Chap. 1. Its importance was barely recognized even in the early twentieth century, but advances in the field spurred interest among Japanese scientists, and it started to be studied in Japan at a deep level as well. As a reference, this appendix presents a brief overview of the early history of Western astrophysics.

Development of Spectroscopy

Astronomical spectroscopy was initiated with an experiment by Joseph von Fraunhofer (1787–1826) in Munich in 1814, who found numerous dark lines across the bright solar spectrum (Fraunhofer lines) (Fig. A.1).

Chemical identification of the Fraunhofer lines was made in 1859 by Gustav Robert Kirchhoff (1824–1887) at Heidelberg University. Kirchhoff showed that sodium, calcium, iron, and other elements exist in the solar atmosphere.

William Huggins (1824–1910) and William Allen Miller (1817–1870) of London took up spectroscopic observations of several stars at Huggins's private Tulse Hill Observatory, and they found that the chemical compositions varied markedly from star to star. Angelo Secchi (1818–1878) of the Vatican Observatory began to classify stars into five types by color and spectral features. He classified over 4000 stars over the course of his life, and his classification was revised by Hermann Vogel (1841–1907) into seven types.

In the USA, John William Draper (1811–1882) and his son Henry Draper (1837–1882), both enthusiastic amateur astronomers, contributed to the photographic observation of the Sun, stars, and nebulae. In 1843, John William Draper found numerous new Fraunhofer lines in the ultraviolet and infrared regions on

T. Kogure, *The History of Modern Astronomy in Japan*, Historical & Cultural Astronomy, https://doi.org/10.1007/978-3-030-57061-3

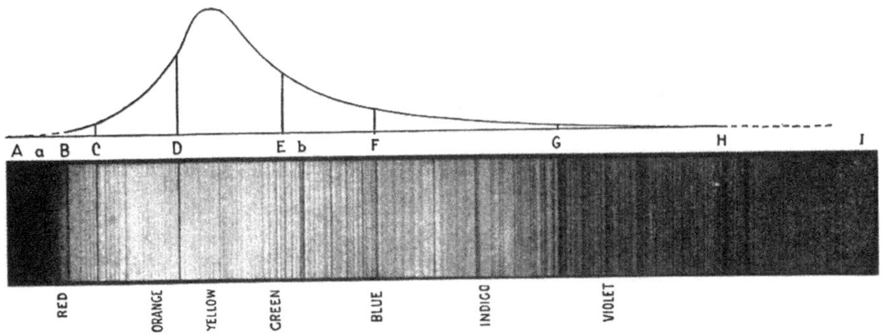

Fig. A.1 Solar spectrum, absorption lines and continuum, sketched by Fraunhofer (King 1955)

photographic spectral plates. In the 1870s, son Henry constructed a private observatory, equipped with a 70-cm reflector, and carried out photographic observations of the Orion nebula and spectroscopic observations of several tens of stars.

Henry's spectroscopic works were followed up by Edward C. Pickering (1846–1919) at the Harvard College Observatory (HCO) in Cambridge, Massachusetts. Pickering organized two big projects: the stellar photometry and the spectral classification of stars, both in a wide field of the sky. For these purposes, Pickering hired many women computers (assistants) at the HCO.

The first woman computer was Williamina P. Fleming (1857–1911), who carried out the spectral classifications of stars based on objective-prism spectrograms. Photographic images made it possible to classify stellar spectra more precisely compared to the classifications of Secchi and Vogel, which were based on sketched spectra. Fleming and Pickering classified spectra using new symbols, A, B, ... M. New classifications for around 10,000 stars were published as the "Henry Draper Memorial Catalogue" under the names Pickering and Fleming in 1890.

Antonia Maury (1866–1952), the second woman computer, promoted her own classification scheme in Pickering's project. Maury paid special attention to the width of spectral lines on her high-dispersion spectra in 1902. She arranged the spectral classes as I, II, III, ..., XXII in the order of stellar effective temperature and the three divisions *a*, *b*, *c* due to the width of the line. Division *c* was defined by sharp absorption lines of hydrogen and helium. The *c* characteristic was soon confirmed by Ejnar Hertzsprung in 1905 as the spectral characteristics of giant stars.

Annie Jump Cannon (1863–1941), the third woman astronomer, took over Fleming's scheme with some amendments. She energetically classified more than 200,000 stars in the northern and southern hemispheres of the sky. The results were published in nine volumes between 1918 and 1924 as the *Henry Draper Catalogue*. Cannon's classification was adopted as the standard system at the General Assembly of the International Astronomical Union (IAU) held in Rome in 1922. Her system, with some modification, is given as follows:

```
            R – N
           /
O – B – A – F – G – K – M
                   \
                    S
```

Stellar photometry was carried out by Pickering himself and the fourth woman computer, Henrietta Leavitt (1868–1921).

Pickering's first project was the establishment of a photometric system of stars, using a visual photometer that he himself had invented. In 1879, he determined, with unprecedented accuracy, the stellar magnitude of 80,000 stars.

The photographic photometry of stars was advanced by Leavitt. In 1909, she began to measure the photographic magnitudes of a sequence of 47 stars near the North Pole. Her work was completed as the North Polar Sequence, which became a standard reference on photographic photometry applicable to any region of the sky. Leavitt also found the period-luminosity relation for Cepheid variables in the Small Magellanic Cloud. Since the distances of these variables were nearly the same, luminosity variation could be regarded as the variation of intrinsic stellar luminosity. Use of the period-luminosity relation became an important technique in measurements of distant objects (Fig. A.2).

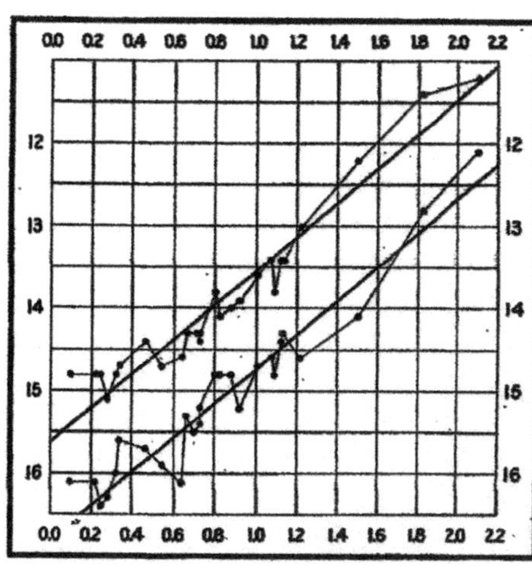

Fig. A.2 Period-luminosity relation, observed by H. Leavitt. The abscissas are the logarithms of the period in days, and the ordinates the apparent magnitudes of stars, which is a ranking logarithm scale. The upper and lower lines denote the magnitudes at light maximum and minimum with the corresponding mean lines (Leavitt and Pickering 1912)

Structure and Evolution of Stars

The idea that the Sun is a gaseous body was inspired by Scottish physicist John James Waterstone (1811–1883), along with German medical doctor Julius Robert von Mayer (1814–1873). Waterstone claimed in 1845 that the Sun was brightening by its own gravitational contraction, while Mayer suggested in the same year that the infalling of small bodies, like comets and asteroids, onto the solar surface represents the origin of solar energy. Waterstone's contraction hypothesis had been advocated by a physiologist and a proponent of energy-conservation law, Hermann von Helmholtz (1821–1894), so the name of Waterstone was replaced by Helmholtz. Helmholtz supposed that a star was born as a bright white star that evolved into a small red star through gravitational contraction. This one-way evolution was widely accepted throughout the nineteenth century.

Jonathan Homer Lane (1819–1880), an examiner at the Patent Office in Washington, D.C., derived the mathematical equations on the internal structure of the Sun in 1870 and numerically solved them in some typical cases, in which he suggested that the central temperature of the Sun should be higher than one million degrees.

In 1907, Robert Emden (1862–1940), a Swiss physicist at the University of Munich, verified Lane's theoretical study on the the stellar structure in a state of hydrostatic and convective equilibrium. Emden paid special attention to the configuration of gas, which satisfies the relation

$$P = K\rho^{(n+1)/n}, \qquad \text{(A.1)}$$

where P and ρ are the gas pressure and density, respectively, and K and n are constants. The constant n is called the polytropic index, and the density gradient decreases as n increases. The internal structure can be analytically solved when $n = 0$, 3, and 5. The stellar radius becomes infinite when n is equal to or greater than 5. Emden numerically solved the spherical gas sphere in many cases of different values of n. The structure of the Sun and the main-sequence stars can be approximately expressed by a polytropic sphere with index $n = 3$.

The idea that the Sun and planets were born from a primordial nebula through its own gravitational attraction had been proposed by Immanuel Kant in 1755 and Pierre-Simon de Laplace in 1797. The physical processes in the gravitational contraction of protostars were first analyzed by August Ritter (1826–1918) at Aachen University in 1898. He showed that when a protostar contracts, its surface temperature increases, while the total luminosity decreases due to the shrinking of the stellar surface. After the star reaches its highest temperature, the energy loss by radiation surpasses the energy gain by contraction and the star gets redder and smaller. Thus, he suggested a two-way evolution of stars under the framework of contraction theory.

Independent of Ritter's theory, Norman Lockyer (1836–1920) of London also proposed a two-way evolution of stars based on his unique classification of stellar

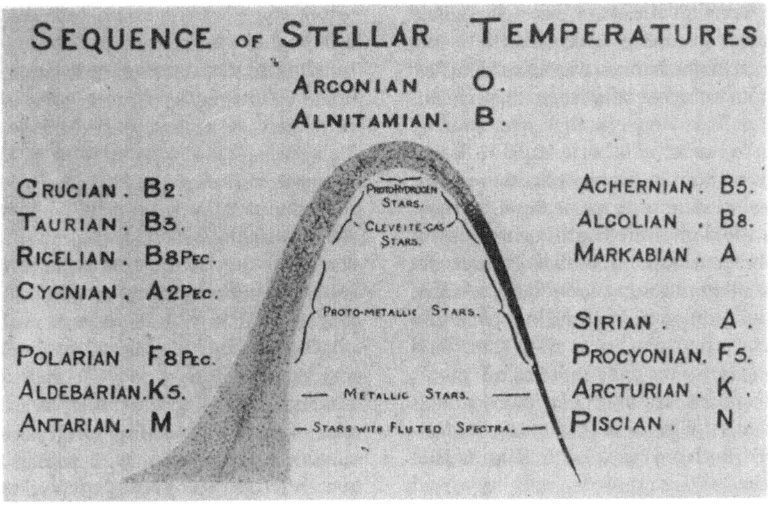

Fig. A.3 Lockyer's arch (Lockyer 1915)

spectra. He arranged the stellar spectral types (Fig. A.3) known as Lockyer's arch, where stars were divided into two temperature sequences, ascending and descending. Stars were born at the base of the ascending branch and evolved into hot stars under the contraction process. After reaching the highest temperature of the O and B types, stars move along the descending branch in the direction of red dwarf stars.

In 1905, Ejnar Hertzsprung (1873–1967) at Urania Observatory in Copenhagen found that late-type stars (G, K, M type) are separated into two groups, based on large and small amounts of parallax (arcsec) and proper motion (arcsec per year), which correspond to nearby stars and distant stars, respectively. Among bright late-type stars, some are at large distances, suggesting that these stars should have intrinsically high luminosity. Hertzsprung thus defined *giant* and *dwarf* stars by their absolute luminosities. A similar relation was also found by Henry Norris Russell (1877–1957) at Princeton University in 1910. The relationship between absolute magnitude and spectral type of stars is shown in Fig. A.4, and it was called the Hertzsprung-Russell diagram (HR diagram).

Russell, like August Ritter, considered the path of stellar evolution on the HR diagram, based on the contraction hypothesis, that a star is born as a red giant and evolved once to an early-type star, and then toward a red dwarf star.

Stellar evolution along the main sequence on the HR diagram was denied by Arthur Stanley Eddington (1882–1944) at Cambridge University Observatory. He discovered the mass-luminosity relation (stellar luminosity is higher in massive stars in the same spectral type). This suggests that a red giant and a red dwarf cannot be the same star. The main sequence does not represent the evolutionary track of stars.

In this way, it became clear that the contraction hypothesis could not explain the path of stellar evolution at all. But then what is the energy source of stellar radiation?

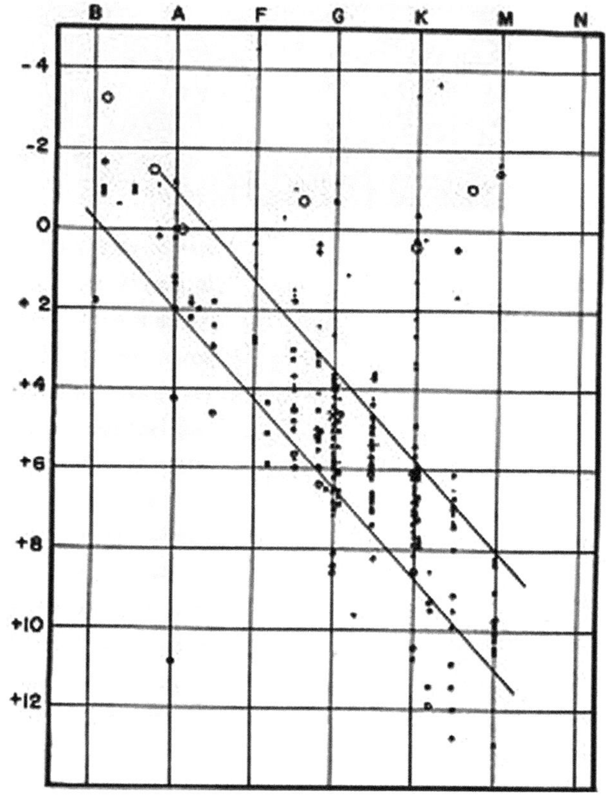

Fig. A.4 HR diagram, constructed by Henry Russel. The abscissas denote spectral type, the ordinates the absolute magnitude (Russell 1914)

In 1926, Eddington suggested the possibility of subatomic processes in some form, such as in the formation of heavy elements from light elements (nuclear fusion), mass annihilation by the collision of electrons with positively charged particles, and the destruction of heavy elements (radioactive elements). Subramanyan Chandrasekhar (1910–1995), who discovered the existence of an upper limit of mass for white dwarf stars, also worked in the early 1930s on the structure and evolution of stars.

Thermonuclear reaction cycles were found by Hans Albrecht Bethe (1906–2005) at Cornell University in Ithaca, New York, in 1937, and by Carl Friedrich von Weizsäcker (1912–2007) at Berlin University in Berlin, Germany, in 1938, independently. These works became the starting points of the modern theory of stellar evolution in the 1940s and thereafter.

Sky Survey and the Galaxy

William Herschel (1738–1822), British but born in Germany, originally a musician, began to observe double stars in a wide field of the sky in around 1776 at his house in Bath, England. By chance he discovered a new planet, Uranus. This made him famous, and he was appointed court astronomer to King George III. Thereafter, he constructed large telescopes (48- and 122-cm reflectors) and carried out survey observations. As a result, he found 2500 nonstellar objects (star clusters and nebulae). William Herschel's surveying activity was taken over by his son John Herschel (1792–1871), who, together with his own observations, collected the data in the *General Catalogue* (GC), which contained 5079 objects and was published in 1864.

John Louis Dreyers (1852–1926) at Armagh Observatory, Northern Ireland, revised the GC and published the *New General Catalogue* (NGC) in 1888, and its supplements, Index Catalogues 1 and 2 by 1910. These catalogs contain 13,196 objects in total.

Using a statistical method for stellar parallax and proper motion, Jacobus Cornelius Kapteyn (1851–1922) of Groningen University proposed a model of the Galaxy in 1921 and 1922. The Galaxy in his model takes an ellipsoidal form with an axial ratio of 5: 1, and its radius in the Galactic plane is around 9000 parsec. The Sun is located near the Galactic Center. A similar model was also proposed by Hugo H. von Seeliger (1849–1924) of the University of Munich in 1920.

In 1918, slightly earlier than the works of Kapteyn and Seeliger, Harlow Shapley (1885–1972) at the Mt. Wilson Observatory was measuring the distance and distribution of globular clusters, based on the period-luminosity relation found by Henrietta Leavitt. For this purpose, he measured the absolute magnitudes of nearby Cepheid variables. The results for six globular clusters indicated that their distance from the Sun were in a range of 6.5 to 18.9 kpc, and their vertical heights from the galactic plane were in a range of 1.8 to 18.8 kpc. These values are one order of magnitude larger than those found by Kapteyn's and Seeliger's models. Moreover, the distribution of globular clusters shows a large concentration toward the constellation Sagittarius. Based on their distribution, Shapley depicted the large-scale structure of the Galaxy as a somewhat flattened stellar system with the diameter of the galactic plane around 10 kpc, with the Sun located far from the center.

On the other hand, Shapley regarded spiral nebulae as small objects distributed around or inside the Galaxy, accepting the claim made in 1916 by A. van Maanen at the Mt. Wilson Observatory about the existence of proper motions in the spiral nebula M 101. If any proper motion were detectable, that object should be a nearby galactic object.

In 1920, Heber Doust Curtis (1872–1942) of Lick Observatory paid attention to the large variety of spiral nebulae with respect to their size and form. He insisted that spiral nebulae are very distant objects outside the Galaxy, though he accepted Kapteyn's rather smaller model of our Galaxy.

Since the views of Shapley and Curtis were so different, the National Academy of Sciences in Washington, D.C., organized an open discussion called the Great Debate in 1920. Both discussions, however, proceeded in parallel without significant progress. In 1934, Shapley and van Maanen accepted the error of measurement of

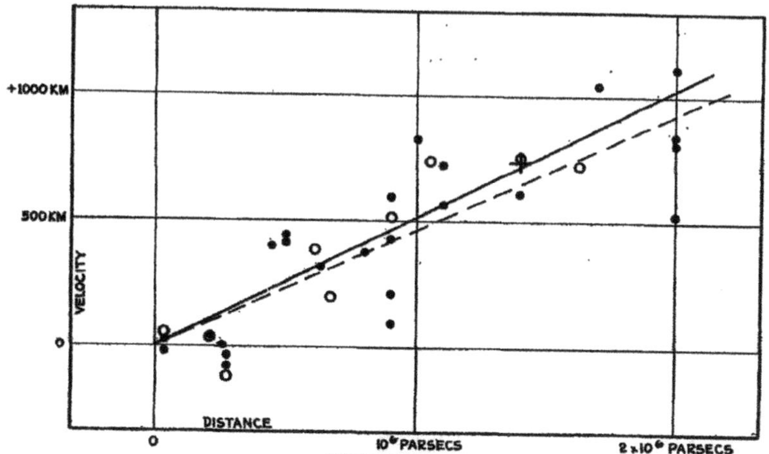

Fig. A.5 Velocity-distance relation of galaxies derived by Hubble. The abscissas denote the distance from the Sun in units of one million parsec, and the ordinates the radial velocity (km s^{-1}), (Hubble 1929)

proper motion in M 101, and the nature of extragalactic nebulae became widely recognized.

In 1915, Vesto Melvin Sligher (1875–1969) at Lowell Observatory in Flagstaff, Arizona, found large line-of-sight velocities for 15 spiral nebulae in a wide range of -300 km s^{-1} (Andromeda Nebula) up to +1100 km s^{-1} (M 104). Inspired by Sligher's observations, Edwin Hubble (1889–1953) measured the radial velocities of spiral nebulae with the 250-cm telescope at the Mt. Wilson Observatory and estimated their distance from the Sun. The radial-velocity distance relation measured by Hubble is shown in Fig. A.5, where he found that the larger the distance, the higher the radial velocity of an extragalactic nebula. This was the discovery of the expanding universe, and Fig. A.5 became a memorial diagram for the new age of extragalactic study and modern cosmology (Chap. 7).

References

Hubble, E. (1929). A relation between distance and radial velocity among extra-galactic nebulae. *Proceedings. National Academy of Sciences, 15*, 168–173.

King, H. S. (1955). *The history of the telescope*. Dover Pub. Inc.

Kogure, T. (2015). *Modern history of astronomy – Origin of astrophysics and pioneers* (現代天文学史—天体物理学の起源と開拓者たち—, 京都大学学術出版社). Kyoto University Press.

Leavitt, H., & Pickering, E. C. (1912). Period of 25 variable stars in the Small Magellanic Cloud. *Harvard College Observatory Circular, 173*, 1–3.

Lockyer, N. (1915). Notes on stellar classification. *Nature, 94*, 618–619.

Russell, H. N. (1914). Relation between the spectra and other characteristics of the stars. II. Brightness and spectral class. *Nature, 93*, 252–258.

Name Index

T. Kogure, *The History of Modern Astronomy in Japan*, Historical & Cultural Astronomy, https://doi.org/10.1007/978-3-030-57061-3

T

Subject Index

T. Kogure, *The History of Modern Astronomy in Japan*, Historical & Cultural Astronomy, https://doi.org/10.1007/978-3-030-57061-3

T

Index of Telescope

T. Kogure, *The History of Modern Astronomy in Japan*, Historical & Cultural Astronomy, https://doi.org/10.1007/978-3-030-57061-3

Index of Research Institute

T. Kogure, *The History of Modern Astronomy in Japan*, Historical & Cultural Astronomy, https://doi.org/10.1007/978-3-030-57061-3

Index of Scientific Satellites and Space Explorers

T. Kogure, *The History of Modern Astronomy in Japan*, Historical & Cultural Astronomy, https://doi.org/10.1007/978-3-030-57061-3

Index of Celestial Objects

T. Kogure, *The History of Modern Astronomy in Japan*, Historical & Cultural Astronomy, https://doi.org/10.1007/978-3-030-57061-3

3. Nova, supernova

4. X-ray source

5. Star clusters and nebulae

6. The Galaxy

7. Galaxies

GPSR Compliance

The European Union's (EU) General Product Safety Regulation (GPSR) is a set of rules that requires consumer products to be safe and our obligations to ensure this.

If you have any concerns about our products, you can contact us on ProductSafety@springernature.com

In case Publisher is established outside the EU, the EU authorized representative is:

Springer Nature Customer Service Center GmbH
Europaplatz 3
69115 Heidelberg, Germany

Batch number: 10732090

Printed by Printforce, the Netherlands